Microbial Ecotoxicology

Cristiana Cravo-Laureau
Christine Cagnon · Béatrice Lauga
Robert Duran
Editors

Microbial Ecotoxicology

Springer

Editors
Cristiana Cravo-Laureau
Equipe Environnement et Microbiologie, IPREM UMR 5254
Université de Pau et des Pays de l'Adour
Pau
France

Béatrice Lauga
Equipe Environnement et Microbiologie, IPREM UMR 5254
Université de Pau et des Pays de l'Adour
Pau
France

Christine Cagnon
Equipe Environnement et Microbiologie, IPREM UMR 5254
Université de Pau et des Pays de l'Adour
Pau
France

Robert Duran
Equipe Environnement et Microbiologie, IPREM UMR 5254
Université de Pau et des Pays de l'Adour
Pau
France

ISBN 978-3-319-61794-7 ISBN 978-3-319-61795-4 (eBook)
DOI 10.1007/978-3-319-61795-4

Library of Congress Control Number: 2017945257

Printed on acid-free paper

This Springer imprint is published by Springer Nature
The registered company is Springer International Publishing AG
The registered company address is: Gewerbestrasse 11, 6330 Cham, Switzerland

Preface

Worldwide human activities are increasingly threatening the environment. Scientists have long been concerned about the effect of pollutants on biota, at the different levels of biological organization, from organisms to ecosystems. In the early 1960s, the emergence of ecotoxicology sensitized society to the harmful effects on the environment of human activities. Recent advances in microbial ecology enrich ecotoxicology with ***Microbial Ecotoxicology***, which aims to investigate the effect of pollutants on microorganisms and, in turn, the role of microorganisms in determining the fate of pollutants. Several initiatives have proposed to promote ***Microbial Ecotoxicology***, organize the scientific community to smooth research exchanges, and facilitate interdisciplinary studies (Ghiglione et al. 2014, 2016; Gu and Wang 2014). Such initiative corresponds to an increasing demand from worldwide politics and society in front of the intense human activities threatening the environmental health. In this context, we believe that it is now time to compile microbiological studies that adopt an ecotoxicological point of view in order to show the multifacets of ***Microbial Ecotoxicology***. The present book is a treatise on ***Microbial Ecotoxicology*** covering the effect of pollutants on microbial ecosystems and the role of microorganisms in ecosystems services. Emphasizing the microbial responses to pollution at different biological levels, this book focuses on metabolic pathways, genetic adaptation, and response at the whole microbial community level. This book also addresses the ecological indicators of ecosystem recovery, as well as microbial biomarkers and biosensors as tools for ***Microbial Ecotoxicology***. In order to cover all these aspects, we have contacted worldwide scientists who have contributed significantly in the advances of ***Microbial Ecotoxicology***. We would like to thank all the authors that have contributed with review chapters. We also thank all colleagues that have accepted to evaluate and comment on the submitted chapters, and their review is essential to guarantee the scientific quality and improve the contributions.

Pau, France

Cristiana Cravo-Laureau
Christine Cagnon
Béatrice Lauga
Robert Duran

References

Ghiglione JF, Martin-Laurent F, Pesce S (2016) Microbial ecotoxicology: an emerging discipline facing contemporary environmental threats. Environ Sci Pollut Res 23:3981–3983. doi:10.1007/s11356-015-5763-1

Ghiglione JF, Martin-Laurent F, Stachowski-Haberkorn S, Pesce S, Vuilleumier S (2014) The coming of age of microbial ecotoxicology: report on the first two meetings in France. Environ Sci Pollut Res Int 21:14241–14245. doi:10.1007/s11356-014-3390-x

Gu JD, Wang Y (2014) Geomicrobial ecotoxicology as a new subject in environmental sciences is proposed. Ecotoxicology 23:1823–1825. doi:10.1007/s10646-014-1359-7

Reviewers

Callaghan Amy, University of Oklahoma, USA
Cao Bin, Nanyang Technological University, Singapore
Cébron Aurélie, Lorraine University, France
Coulon Frédéric, Cranfield University, UK
Gieg Lisa, University of Calgary, Canada
Giloteaux Ludovic, Cornell University, USA
Griffiths Bryan, Scotland's Rural College, UK
Gutiérrez Juan Carlos, Universidad Complutense de Madrid, Spain
Head Ian, University of Newcastle, UK
Hickey William, University of Wisconsin, USA
Jouanneau Sulivan, Nantes University, France
Kirchman David, University of Delaware, USA
Kunze Gotthard, ipk-Gatersleben, Germany
Lambreva Maya, CNR, Italy
Morin Soizic, IRSTEA, France
Pesce Stéphane, IRSTEA, France
Peyretaillade Eric, Auvergne University, France
Richaume Agnès, Lyon University, France
Sabater Sergi, Girona University, Spain
Simonet Pascal, Ecole Centrale Lyon, France
Tang Jingchun, Nankai University, China
Xu Tingting, University of Tennessee, USA
Yarwood Stephanie, University of Maryland, USA

Contents

Editors and Contributors

About the Editors

Cristiana Cravo-Laureau is Associate Professor in Environmental Microbiology at Pau University (France). She is member of the Institute for Analytical Sciences and Physico-Chemistry for Environment and Materials (IPREM), a CNRS-UPPA (University of Pau and Pays de l'Adour) joined unit. She earned her Ph.D. in Environmental Science from the Aix-Marseille University (France) in 2003, completed a postdoc at IRD (France), and joined the Pau University in 2004. Her research focuses on microbial ecology and ecotoxicology. She has special interest in investigating the involvement of microorganisms in biogeochemical cycles and their role in pollutants degradation in marine environments, particularly in coastal sediments. She has an expertise in anoxic environments and anaerobic microorganisms, from community scale to pure strains. In addition to examine natural environments, she develops experimental ecology approaches. Her research also deals with the impact of oil pollution on aerobic and anaerobic prokaryotic communities. She also expanded her research interests to the impact of oxic/anoxic oscillations on microbial communities and organic pollutants degradation.

Christine Cagnon is Associate Professor in Environmental Molecular Microbiology at Pau University (France). She is member of the Institute for Analytical Sciences and Physico-Chemistry for Environment and Materials (IPREM), a CNRS-UPPA (University of Pau and Pays de l'Adour) joined unit. She received a Ph.D. in Microbiology from the National Institute of Applied Sciences (INSA) of Toulouse (France) and completed her postdoc at Protein Engineering Research Institute (PERI) in Osaka (Japan) before joining Pau University. She first used her expertise in molecular biology, bacterial genetics, and biochemistry to protein engineering. At Pau University, she developed research on molecular responses of organisms to chemical stresses with special references to community structure, molecular markers, and genomic plasticity. She focuses since 11 years on bacterial community adaptation process and on the implication of mobile genetic elements. She targets especially integrons and recently highlighted the relation between environmental integron diversity and pollution in natural environments.

Béatrice Lauga completed a Ph.D. in Plant Biology and Genetics at the University of Pau and Pays de l'Adour, France. She was a postdoctoral fellow at the ICAPB (Institute of Cell, Animal and Population Biology) in Profs B. et D. Charlesworth's Lab, Scotland, and then at the University of Pau and Pays de l'Adour, where she started working with microbes in the environment. She is now Professor in Microbial Ecology at the University of Pau and Pays de l'Adour and the Head of the PREMICE platform (a core facility dedicated to the investigation of microbial diversity). Her research interests focus on pattern of microbial diversity with special emphasis on its role on the functioning of ecosystems. Taxonomic and functional diversities are both investigated. She explores a variety of environments to unravel forces that shape the assemblage of microorganisms and the structure of natural microbial communities. Most of the environments she studies are imprinted by contaminants originating from anthropogenic activities. She develops a range of tools and methods to investigate highly contaminated or complex samples. Her approaches are clearly influenced by her past experiences in population genetics and their links to community ecology. As an example, she recently deciphers the role of dispersal versus local contamination on community assembly in acid mine drainages and in tropical rivers highly contaminated with pesticides.

Robert Duran is Professor in Environmental Molecular Microbiology at Pau University (France). He is member of the Institute for Analytical Sciences and Physico-Chemistry for Environment and Materials (IPREM), a CNRS-UPPA (University of Pau and Pays de l'Adour) joined unit. He received a Ph.D. in Biochemistry, Molecular and Cell Biology from Montpellier University (France) and completed his postdoc at Tokyo University (Japan) before joining Pau University. He has been working for 20 years in environmental microbiology and molecular microbiology research. Professor Duran has been a member of several scientific boards committees, and he is presently Editor for Environmental Science and Pollution Research (Springer). He has successfully completed more than 20 projects focusing on microbial ecology and diversity, with an emphasis on extreme, polluted, and estuarine and coastal ecosystems. The main research objectives are to understand the impact of pollutants on microbial communities and, in turn, to characterize the role of microorganisms in determining the fate of pollutants. Relevant recent discoveries include the characterization of the taxonomic and functional microbial diversity in extreme and contaminated ecosystems: behavior of sulfate reducing bacteria populations in acid mine drainage and microbial mats according to the fluctuation of environmental parameters, the diversity of ring-hydroxylating dioxygenases in polluted microbial mats, and the organization of hydrocarbon-degrading prokaryotic communities in mudflat sediments.

Contributors

Arturo Aburto-Medina School of Science, Centre for Environmental Sustainability and Remediation, RMIT University, Bundoora, VIC, Australia

Francisco Amaro Dpto. Microbiología-III, Facultad de Biología, Universidad Complutense (UCM), Madrid, Spain

Ali Assaf University of Nantes, La Roche sur Yon, France

Andrew S. Ball School of Science, Centre for Environmental Sustainability and Remediation, RMIT University, Bundoora, VIC, Australia

Laura Barral Institut d'Ecologia Aquàtica i Departament de Ciències Ambientals, Girona, Spain

Simone C. Birrer Evolution and Ecology Research Centre, School of Biological, Earth and Environmental Sciences, University of New South Wales, Sydney, NSW, Australia; The Sydney Institute of Marine Science, Mosman, NSW, Australia

Marine Bittel University of Nantes, La Roche sur Yon, France; TRONICO Vigicell, La Roche sur Yon, France

Berta Bonet Institut d'Ecologia Aquàtica i Departament de Ciències Ambientals, Girona, Spain; School of Geography, Earth and Environmental Sciences, University of Birmingham, Birmingham, UK

Chloé Bonnineau Irstea, UR MALY, Centre de Lyon-Villeurbanne, Villeurbanne Cedex, France

Christine Cagnon Equipe Environnement et Microbiologie, IPREM UMR 5254, MELODY Group, Université de Pau et Des Pays de L'Adour, Pau, France

Frédéric Coulon School of Water, Energy and Environment, Cranfield University, Cranfield, UK

Cristiana Cravo-Laureau Equipe Environnement et Microbiologie, IPREM UMR 5254, MELODY Group, Université de Pau et Des Pays de L'Adour, Pau, France

Katherine A. Dafforn Evolution and Ecology Research Centre, School of Biological, Earth and Environmental Sciences, University of New South Wales, Sydney, NSW, Australia; The Sydney Institute of Marine Science, Mosman, NSW, Australia

Vidya de Gannes Department of Food Production, University of the West Indies, St. Augustine, Trinidad and Tobago

Robert Duran Equipe Environnement et Microbiologie, IPREM UMR 5254, MELODY Group, Université de Pau et Des Pays de L'Adour, Pau, France

Marie-José Durand University of Nantes, La Roche sur Yon, France

Samuel Eduok School of Water, Energy and Environment, Cranfield University, Cranfield, UK; Department of Microbiology, University of Uyo, Uyo, Nigeria

Philippe Garrigues Institut des Sciences Moléculaires, UMR 5255 CNRS/Université de Bordeaux, Talence, France

Jean-François Ghiglione Laboratoire d'Océanographie Microbienne, Observatoire Océanologique, Sorbonne Universités, CNRS, UPMC Univ Paris 06, UMR 7621, Banyuls-sur-Mer, France

Helena Guasch Institut d'Ecologia Aquàtica i Departament de Ciències Ambientals, Girona, Spain

Juan Carlos Gutiérrez Dpto. Microbiología-III, Facultad de Biología, Universidad Complutense (UCM), Madrid, Spain

William J. Hickey O.N. Allen Laboratory for Soil Microbiology, Department of Soil Science, University of Wisconsin, Madison, WI, USA

Emma L. Johnston Evolution and Ecology Research Centre, School of Biological, Earth and Environmental Sciences, University of New South Wales, Sydney, NSW, Australia; The Sydney Institute of Marine Science, Mosman, NSW, Australia

Sulivan Jouanneau University of Nantes, La Roche sur Yon, France

Leadin Salah Khudur School of Science, Centre for Environmental Sustainability and Remediation, RMIT University, Bundoora, VIC, Australia

Anna I. Koukkou Sector of Organic Chemistry and Biochemistry, Department of Chemistry, University of Ioannina, Ioannina, Greece

Béatrice Lauga Equipe Environnement et Microbiologie, IPREM UMR 5254, MELODY Group, Université de Pau et Des Pays de L'Adour, Pau, France

Fabrice Martin-Laurent Agroécologie, AgroSup Dijon, INRA, Université de Bourgogne Franche-Comté, Dijon, France

Ana Martín-González Dpto. Microbiología-III, Facultad de Biología, Universidad Complutense (UCM), Madrid, Spain

Geneviève Massard-Guilbaud Environmental, Economic and Social History, EHESS, Paris, France

Charles-François Mathis Modern History, Université Bordeaux-Montaigne, UFR Humanités, Pessac, France

Stéphane Pesce Irstea, UR MALY, Centre de Lyon-Villeurbanne, Villeurbanne Cedex, France

Suja Rajan Department of Biological Sciences, University of Alabama, Tuscaloosa, AL, USA

Sergi Sabater Catalan Institute for Water Research (ICRA), GRECO-IEA, University of Girona, Girona, Spain

Esmaeil Shahsavari School of Science, Centre for Environmental Sustainability and Remediation, RMIT University, Bundoora, VIC, Australia

Patricia A. Sobecky Department of Biological Sciences, University of Alabama, Tuscaloosa, AL, USA

Mohamed Taha School of Science, Centre for Environmental Sustainability and Remediation, RMIT University, Bundoora, VIC, Australia; Department of Biochemistry, Faculty of Agriculture, Benha University, Moshtohor, Toukh, Egypt

Gérald Thouand University of Nantes, La Roche sur Yon, France

Elpiniki Vandera Sector of Organic Chemistry and Biochemistry, Department of Chemistry, University of Ioannina, Ioannina, Greece

Part I
Introduction

Chapter 1
A Brief Introduction to the History of Pollution: From Local to Global

Geneviève Massard-Guilbaud and Charles-François Mathis

Abstract The way pollution can be defined has greatly varied in the course of history, according to times and places. It has long retained a moral and religious dimension. That some substances were "out of place" is attested since at least ancient civilisations. In France, as in most European countries, this pollution has been dealt with in the Middle Ages and in the early Modern period by removing dirt and dirt-producing activities onto the fringe of cities. This began to change at the end of the 18th century when states started to promote industry as a means of power. New pollutions also appeared with the birth of the modern chemical industry. But in the 20th century the scale of pollution changed dramatically: it became more intensive, with a growing number of pollutants discharged in the biosphere; more global, affecting all places on earth; more lasting, with disastrous cumulative effects.

Keywords Nuisance · Pollution · Environment · History

1.1 Pollution: What Are We Talking About?

Whoever wants to talk about pollution history must first precise *what* he is talking about. Indeed, the notion of pollution can cover very different things according to times and places. Today's pollution is not yesterday's, even less that of the eighteenth century or of the Middle Ages. This is not mainly due to the fact that "they did not know" that this or that product was a pollutant, but to the fact that each society defines pollution according to its own criteria, beliefs, perceptions and feelings. Each society has its own conception of what is clean and unclean, healthy

G. Massard-Guilbaud
Environmental, Economic and Social History, EHESS, Paris, France

C.-F. Mathis (✉)
Modern History, Université Bordeaux-Montaigne, UFR Humanités, Domaine Universitaire, 33607 Pessac, France
e-mail: cfmathis@hotmail.com

C. Cravo-Laureau et al. (eds.), *Microbial Ecotoxicology*,
DOI 10.1007/978-3-319-61795-4_1

and unhealthy and, as anthropologist Mary Douglas put it (Douglas 1966), its specific conception of order. In this order, pollution is just what is "out of place". Indeed, no product is a pollutant in itself, but it can become one in specific circumstances (when it is introduced where it should not be) or according to other products it is associated or mixed with. Easy examples: coal or oil, as long as they remain in the earth where they formed over the millennia, are no pollutants. But they become so when we burn them; arsenic is not a pollutant, as long as it is not ingested in quantity by humans or animals. It can even become a medicine when used by homoeopaths in infinitesimal doses. The experts working, today, on the REACH classification are confronted to this issue: pollution does not depend on the product, but of what you make of it.

It is interesting to have a look at the origins of the word "pollution" itself. We'll do so for the French language. The word pollution comes from the Latin *pollutio* which meant dirt and appeared in French in the twelfth century. It has, at that time, lost its Latin sense and means stain or impurity, with both moral and religious connotations. In the seventeenth and eighteenth centuries, French dictionaries provide two different definitions for the word pollution: (1) profanation of a religious sanctuary; (2) obscenity committed on one's body by indecent touching. The first occurrence of the term pollution close to the meaning we give it today is to be found in the 1874 edition of the Littré Dictionary: "action of dirtying with litter". The example provided to illustrate this definition was drawn from the *Journal Officiel de la République française* [official publication of the French Republic] issued on the twentieth of November of the same year. Historians who read the records confirm that the word do appear with this meaning in the 1860s and 1870s, though it was used for water only. It would extend to air and soil at the end of the century. But the Littré is the only dictionary to note its appearance with this meaning as early as the last third of the nineteenth century, which proves that it was not so common yet. The 1930s edition of the Académie française dictionary still sticks to the old meanings. And the 2010 edition of the reference dictionary *Le Robert* gets the wrong century, dating the modern meaning from the 1960s!

This digression by the evolution of the word attests how strongly and how long the word pollution kept its old religious or moral meaning, and the relationship long maintained between dirt and impurity. It also explains why historians studying the nineteenth century or beginning of the twentieth century records find so often pollution called "the demon", or the "modern demon"—in French as well as in English (Bernhardt and Massard-Guilbaud 2002, p. 16). Incidentally, the monks were the first to try hard to clean and sanitize their environment, bringing in their monasteries large amounts of water and draining used waters away, as early as medieval times. Clean water was seen as the main remedy against material and moral stain. In medieval and early modern painting, running waters and fountains are metaphors of moral purity (Fournier 2002).

The fact the word pollution was not used in medieval and early modern French to talk about *material* pollution does not however mean that the nuisances created by crafts or the insufficient sanitation did not exist or that they were not sensed. The existence of pollution of anthropic origin is evidenced as early as ancient civilisation. The analysis of samples taken off Greenland ice show that the concentration of copper began to raise above their natural level as early as 2500 BC. In the Roman period, the concentration was twice what it was before man began to mine it (Hong et al. 1996). Researchers have also shown that in the middle of the eighteenth century, the concentration of lead in the ice was ten times higher than its natural level (Boutron et al. 1993).

From the Middle Ages onto the eighteenth century, many other words were used to think and talk about pollution: the substantives dirt, stench, nuisance … the adjectives unhealthy, infectious, corrupted, soiled, rotten … Not only were these nuisances named, but they were also regulated. In all the countries where this has been studied (Poulussen 1991; Le Roux 2011; Parmentier 2008; Cavert 2016), nuisances were a juridical category. In England, the law rested on the maxim "*Sic utere tuo ut alienum non laedas*", meaning that one should not disturb others while enjoying one's own possessions. In England, the Common Law of nuisances went back to the Middle Ages and had its own jurisdiction, the Nuisance Assize. Contraveners could be fined, see their business closed or incarcerated (Cokayne 2007). In France, nuisance regulation rested on a very large corpus of edicts, arrests, orders and decrees emanating from various authorities (parliaments, police, municipalities …), formed since the Middle Ages. The case of Paris in the eighteenth century has been well studied by Thomas Le Roux (Le Roux 2011). The police dealt with nuisances according to the common law of servitudes and neighbourhood. But they also had a specific action against nuisances and the protection of public health was one of their duties. The main principle to protect the city dwellers against nuisances was their removal. All the trades dealing with animal raw materials and not least tanneries but also dying and many others had to settle outside the city, in areas that were sort of sacrificed to industry.

This was also the case for all industries using putrefaction as a means to transform materials, due to the stench they produce. The neo-Hippocratic medicine of that time, which saw in miasmas the origins of illnesses, granted a specific interest to three factors: climate, dangerous substances contained by air, water and the soil, and the mephitic vapours likely to contaminate and corrupt organisms. Those which emanated from marshes or putrefying matters were feared. The odour was then a strong indicator of pollution, and the trades that produced stench particularly dreaded. Those likely to cause fire or important noise were also under close surveillance.

This moving away, however, did not apply to all noxious trades. Tallow foundries, candle makers, some ovens, known for the nuisances they created, were allowed to stay in the city. In this case, the practice was a pragmatic mix of prevention, dialogue and surveillance, and the decisions made on the case by case basis (Le Roux 2011). Amongst the measures of prevention was the so-called survey of *commodo et incommodo*. Formalised in 1749 for the workshops which killed butchery animals, it extended then to other trades. This procedure included information, prevention and dialogue. The role of the police was to conciliate as far as possible all vested interests. Thanks to this sophisticated process, nuisances were more or less under control, in Paris, by the middle of the eighteenth century.

Complaints, that were so many at the beginning of the century, had nearly disappeared.

Although heavy modern industry had developed earlier in the United Kingdom, France was the country where this old mode of regulation, so far more or less common to all industrialised areas, was first to be upset. For this reason, we will concentrate on France to give the reader an idea of the way authorities reacted to the new pollution created by the industrial revolution, before coming back to a more global view.

1.2 Pollution and Industrial Revolution: The French Case

The birth of a new chemistry and especially the making on an industrial scale of what was then called the "strong acids" was to challenge the former order of regulation. 1773 can be considered as a milestone. In that year began in Rouen a trial opposing the owner of a large chemical factory (actually the first industrial chemical factory of such a large size in the country), John Holker, to his neighbours who complained about the terrible effects, for the surroundings, of his new production: sulphuric acid made in lead chambers. This trial was emblematic of the turn that was rapidly going to extend to all the country and for this reason deserves a few lines here. Holker, an Englishman who had fled his country and been naturalized French, had "borrowed" (actually stolen by the way of industrial spying) this making process to his compatriot Roebuck, who implemented it on the banks of the Firth of Forth. The lead chambers process made it possible to make sulphuric acid on a large scale and at a much lower cost than the methods implemented so far. Sulphuric acid was a fundamental product to many industries, including textile. Holker was at that time the only one in France mastering the lead chamber method —as far as we can say that, the process being far from finalised. For this reason, the king was ready to protect him, whatever the nuisances he created. And this is what happened. Without entering the details of the trial, let us say that the case ended at the top of the state as an arm-wrestling contest between the Police minister (who supported the action of the police in favour of the neighbours) and the Commerce minister (who supported the industrialist). The Conseil du Roi ruled in the latter's favour, announcing times when an industrial production could be declared of national interest, whatever the consequences for the neighbours in terms of health and the environment.

But this was only a beginning, and sulphuric acid was not the only product concerned. Coal provides another example. So far, coal had been used in France by a limited number of trades. This number was now growing, not least due to the new steam engines. Although the problems caused by the carbon dioxide they gave off were well known, the government encouraged their use, partly because of wood scarcity, partly because of "the necessities of modernisation", according to the terminology of the time. The production of soda from marine salt (Leblanc process) was another industry which was strongly encouraged, in spite of the massive release, in the atmosphere, of calcium sulphur and muriatic acid it involved. Industries using traditional processes were still regulated by old rules while the new and flourishing ones, generally using mineral (instead of organic) materials and fire (instead of putrefaction) as a tool for transforming them, were encouraged on behalf of economic modernization. A tendency had appeared to treat the pollution they emitted as a political and administrative issue rather than as a sanitary and judiciary issue.

At that time, however, the police and the civil courts still had the right to order the move or even the closing of polluting factories. They used it with more or less rigour according to the place. Chemists (who were usually both scientists *and* industrialists) then complained about what they called an "intolerable arbitrariness" (Massard-Guilbaud 2010). In the meantime, the revolutionary rulers gave these chemists a new political weight in calling for their help during the wars against the combined forces of the European monarchies (Guillerme et al. 2004). They would use this new influence to get the tendency just described formalized. The way pollution was regulated dramatically changed. The powers of police and courts were strongly reduced in favour of the administration.

This happened in France, and beyond France, in the large part of Western Europe occupied by the French, in 1810. The decree of 15 October 1810 that organised the new deal was taken after two successive reports had been asked to the chemistry class of the French Academy by the famous chemist Jean-Antoine Chaptal, also Minister of the Interior. The scientists who sat in the Academy were obviously juge and jury: scientists called to say what was or was not polluting, they were also, as industrialists, the main polluters of the country. As we can guess from this configuration, the 1810 decree was not made to protect the health of the population nor the environment. It was made to confirm that from then on, polluting industry deserved the protection of administration against their neighbours' prosecution. Or, to tell it differently, as all kinds of private property were important to the French government issued from the Revolution, to arbitrate between two forms of property and the rights associated to them: the right, for land owners, to enjoy their property without being polluted by a factory, and on the other hand the right to practise new and profitable trades without being threatened by neighbours, police and the courts. Pollution itself was not condemned, just treated as a secondary effect of trades deemed by the middle class valuable both for the economy and the working class who needed work.

When in 2010 the history committee of the ministry of Ecology decided to celebrate the bicentenary of this decree, it was difficult for historians to persuade those working for its administration that the 1810 decree was definitely not the first ever law adopted to protect the environment, as many still believe. But all historians agree that it really meant to protect industry, whatever the damages the latter could cause.

The decree classified industry identified as polluting in three classes going from the less to the more so. Those of the two first classes would have to get an authorisation before being practiced, and the most polluting should also be settled far from any habitation. All would have to implement the safety and sanitary measures imposed upon them by the administration. On the other hand, once they had been authorised, nobody could pretend to see them moved or closed. The civil courts, which in the past used this ability, could from then on only grant financial compensations to those who would suffer pollution. Pollution had become a political and administrative issue (in the sense that it was ruled by the central state or its local representatives, the *préfets*) instead of the health and sanitary problem dealt with by the police that it was in the past.

The philosophy of this decree was to remain the basis of the modern regulation, though the moving away was renounced to in favour of technical measures supposedly reducing the nuisances, and the nomenclature was modified on several occasions before being fundamentally reformed in the 1970s. People tried to use in their own favour the 1810 decree, that they often called "the decree that protects us from pollution". But this was most often to no avail, for a number of reasons. First of all, the nomenclature that classified industries according to the degree of pollution they created was done by a committee composed of industrialists and engineers, little inclined to thwart their peers. Secondly, even when appropriate measures were ordered, they usually went unheeded. Only a few cities appointed inspectors to control their implementation. The central state, on its side, did not create inspectors before 1917. Ill trained for this task, too few and without real possibility of sanction, they were rather helpless. Finally, pollution was very little analysed, measured and understood, in a time when, at the end of the century, another industrial revolution brought on the markets an increasing number of new chemical products. While a growing number of physicians understood and advertised the effects of various pollutions on health, their voice remained little heard until well into the twentieth century.

1.3 The Twentieth Century: Changes in Nature and Scale

In the 20th century, the type, level and scale of pollution changed dramatically. The historian John McNeill (2000) argues that never before were environmental changes as intensive, and to such an extent triggered by human action. In the rest of this text, we will therefore look at pollution on a more global scale.

The great wars of that century exemplify this tendency (Hupy 2008). The appalling destructions they caused, especially in the landscapes (mainly by the destruction of forests by bombings) are well recorded. But they also polluted the environments where they took place. During the First World War, chemical weapons were used massively for the first time: about 65 million shells were fired on the Western front. The consequences on the French, Belgian and Dutch ecosystems are still to be specified. The existing studies show that the concentration of lead, zinc, copper, sometimes arsenic, in the soils, are higher than they should be. The most dramatic case is that of the "Gas Place", a 70-meter-wide circle North-East of Verdun where 200,000 chemical shells were incinerated before 1928 and where nothing has been able to grow ever since. In 2012, the consumption of water in 544 municipalities in the North of France was restricted because of levels of pollutions certainly provoked by weapons and munitions abandoned in their territories during the Great War (Masson-Loodts 2014).

Surprisingly, the most devastating polluting effect of the Second World War does not come from the radiations of the nuclear explosions of Hiroshima and Nagasaki: recent studies tend to show that the ecosystems were resilient and quickly recovered (Tsutsui 2003). More damaging seem to have been for instance the 10,000 tons of phosphorus bombs that were dropped on Hamburg in July 1943, or, more generally, the intense mining and industrialisation of the fighting countries (Hamblin 2013).

During the Vietnam war, about 70 million litres of the "agent orange" herbicide were spread over the country between 1961 and 1971, with dreadful consequences for the population (who still suffers from cancers and malformations) and the environment: 40% of the arable lands have been contaminated (Stellman et al. 2003).

Beyond the case of these major wars, most researchers agree that the period post-1945 was a turning point in terms of pollution. In *The Earth as transformed by human action*, Turner (1990) uses 10 indicators to estimate the global and regional changes in the biosphere from 10,000 years ago to the mid-1980s, among which carbon, sulphur, nitrogen, phosphorus, lead and carbon tetrachloride releases. Except for carbon, they all reach their first quartile of their 1985 total change after 1900, and, except lead, after 1945: the releases of these chemical components have therefore been far more intensive in the last half-century than in the whole of human history. Steffen (2005), in *Global Change and the Earth System*, also depicts a trajectory of deep global changes that take off around 1800 and speed up after the Second World War. As for Christian Pfister (1995), he argues that it was during the 1950s that the global threats to the earth really began, mainly because declining oil prices led to a wasteful organisation and exploitation of the world. Finally, Donald Hughes (2009) claims that "the kinds of changes inflicted by industry on ecosystems since the Second World War include some that had not been know during previous centuries. Plutonium and other radioactive wastes, non-biodegradable insecticides, chlorofluorocarbons, plastics, artificial pheromones and hormones, and many of the rest of the tens of thousands of industrial chemicals in use either did not exist or were not disseminated in major quantities until recently". All these authors emphasize the main features of the post-1945 environmental pollution, namely its intensification, its globalisation, its invisibility and its long-lasting effects.

1.4 Intensification of Pollution

First of all, environmental pollution considerably intensified after the Second World War. Several reasons can explain this, the first one being related to changes in agriculture. The use of phosphate-enriched fertilizers in agriculture started in the 19th century, and the Haber-Bosch process to create nitrate was invented in 1913. The possibility to add phosphorus and nitrogen to the soil liberated agriculture from its previous limits and from the metabolic cycle that had been preserved so far. Indeed, up to this moment, it was necessary to restore the nutriments of the soil by the use of manure, crop rotations or the import of guano from Peru. This was no longer necessary; recycling waste from the cities, for instance, which had been a concern throughout most of the 19th century, but was becoming more and more difficult with the use of water-flowing devices, was gradually abandoned. But this took time, and it is only after 1945 that the agrarian revolution really took off, with monoculture production based on mechanisation and the massive use of both fertilizers and pesticides; it was first implemented in developed countries, and, through the "green revolution", in developing countries, mainly from the 1970s onwards. In 1940, 4 million tons of fertilizers were used in the world; it reached 40 million tons in 1965 and 150 million in 1990 (McNeill 2000, p. 54). As a consequence, nowadays, the flux of nitrogen produced by mankind is twice as important as the natural, while, for the flux of phosphate, the ratio is of 8 to 1! About half of these products ends up in waters, whether ground water, rivers, or, *in fine*, oceans. 9 million new tons of phosphate are thus accumulating every year in the oceans. The consequences are well known: a lower oxygen content of waters, and a eutrophication of rivers and estuaries (Bonneuil and Fressoz 2013, p. 23).

This more intensive agriculture has enabled a considerable growth of the human population: from 1 billion in 1800, it grew to more than 2 billion in 1945 to reach 7 billion today. Turner (1990) considers the increase of population as the major force of environmental change, especially in the 20th century. Such a growth was accompanied by a redistribution of the population and a massive urbanisation: in 2009, more human beings lived in cities than in the country; this can be considered as a decisive turning point in the history of mankind. The problems are roughly the same as in the 19th century, but greatly increased: the mere concentration of people induces massive discharges in the soils, waters and in the atmosphere, which durably affect the close and even remote environments of the towns (concentrations of metals in urban soils are between 10 and 100 times higher than in natural conditions). Mexico City or Beijing, both approaching 20 million people, are telling examples of such issues, with their legendary atmospheric pollution, the contamination of water, the high levels of heavy metals in their soils and the replacing of natural ecosystems by built land. The fact that a growing part of mankind is adopting a non-sustainable way of living, based on automobile, mass consumerism and its wasteful mentality, makes matters even worse. The world consumption of energy has considerably risen since 1800.[1] This has led, among other things, to a dramatic increase in the emission of greenhouse gases, which has reached 14 Giga tons equivalent carbon in 2000. The production of energy is responsible for 38% of these emissions, industries for 25%, transports for 24%, and heating and other uses

[1]The rise is certain, but estimates vary: McNeill believes that it has been multiplied by 40 between 1800 and 1990, and by 12 since 1900 (p. 41); Jancovici gives a factor of 150 since 1850 and 30 since 1900 (p. 19). Of course these figures hide considerable variations between countries according to their economic growth and development.

for the remaining 13%. Such levels are unknown in the history of mankind and their effects on ecosystems still to be fully measured (Jancovici 2002, pp. 150–158).

The last major fact that contributed to the huge increase of pollution in the 20th century, and mainly in its second half, is the industrialisation of the world and the growing number of heavy metals and chemicals discharged in the biosphere. The production of chemicals, that started, as we have seen, in the 19th century, has greatly increased in the course of the 20th century. About 10 millions have been synthesised since 1900, and around 150,000 have been marketed. Lead and cadmium emissions have been multiplied by 20 between 1875 and 1975, with considerable damages: in 2013 in China, the magazine *China Weekly*, revealed that nearly half of the rice sold in Guangzhou was strongly contaminated by water polluted by cadmium, recalling what happened in Japan, where, in 1980, 10% of the rice was inedible for the same reason (McNeill 2000, pp. 57–58). Lake Baïkal under the former USSR is another famous example of an ecosystem degradation by industry. Considered as one of the clearest fresh water lakes in the world, it provided untreated clean water to its inhabitants until cellulose plants were established on its shores in 1958 and discharged their effluents. Even though they closed in 2013, the addition of waste dumped into the waters and of pollution by transports have deeply altered this fragile and unique ecosystem (Hughes 2009, p. 191)

1.5 A Global and Cumulative Pollution

The second main feature of post-1945 pollution is its globalisation. This is due partly to the intensification, but also to the diffusion of industry, chemicals, and economic growth all over the world. Up to the middle of the 19th century, pollution was mainly concentrated in some urban or industrial regions, even though it was less and less contained and started to cover larger areas. It now became worldwide: no place on earth, no ecosystem remained undisturbed. Lead coming from petrol has been found in Greenland ices, and snowflakes on the South Pole have revealed traces of insecticides (Ramade 2000). The globalisation in the depletion of the ozone layer by chlorofluorocarbons, for instance, was dramatically revealed in 1985 when British scientists announced the existence of a hole over the Antarctic. This led to the Montreal Treaty in 1987, that planned to reduce by 50% in ten years the emissions of CFC. But despite its implementation and further strengthening of the legislation, the weakening of the ozone layer continues. The omnipresence of plastics, whose use has increased 20-fold in the past half century, is another sign of this globalisation: an estimated 150 million tons of plastic can already be found in all the oceans, and, at the current rate, this amount could double by 2050 (Ellen MacArthur Foundation 2016).

Moreover, what is now better perceived is the cumulative effect of pollution. As such, it is of course not new: the pollution by alkali fumes around St Helens in Great Britain in the 19th century did accumulate in the soil and along the food chain. But this is far better understood in the 20th century, mainly because of the spread of pollution in most ecosystems. *Silent Spring*, published with great success in 1962 by Rachel Carson, is probably one of the earliest (and certainly one the most famous) studies of this phenomenon. Carson indeed demonstrates how the widespread use of DDT in agriculture, but also to destroy mosquitoes and flies in cities, was responsible for the dissemination of long-lasting pollutants, that would accumulate in the species situated high on the food chain. By the time the book was published, DDT was found in mothers' milk but also in penguins in the Antarctic; birds, especially, were threatened (the reproduction of pelicans or ospreys was put at risk), leading to a possible spring deprived of their songs… Even though the use of DDT has been banned in the United-States and in Europe, the former were producing pesticides in the 1990s at a rate thirteen thousand times faster than at the beginning of the 1960s (Radkau 2008, p. 295).

One of the most common long-lasting pollutions of the environment comes from metals such as lead or mercury, emanating from industries, dispersed in the air or water, and accumulating in the biosphere. The case of Minamata bay, in Japan, is a particularly well-documented example. In 1910, the Nippon Chisso company established there a chemical factory, that started rejecting mercury in the water in 1932. About a decade later, strange things started to happen to animals: fish died by hundreds, seagulls collided with walls and electric wires, and cats had an erratic behaviour, contracting the "disease of the dancing cats" before dying. Soon afterwards, around 1956, humans started to be affected: children suffered from brain lesions and were born deformed. It became clear as early as the late 1950s that mercury was at the origin of these disasters, but the company was so important for the economy of the area that it was difficult to fight it. Mercury only ceased to be discharged at the end of the 1960s, while the company was convicted a decade later. In 1990, about a hundred people had died from the "Minamata disease" while between 4 and 5 thousand were affected (Bouguerra 1997). One could add examples of nuclear pollution provoked by catastrophes such as those of Chernobyl or Fukushima.

Finally, this cumulative effect can be explained by the invisibility of most pollutions nowadays. Once again, as we have seen, they existed in the past. But then, the attention was focused on more visible nuisances: fumes, stinking waters, etc. In most developed societies today, these nuisances have been mainly displaced or covered. The population is now more and more concerned with other and more insidious forms of pollution, that are sometimes revealed belatedly. According to Herzlich and Pierret (1984), in France in the 1960s, the fear of epidemics and infectious diseases gave way to fears about the consequences of civilisation. Joachim Radkau agrees with such a statement, and believes that Western societies in general developed in the course of the 1950s–1960s a renewed concern for environmentally induced diseases, that was partially focused on cancer and the effects of nuclear technology (Radkau 2008, pp. 266–268). One has nonetheless to wait for the 1970s for a general concern over invisible pollutions, giving birth to environmental movements. In that respect, the cases of DDT and Minamata are two turning points for the United-States and Japan. In Germany, a particularly telling case is the issue of the *Waldsterben* (death of forests), which made the newspapers headlines in 1981 and was decisive in the political organisation of ecologists in the country. As early as the 1850s, scientists had analysed the effects on forests of sulphur dioxide discharged in the atmosphere by iron mills in the vicinity of Dresden. Experiments were made on plants and cows to demonstrate the deleterious effect of such pollution—but these very innovative researches stumbled on the impossibility to determine a threshold of harmfulness. The issue reappeared now and then, but never with the intensity of the 1980s: by then, the pollution was said to be global; forests were dying because of sulphur dioxide emissions not only in Germany, but also all over the world. It was one of the first times when public opinion became aware of the possibility of global ecological catastrophes (Brüggemeier 2002).

That mankind was now capable of disrupting some of the basic cycles of nature and of transforming it on a global scale, suggested the coinage of the word "Anthropocene" in 2000 by Paul Crutzen, Nobel prize in chemistry: it described a new geological era where mankind could be considered as a geological agent per se. This notion, which is still not accepted by the international associations of geology, has led to multiple controversies. It is first difficult to determine a date when it would begin, even though a slight consensus tends towards the beginning of the industrial era at the end of the 18th century. By considering mankind as a whole, the notion also tends to disregard regional differences in the impact people have on the earth. It may also give too much prominence to economic and technological factors, while neglecting political and ideological decisions (Bonneuil and Fressoz 2013). It has nonetheless the benefit of alerting on the scale of the human impact on the earth as a whole, and on its acceleration in the last half-century or so.

Strong difficulties are thus to be faced by mankind. As Turner (1990) has shown, the three main forces of environmental change are population growth, technological innovation and socio-cultural organisation. None of them is easy to change and adapt to reach a sustainable relation with our planet. And pollution is only one of the many anthropic impacts on the environment of the earth: water control, deforestation, waste treatment, the building of lands and the destruction of biodiversity, among others, are issues that need to be addressed. But this rather bleak prospect should not obscure the efforts made, and the successes encountered, in the fight against environmental degradation. The 2005 Millennium Ecosystem Assessment, commissioned by the United Nations, offers in that sense some hope and prospects for the future: without hiding the complex implementation of solutions, it suggests some scenarios that could prevent further degradation or even allow a partial recovery of some damaged ecosystems (Millennium Ecosystem Assessment 2005).

References

Bernhardt C, Massard-Guilbaud G (eds) (2002) Le Démon moderne. La pollution dans les sociétés urbaines et industrielles d'Europe. Presses Universitaires Blaise-Pascal, Clermont-Ferrand

Bonneuil C, Fressoz JB (2013) L'événement anthropocène. Seuil, Paris
Bouguerra ML (1997) La pollution invisible. PUF, Paris
Boutron C, Candelone JP, Hong S (1993) Le plomb dans les neiges et glaces du Groënland. Pollution atmosphérique 139. http://lodel.irevues.inist.fr/pollution-atmospherique/index.php?id=4271. Accessed 30 May 2016
Brüggemeier FJ (2002) Le dépérissement de la forêt (*Waldsterben*): construction et deconstruction d'un problème d'environnement. In: Bernhardt C, Massard-Guilbaud G (eds) Le Démon moderne. La pollution dans les sociétés urbaines et industrielles d'Europe. Presses Universitaires Blaise-Pascal, Clermont-Ferrand, pp 75–90
Carson R (1962) Silent spring. Houghton Mifflin, Boston
Cavert WM (2016) The smoke of London, energy and environment in early modern city. Cambridge University Press, Cambridge
Cokayne E (2007) Hubbub, filth, noise and stench in England 1600–1770. Yale University Press, New Haven
Douglas M (1966) De la souillure. Essai sur les notions de pollution et de tabou. La Découverte, Paris [original English edition 1967, French edition 1966]
Ellen MacArthur Foundation (2016) The new plastics economy. World Economic Forum. http://www3.weforum.org/docs/WEF_The_New_Plastics_Economy.pdf
Fournier P (2002) De la souillure à la pollution. Un essai d'interprétation des origines de l'idée de pollution. In: Bernhardt C, Massard-Guilbaud G (eds) Le Démon moderne. La pollution dans les sociétiés urbaines et industrielles d'Europe. Presses Universitaires Blaise-Pascal, Clermont-Ferrand, pp 33–56
Guillerme A, Lefort AC, Jigaudon G (2004) Dangereux, insalubres et incommodes. Paysages industriels en banlieue parisienne, XIXe-XXe siècles. Champvallon, Seyssel
Hamblin JD (2013) Environmental dimensions of World War II. In: Zeiler TW (ed) A companion to world war II. Wiley-Blackwell, Malden (Mass.), pp 698–716
Herzlich C, Pierret J (1984) Malades d'hier, malades d'aujourd'hui. Payot, Paris
Hong S, Candelone JP, Patterson CC et al (1996) History of ancient copper smelting pollution during roman and medieval time recorded in Greenland ice. Science 272:246–249
Hughes JD (2009) An Environmental History of the World. Humankind's changing role in the community of life. Routledge, London
Hupy J (2008) The environmental footprint of war. Environ Hist 14:405–421
Jancovici JM (2002) L'avenir climatique. Seuil, Paris
Le Roux T (2011) Le Laboratoire des pollutions industrielles. Paris 1770–1830. Albin Michel, Paris
Massard-Guilbaud G (2010) Histoire de la pollution industrielle, France, 1789–1914. Éditions de l'École des Hautes Études en Sciences Sociales, Paris
Masson-Loodts I (2014) Paysages en bataille: les séquelles environnementales de la Grande guerre. Nevicata, Bruxelles
McNeill J (2000) Something New Under the Sun. Norton, New York
Millennium Ecosystem Assessment (2005) Ecosystems and Human Well-being: Synthesis. Island Press, Washington
Parmentier I (2008) Histoire de l'environnement en pays de Charleroi, 1730–1830. Pollution et nuisances dans un paysage en voie d'industrialisation. Académie Royale de Belgique, Bruxelles
Pfister C (1995) Das 1950er Syndrom: Der Weg in die Konsumgesellschaft. Haupt, Bern
Poulussen P (1991) Ennuis de voisinage et pollution de l'environnement. In: van der Stock Jan (ed) La Ville en Flandres Culture et Société 1477–1787. Crédit communal, Bruxelles
Radkau J (2008) Nature and Power. Cambridge University Press, Cambridge
Ramade F (2000) Dictionnaire encyclopédique des pollutions: les polluants, de l'environnement à l'homme. Ediscience international, Paris

Steffen W (2005) Global change and the earth system: a planet under pressure. Springer, New York

Stellman JM, Stellman SD, Richard C et al (2003) The extent and patterns of usage of agent orange and other herbicides in Vietnam. Nature 422:681–687

Tsutsui W (2003) Landscapes in the dark valley: toward an environmental history of wartime Japan. Environ Hist 8(2):294–311

Turner BL (1990) The Earth as transformed by human action. Global and regional changes in the Biosphere over the past 300 years. Cambridge University Press, Cambridge

Chapter 2
From Microbial Ecology to Microbial Ecotoxicology

Esmaeil Shahsavari, Arturo Aburto-Medina, Leadin Salah Khudur, Mohamed Taha and Andrew S. Ball

Abstract Because of their ubiquity, abundance and metabolic activities, microorganisms play a crucial role in the biogeochemical cycling of elements in the environment. Any perturbations in the activity and diversity of the microbial community are likely to lead to significant impacts in terms not only of biogeochemical cycling but also ecosystem resilience. Human activities and industrialization have resulted in the release of millions of tonnes of chemicals and pollutants into the environments; some of these are toxic to living organisms. However there is a lack of information about the toxicity effects at the ecosystem level; where toxicity tests have been included in studies the basis has been the use of target species including plants (e.g. radish germination), worms (e.g. earthworm survival) and microbes (e.g. the Microtox bioassay test) to evaluate the effect of the pollutant on the target organisms. Microbial ecotoxicology represents an emerging discipline that encompasses microbial ecology, microbial toxicology, chemistry and physics and that offers great potential in the assessment of the fate and impact of environmental pollutants at the ecosystem level. In this introduction we discuss the importance of microbial ecology together with some of the advantages of the application of the recently established microbial ecotoxicology discipline in order to reliably assess the impact of contamination on the resilience and the functionality of the microbial community.

Keywords Biosensors · Bioreporters · Bioassays · Microorganisms · Ecotoxicology · Next generation sequencing · Environmental pollution

E. Shahsavari · A. Aburto-Medina · L.S. Khudur · M. Taha · A.S. Ball (✉)
School of Science, Centre for Environmental Sustainability and Remediation,
RMIT University, Bundoora, VIC 3083, Australia
e-mail: andy.ball@rmit.edu.au

M. Taha
Department of Biochemistry, Faculty of Agriculture, Benha University,
Moshtohor, Toukh 13736, Egypt

C. Cravo-Laureau et al. (eds.), *Microbial Ecotoxicology*,
DOI 10.1007/978-3-319-61795-4_2

2.1 Introduction

Microorganisms are the most abundant organisms on the planet, with recent estimates suggesting that the total numbers of microbes on the planet is between 9.2×10^{29} and 31.7×10^{29} (Kallmeyer et al. 2012). Bacteria represent the most diverse group of organisms; estimates suggest the number of bacterial species in the world range from 10^7 to 10^9 (Dykhuizen 1998; Curtis et al. 2002). Given these facts it is not surprising that microbes collectively exhibit the greatest metabolic diversity. Because of their ubiquity, abundance and metabolic activities, microorganisms play a crucial role in the biogeochemical cycling of elements in the environment. Understanding the role of microorganisms in the biogeochemical cycling is complex, requiring research into not only the structure and the function of microbial communities but also their synecological interactions. Historically this has taken place through a range of culture-dependent methods, which led to the recognition of the extraordinary diversity of microbial life. However, with the application of molecular microbial techniques, estimates of microbial diversity have increased dramatically. In particular, the advent of low cost, next generation sequencing technologies has led to an exponential increase in sequence-based microbial community studies investigating taxon diversity and community structure (e.g. via rRNA gene analysis) and/or microbial function via metagenomics of uncultivated microorganisms present within the environment. Such approaches have revealed a diverse wealth of hitherto unknown microbial taxa and provided new understanding of the ecological and biological functions and adaptations of environmental microbes. What is required now is to link this understanding of microbial diversity and complexity to ecosystem function. In natural environments, microorganisms interact with both biotic and abiotic components of their ecosystems; these interactions are essential for ecosystem function with key specific functions including biogeochemical cycling, biodegradation of pollutants and the impacts of microbes upon the activity and health of plants and animals, including humans. Defining the specific role of individual microorganisms in the environment is complex, due in part to the metabolic flexibility and diversity within individual species, and additionally by functional redundancy whereby diverse species can carry out the same biological activity.

In summary, microbial communities play a central role in ecosystem functioning. Any perturbations in the activity and diversity of the microbial community are likely to lead to significant impacts in terms of not only biogeochemical cycling but also ecosystem resilience. It is therefore not surprising that the ecological impacts of pollutants at the microbial scale and on the various functions that they carry out in the ecosystem together with the role of microbial communities in the ecodynamics of the pollutant-impacted ecosystem have formed the basis of microbial ecotoxicology (Ghiglione et al. 2016), a discipline that has emerged from microbial ecology. Here we discuss the importance of microbial ecology together with some of the advantages of the application of the recently established microbial ecotoxicology discipline in terms of assessing changes in the diversity and functionality of the microbial community in pollutant-impacted environments. Specifically, we examine environmental pollution and the role of microorganisms in the removal of these pollutants. Following this we briefly review the techniques available for microbial ecology before looking at the development of microbial ecotoxicology and the use of biosensors before examining future prospects for this emerging area of research.

2.2 Tools Used in Microbial Ecology

A variety of methods have been developed to study microbial ecology and in particular to investigate microbial diversity and functions. Traditional methods involving the culturing of indigenous microflora to study microbial processes (e.g. plate counting) have been widely employed; however it is now well known that less than 1% of all environmental microorganisms can be cultured (Amann et al. 1995). Recent developments in molecular microbial ecology have led to new insights into microbial ecology and this has led to a far better understanding of the ecology of microbial communities in different matrices. Characterization of the activity and diversity of microorganisms in the environment has been an active area of research due to the crucial role of microbial taxonomy and functional diversity in the ecosystem. Many methods and approaches have been developed in order to allow microbiologists to better assess microbial diversity in natural ecosystems. Among these approaches are a number of classic culture-based techniques, including:

- Dilution plating and culturing methods developed using a variety of culturing techniques and culture media designed to increase the growth of certain microbial species (Boehm et al. 1997; Hill et al. 2000).
- Community-level physiological profiling, which can be performed by the BIOLOG® system. This has been used widely to analyse microbial communities based on the ability of microbes to utilize different carbon sources (Lehman et al. 1995; Hill et al. 2000).

Alternative, culture-independent techniques based either on the biochemical or nucleic acid composition of microorganisms or fluorescent microscopic approaches have also been developed and successfully applied in microbial ecology. These techniques include:-

- Phospholipid fatty acid (PLFA) analysis. Phospholipid fatty acids are essential components of the cell membrane of microorganisms which break down rapidly after cell death, serving as an indicator to distinguish between living and dead organisms. In addition, different PLFA profiles vary according to the composition of the microbial community, making this technique useful in distinguishing between microbial communities (Hill et al. 2000).
- Nucleic acid approaches which are the most widely used and arguably the most useful tools for studying microbial ecology as nucleic acids are present in all

forms of life (Woese et al. 1990) and contain unique and highly conserved regions. For example, fluorescent in situ hybridization (FISH), has been used to study microbial communities due to the ability of this technique to determine and quantify specific microbial groups (Watanabe 2001).

Nucleic acids approaches (mostly DNA, but also RNA) have often been used to underpin microbial community analysis. Some of the methods developed, such as FISH, microarray and whole metagenome sequence analysis (metagenomics) evaluate nucleic acids directly, while others, such as denaturing gradient gel electrophoresis (DGGE) terminal restriction fragment length polymorphism (T-RFLP), ribosomal intergenic spacer analysis (RISA) or automated RISA (ARISA) and sequence analysis of 16 rDNA gene libraries require PCR to increase copies of a target gene for easier detection (Nakatsu 2007) (Fig. 2.1). Quantitative PCR (qPCR) is another useful, cost effective, sensitive and high throughput tool to study the abundance and expression of taxonomic or functional genes. Unlike other approaches, in this method the target genes can be fully quantified (Shahsavari et al. 2016).

Another technique used to establish the function of microorganisms in ecosystems is stable isotope probing (SIP); for example this technique has been used to identify the microorganisms responsible for the aerobic degradation of phenol (Manefield et al. 2002), the consumption of methanol by microorganisms (Radajewski et al. 2000) and the anaerobic degradation of benzene and toluene

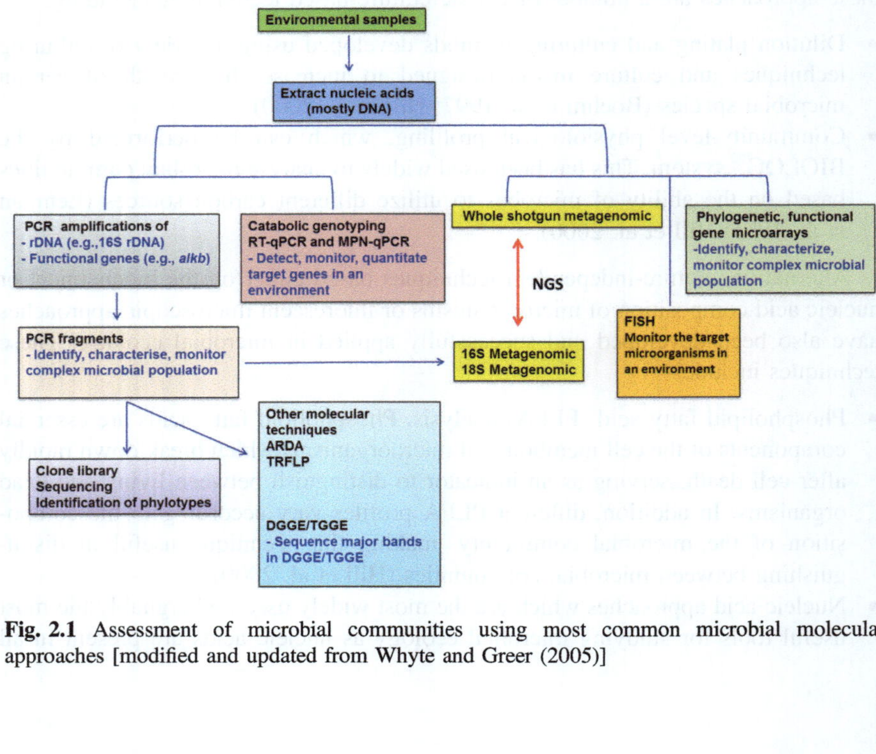

Fig. 2.1 Assessment of microbial communities using most common microbial molecular approaches [modified and updated from Whyte and Greer (2005)]

Table 2.1 The main molecular tools used to assess microbial communities in polluted environments (Kasai 2011; Hirsch et al. 2010; Malik et al. 2008)

Method	Advantages	Disadvantages
RISA/ARISA	Rapid and simple rRNA fingerprinting Highly reproducible	Limited database for ribosomal intergenic spacer sequences
ARDRA	Simple Highly reproducible	Limited resolution Sequence information unavailable
Clone library and sequencing	Possible detection of species	Time consuming Expensive
DGGE/TGGE	Inexpensive Possible to sequencing the band to determine the related species High resolution (1 bp)	Overestimates diversity Several bands may come from one species Assessment of only 1–2% of the microbial population
TRFLP	Simple High reproducibility	Requires expensive equipment. Requires multiple restriction enzymes Sequence information unavailable Several peaks form from one species
FISH	DNA isolation and PCR bias independent	Limited number of probes available (about 3) Background fluorescence interferes with detection of organisms Probe permeability
Microarray technologies	High throughput PCR bias independent	Expensive Non-specific hybridisation Applicable to known sequences only
Metagenomic analysis	Assessment of whole microbial community Much more information provided	Expensive Generates huge amount of data Requires high-performance computing and automated software

(Aburto and Ball 2009; Cupples 2011; Herrmann et al. 2010; Kunapuli et al. 2007; Oka et al. 2008) among several other contaminants.

However, whether traditional, molecular approaches or NGS are to be employed, consideration must be given to the fact that each method has advantages and disadvantages (Table 2.1). Overall, due to the limitations associated with both traditional (culture-based methods) and molecular methods, a combination of both approaches is most often desirable in microbial ecological studies (Zhang et al. 2012).

Among the different molecular methods used, some have become widely used in microbial ecology; for example DGGE has been used over the past 20 years as a community fingerprinting technique, while over the past 5 years metagenomics has become the dominant analytical protocol. Table 2.1 provides details of these various approaches. According to Whyte and Greer (2005), more than 1000 papers have been published by authors that have used DGGE for the analysis of various environmental microbial communities over a 10 year period. The author's estimation using Google Scholar and the search terms "environmental pollution" and "DGGE" resulted in 16,100 records by April 2016. DGGE has been used extensively since its introduction (Muyzer et al. 1993) to assess the effect of different chemical compounds such as petroleum hydrocarbons (Aburto-Medina et al. 2012; Adetutu et al. 2013; Shahsavari et al. 2013; Simons et al. 2013), tetrachloroethene (Patil et al. 2013) and metals on natural microbial communities (Reith et al. 2016).

DGGE and other gradient gel electrophoresis methods separate PCR-amplified DNA fragments from environmental samples (e.g. soil) based on differences in the GC content of the amplified gene (e.g. 16S rDNA). Sequences with different GC content show differential mobility through a DNA-denaturing gel (Whyte and Greer 2005). Denaturing conditions can be made with the use of urea and formamide in DGGE or by temperature in the case of TGGE. The main advantages of DGGE/TGGE are the ability to excise and directly sequence bands of interest which can then be compared with available sequences in online databases (GenBank or EMBL) to identify the putative microorganism. Also, DGGE is inexpensive and many samples can be run at the same time (for details about full DGGE's protocol see Green et al. 2010; Whyte and Greer 2005). Like other molecular microbial tools, DGGE has some disadvantages (Table 2.1) and the limitations of using this method should be considered when employing this technology for the evaluation of microbial communities.

The latest technique used by microbial ecologists is metagenomics. This is a preferred technique as it provides vast information that cannot be obtained with any other technique. The importance and ubiquity of this technique is such that it has been applied in many disciplines such as medicine, agriculture, energy production, bioforensics and bioremediation.

Metagenomics is defined as the analysis of DNA from microbial communities in environmental samples without the requirement for microbial culturing (Oulas et al. 2015). Historically, it was categorized as PCR independent analysis; however metagenomic analysis can be carried out following PCR amplification of certain genes of interest (e.g. 16S rDNA or 18S rDNA). Recently whole shotgun metagenomics has been applied in different environments including those impacted by pollution. Perhaps the greatest advantage of metagenomics is that it also provides information on the functional gene composition of microbial communities rather than merely phylogenetic surveys. It therefore provides the opportunity to address two key microbial ecological questions: which microorganisms are present and what are their roles? Thus, the information relating to functional genes may potentially reveal novel enzymes and biocatalysts (Thomas et al. 2012).

Recent advances in high-throughput sequencing methods, also called next generation sequencing (as opposed to Sanger sequencing) have revolutionized metagenomic studies. The first NGS platform was developed by Roche (formerly Life Sciences), 454 pyrosequencing in 2005. Later, many sequencing platforms such as Illumina, Ion Torrent, PacBio and SOLiD were developed by different vendors (van Dijk et al. 2014); however, the Illumina platforms (e.g. HiSeq or MiSeq) have become the most popular among researchers.

NGS technologies have been widely used recently to evaluate the effect of pollutants on microbial communities and their interactions, making them essential tools in microbial ecology (Adetutu et al. 2015; Costa et al. 2015; Shahi et al. 2016; Yergeau et al. 2012; Fang et al. 2013, 2014; Kao et al. 2016; Gołębiewski et al. 2014).

In summary, it can be seen that microbial ecology represents a key discipline in the field of contaminated environments. It is also apparent that the development of culture independent approaches has led to a much greater comprehension of microbial activity and diversity in the environment and the impact that pollutants have on both the activity and diversity of the microflora.

2.3 Environmental Pollution

Environmental pollution is a term which refers to the release, discharge or disposal of various pollutants and contaminants (solid, liquid or gases; physical, chemical and biological) as a consequence of pollution diffusion and industrial processes without appropriate treatment to meet the standard regulations required prior to discharge into the environment (water, air and soil), causing a threat to humans, animals, plants and to the entire environment (Wu et al. 2010). The current scale of environmental pollution is unprecedented; a recent Cornell University research survey concluded that about 40% of human deaths worldwide were caused by different types of pollution in the different environmental compartments (water, air and soil) (Pimentel et al. 2010). The situation is worse in developing countries. Unfortunately, approximately 90–95% of untreated urban sewage, toxic and carcinogenic chemicals generated are discharged directly into running waters in developing countries without appropriate treatment. Pollution is also a serious threat to water quality in developed countries, including the US, U.K. and Australia. In the UK it has been estimated that 110 tonnes of benzene and over one million tonnes of petroleum hydrocarbons are spilled into terrestrial ecosystems every year, mainly through vehicle emissions and from the chemical industry (Fahy et al. 2006). In Australia it is estimated that around 27 million litres of crude oil waste are illegally dumped annually without appropriate treatment (Aleer et al. 2011). Polluted water, air and soil all impact the biosphere, including humans due to the dependency on safe environmental conditions of the food-chain. Physical, chemical and biological contaminants that are directly or indirectly discharged into the environment will have an impact on living organisms.

The removal of these contaminants from the environment represents an essential step in restoring ecosystem functions and services that we depend on to maintain both ecosystem- and human-health. Biological systems and microorganisms in particular are playing an increasing role in the clean-up or remediation of contaminated sites (Arias et al. 2005). Bioremediation is already widely practiced as a safe, environmentally friendly approach for the clean-up of various pollutants such as petroleum hydrocarbons (Lors et al. 2012) heavy metals (Mejáre and Bülow 2001), pesticides (Scott et al. 2008), and waste products from the dye (Senan and Abraham 2004), and pulp and paper industry (Paszczynski and Crawford 1995).

Microorganisms in particular are useful in the degradation of environmental contaminants in two general ways:

- Many microorganisms, including those indigenous to the particular environment can degrade the pollutants or transform them into less toxic compounds, reducing the concentration of pollutants and importantly reducing their toxic effects in the environment.
- Microorganisms themselves can serve as important indicators of environmental pollution since any environmental impact is translated in community changes that can be recorded. Moreover, since microorganisms are ubiquitous in the environment, a community response to environmental pollutants can be observed.

2.4 Ecotoxicological Evaluation

Traditional chemical analyses are the usual way to evaluate the degradation of pollutants and monitor the contamination levels as well as to assess the impacts on various organisms. However, these techniques (gas chromatography-mass spectrometry, GCMS for hydrocarbons and inductively coupled plasma mass spectrometry, ICPMS for metals) are not effective in assessing the efficacy of the clean-up method in terms of toxicity reduction and bioavailability, or to evaluate the ecotoxicological outcomes of the process on the ecosystem at the treated sites (Molina-Barahona et al. 2005). A reduction in the concentration of contaminants does not necessarily mean a reduction in their toxicity due to many possible reasons as has been shown previously (Khudur et al. 2015). Therefore, integration between chemical analytical data, ecotoxicological assessments and the evaluation of the microbial community is required, though seldom carried out, to evaluate the efficiency of the approach used to treat the contaminated area and the resultant environmental outcomes in terms of its effects on ecosystems and humans.

The key indicators of the risk posed by chemicals to human health and other organisms in the environment is related to the bioavailability of the contaminant, which can be defined as the difference between the amount of the contaminant to which an organism is exposed and the actual dose of the substance the organism receives (Naidu 2011). Many factors determine the environmental bioavailability of a contaminant. These factors might include (a) the solubility of the contaminant in the soil solution (Lanno et al. 2004), (b) the chemical mass transfer rates, specifically the molecular weight of the desorbed fraction (Cornelissen et al. 1998), and (c) the type of organism exposed to the chemicals (Reid et al. 2000).

To examine the impact of chemicals on the biota, a variety of toxicity test methods and procedures have been proposed (Dutka and Kwan 1981). Inhibition of natural bacterial bioluminescence, which is known as the Microtox test, has been developed as a cost-effective, easy pre-screening test, which is based on measuring the inhibition in light emitted by a marine bacterial species, *Aliivibrio fischeri* (also known as *Vibrio fischeri* and formerly known as *Photobacterium phosphoreum*) (Kamlet et al. 1986). The bioluminescence results from a complex set of energy-producing reactions controlled by the expression of six genes with light output induced at high cell density. Therefore, inhibition of the enzymes by the pollutants alters the rate of gene expression and subsequently the amount of bioluminescence emitted.

The Microtox test is used to determine the toxicity of a contaminant and is commonly applied as a quick bioassay test, in an attempt to alert monitoring agencies of potential toxic conditions, as well as the rapid assessment of the changes in environmental quality. The endpoint of this test is the determination of the concentration of a contaminant which causes a reduction in the bioluminescence by 50% after a certain time, usually 5, 10 and 15 min. This is referred to as Effective Concentration 50 (EC_{50}) (Kamlet et al. 1986). On the other hand, a chronic toxicity test aims to evaluate the toxicity of a substance by detecting changes in physiological functions and usually performed on macroscopic organisms.

2.5 Microbial Ecotoxicology as a New Discipline

Microbial ecotoxicology is a new discipline that encompasses microbial ecology, microbial toxicology, and chemistry and physics. This new discipline may be distinguished from microbial toxicology since the latter has usually looked at the toxic effects of compounds on a certain microorganism and infer that toxicity to the whole community while microbial ecotoxicology aims to compile all the different approaches such as analytical methods, enzymatic measurements, toxicity measurements and culture-independent methods among others in order to have a more accurate assessment of the toxic compounds in the whole community. This is a significant development from simple microbial toxicology work, requiring the study of ecotoxicological approaches at the community level. However the outcome of microbial ecotoxicology is of ecological relevance. Consequently this discipline has recently been proposed and is currently being consolidated (Ghiglione et al. 2016, 2014; Gu and Wang 2014).

For some time microbial ecologists have been conducting and reporting successful studies on the bioremediation of toxic compounds (Aburto et al. 2009; Aburto-Medina et al. 2012; Shahsavari et al. 2013, 2015a, b). Most of these studies have relied on the use of analytical methods such as gas chromatography-mass spectrometry to confirm the degradation of the contaminants; however there is also a need to assess the toxicity of the remaining compounds following remediation with several toxicity methods such as the Microtox bioassay and the

Toxi-chromoPad test (Ahtiainen et al. 2002). These tests have been used for over twenty years (van Beelen and Doelman 1997) and can be classified as single species tests, carbon and nitrogen transformations, enzymatic tests (Margesin et al. 2000), biomass measurements (Margesin et al. 2000) and tests assessing changes in microbial diversity (Khudur et al. 2015).

Furthermore, the tests have been used to assess the toxicity of different contaminants such as herbicides (Bonnet et al. 2007, 2008), hydrocarbons (Płaza et al. 2009), dyes (Ogawa et al. 1988), wastewater (Sazykin et al. 2016) and heavy metals (Preston et al. 2000) among others. Some of the advantages of microbial toxicology tests are:

- They are generally simple to carry out.
- Some assays are almost as accurate as the chemical methods.
- Importantly they provide information on the biological effects of contaminants.
- This approach allows the establishment of toxicologically safe endpoints.

In parallel, other tests that have been carried out may provide a better representation of the status of the whole community and they include the measurement of growth, basal respiration and enzymatic potential in order to evaluate the effect of contaminants on the microorganisms in the ecosystem. Specifically, previous studies have reported the use of microbial biomass (Ingham et al. 1986; Muñoz-Leoz et al. 2011), photosynthesis (Sabater et al. 2007) basal respiration (Gong et al. 2001; Kumpiene et al. 2009; Muñoz-Leoz et al. 2011), enzyme activities (Gong et al. 2001; Kumpiene et al. 2009; Liu et al. 2009; Mora et al. 2005; Muñoz-Leoz et al. 2011; Renella et al. 2008; Tscherko and Kandeler 1997) and nitrification (Muñoz-Leoz et al. 2011; Sverdrup et al. 2002) as indicators of the effects of contaminants such as hydrocarbons, metals, pesticides and explosives on the microbial community. Some of the enzymes that have been used as indicators of the microbial community status include dehydrogenases, ureases, arylsulfatases, phosphatases and β-glucosidases.

The measurement of enzyme activities is a good parameter to determine the toxicity of the matrix. Arylsulfatase, β-glucosidase and dehydrogenase were used to evaluate the toxicity of a heavy metal contaminated soil treated with organic and inorganic amendments (Mora et al. 2005). Similarly, Kumpiene and colleagues used enzyme activities as indicators in the phytostabilisaton of a Pb and Cu contaminated soil. Phosphatase, glycosidase, sulfatase and urease were measured and their increased activity indicated sustainable management of the treated soils since these enzymes are involved in organic matter decomposition and the biogeochemical cycle of macronutrients (Kumpiene et al. 2009). Another study also confirmed that the ratio of arylsulphatase to microbial biomass is a sensitive index to evaluate contamination (Tscherko and Kandeler 1997).

Soil basal respiration and substrate-induced respiration indicate the actual respiratory microbial activity and the maximum potential respiratory microbial activity respectively. These are also good indicators of soil microbial activity and have been measured to assess the effect of the fungicide tebuconazole (Muñoz-Leoz et al. 2011) and the explosive compound RDX (Gong et al. 2001) on the microbial community.

The Microtox, Microtox solid Phase test assay, the P450 reporter gene system and the Toxi-chromo Pad have also been very useful in evaluating the toxicity of hydrocarbon contaminated sediments (Mueller et al. 2003) and creosote contaminated soil (Ahtiainen et al. 2002). Although the Microtox result is based on a single microorganism, it has been extrapolated to the whole community, in order to overcome this drawback, biofilms have also recently been used as indicators of the effects of chemicals on the microbial community (Sabater et al. 2007) or to assess the quality of riverine systems (Burns and Ryder 2001).

Thus, the evaluation of the Microtox, the biogeochemical cycle parameters and biofilms is a good indication of the toxic effects of contaminants on the microbial community and confirms there is a strong link between microbial ecotoxicological studies and the microbial ecology of the system.

In summary, assessment of the activity of key microbial enzymes or processes within a biogeochemical cycle provides a good indication of the toxic effects of contaminants on the microbial community. For example nitrification has been used for microbial toxicological assessment of the impact of a contaminant. Smolders et al. (2001) used potential nitrification rate tests (PNR) to evaluate metal toxicity in metal salt–spiked soils, uncontaminated soils and field soil contaminated with metals from previous smelting activities. The authors concluded that nitrification was sensitive to metal, although they stated that overall it was not a useful soil assay. In another study Broos et al. (2007) measured substrate-induced nitrification (SIN) and substrate-induced respiration (SIR) in the top soils of 12 Australian soils amended with $ZnSO_4$ or $CuSO_4$. The median effect concentration (EC_{50}) values for Zn and Cu based on total metal concentrations varied between 107 and 8298 mg kg^{-1} for Zn and 108 and 2155 mg kg^{-1} respectively among soils. The results of this study showed significant relationships between the EC50 values for SIR and background Zn concentrations and the cation exchange capacity (CEC0 for Zn, and the presence of clay and log CEC for Cu.

It is also important to point out that molecular tools can detect changes in the microbial community that can be missed by the measurement of only community level end points (Widenfalk et al. 2008). Therefore, an ideal ecotoxicology study should include the assessment of the biogeochemical parameters and the use of molecular tools, especially with the advent of the next generation sequencing, which provides the largest amount of data (up to one billion short reads per run) with a relatively low cost (Metzker 2010; Schuster 2007).

Biological tools such as biosensors and biomarkers are extremely helpful to provide signals for potential damage in the environment and have been widely used in microbial toxicology (Hansen 2008; Hansen and Usedom 1997). Biosensors act as sensing systems and biomarkers, providing biochemical responses that indicate the initial level of damage and provide information in order to take precautionary action. Thus, the recognition of an early sign of environmental damage will prevent eventual larger damage to the environment.

Genetic engineering has provided the tools to introduce genes encoding enzymes and luminescent proteins into bacterial species that act as reporter genes (van der Meer and Belkin 2010). Some of these genes include the *lux* genes of bioluminescent bacteria such as *Aliivibrio fischeri*, *Vibrio harveyi* and *Photorhabdus luminescens* (formerly *Xenorhabdus luminescens*). Other reporter genes are the lucFF from the firefly luciferase and the *gfp* genes encoding the green fluorescent protein (GFP) from the jellyfish *Aequorea victoria* among others. An excellent review on the use of reporter and sensor proteins used in ecotoxicology studies has been prepared by Van der Meer and Belkin (van der Meer and Belkin 2010).

These recombinant microorganisms are usually termed bioreporters and they allow the quantification of bioavailability which is a direct indicator of the contaminants effect on living organisms (Farré and Barceló 2009). *Escherichia coli* has been extensively used as a bioreporter (Robbens et al. 2010), used to detect bioavailable iron (Bachmann 2003) ionic mercury (Xu et al. 2015), arsenic (Yoon et al. 2015) and zinc (Maderova et al. 2011) among other contaminants. Other engineered microorganisms include *Pseudomonas fluorescens* and *Pseudomonas putida* which have been used for the detection of halogenated compounds and heavy metals (Preston et al. 2000; Sütterlin et al. 2008; Weitz et al. 2001). A recent review also indicates the use of cyanobacteria as bioreporters (Mateo et al. 2015). Microalgae such as *Chlorella vulgaris, Chlorella fusca*, (formerly *Selenastrum capricornotum*) and *Dunaliella salina* are sensitive to pesticides and have also been used as toxicity indicators. Several reviews have compiled the advances of biosensors at different time points (D'Souza 2001; Rodriguez-Mozaz et al. 2003; Su et al. 2011; van der Meer and Belkin 2010). Furthermore biosensors have been recently used to assess the eco and genotoxicity of airborne emissions (Kováts and Horváth 2016) and to compare the ecotoxicological contamination of wastewaters in Russia and Germany (Sazykin et al. 2016). Some of the most common microorganisms used as biosensors to different contaminants are listed in Table 2.2.

Thus, another challenge of microbial ecotoxicology as an emerging discipline will be to merge and perform in parallel all the methods mentioned above (targeting single species and the whole community). This will provide a better assessment of the site toxicity and in turn provide crucial information about the success of any remediation intervention.

2.6 Future Aspects

So far, microbial ecotoxicology approaches include the measurement of biogeochemical parameters (photosynthesis, respiration, denitrification, decomposition, and enzymatic potential), the Microtox, biosensors and bioreporters in order to evaluate the effects of contaminants on the microbial community.

However, there is a requirement for a better toxicity indicator for the entire microbial community in the system in order to relate specific functions (e.g. enzyme activities) to certain groups of organisms and their abundance. With the advent of next generation sequencing, we are able to detect differences in functional genes at the community level *alkB* genes involved in degradation alkane and if we couple

Table 2.2 Microorganisms used as biosensors for a wide range of contaminants

Microorganism	Contaminant	Reference
Achromobacter	Surfactant detection	Taranova et al. (2002)
Acinetobacter	Phenol	Abd-El-Haleem et al. (2002)
Aliivibrio fischeri (formerly *Vibrio fischeri*)	Landfill leachate Oil polluted soil Insecticides: Ametryn	Thomas et al. (2009), Bundy et al. (2004), Farré et al. (2014)
Bacillus subtilis	Dyes Zinc, Pb and Cd	Ogawa et al. (1988), Kahru et al. (2005)
Chlorella vulgaris	Heavy metals Atrazine, simazine and diuron Herbicides	Durrieu and Tran-Minh (2002), Naessens et al. (2000), González-Barreiro et al. (2006)
Escherichia coli	Heavy metals Pollutants Zinc, Cu and Cd Aluminium Dioxins Wastewater Zinc bioavailability in soils PAHs Dibenzo-p-dioxins Arsenic Mercury	Vollmer et al. (1997), Bechor et al. (2002), Preston et al. (2000, Guzzo et al. (1992), Min et al. (2003), Sazykin et al. (2016), Maderova and Paton (2013), Gu and Chang (2001), Yoon et al. (2015), Xu et al. (2015)
Janthinobacterium lividum	Organic compounds and heavy metals	Cho et al. (2004)
Photobacterium leiognathi	Metals, pesticides, PAHs	Ulitzur et al. (2002)
Pseudomonas fluorescens	Zinc, Cu and Cd Olive mill wastewater Cu, Zn and 3,5-DCP Cu in soils Naphthalene	Preston et al. (2000), Mekki et al. (2008), Weitz et al. (2001), Maderova et al. (2011), King et al. (1990)
Pseudomonas putida	Benzalkonium chloride Cu, Zn and 3,5-DCP	Sütterlin et al. (2008), Weitz et al. (2001)
Raphidocelis subcapitata (formerly *Selenastrum capricornotum*)	Mercury Zinc, Pb and Cd	Juneau and Popovic (1999), Kahru et al. (2005)
Synechococus (cyanobacteria)	Marine oil spills	Brussaard et al. (2016)
Synechocystis	Herbicides	Shao et al. (2002)
Sphingomonas yanoikuyae	Fluorene	Bastiaens et al. (2001)
Spirostomum ambiguum (ciliate)	Heavy metals & hydrocarbons	Płaza et al. (2009)
Spirulina subsalsa	Heavy metals, triazinic herbicides, carbamate insecticides	Campanella et al. (2001)

(continued)

Table 2.2 (continued)

Microorganism	Contaminant	Reference
Symbiodinium	Herbicides: diuron and atrazine	Jones et al. (2003)
Tetrahymena pyriformis (ciliate)	Herbicides	Bonnet et al. (2007), (2008)
Tetrahymena termofila (ciliate)	Micotoxins	Benitez et al. (1994)
Tetrahymena pyriformis (ciliate)	Heavy metals	Gutiérrez et al. (2003)
Trichosporum cutaneum (fungi)	Wastewater BOD sensor Alkylbenzene sulfonate	Marty et al. (1997), Nomura et al. (1998)
Vibrio harveyi	Aluminium	Guzzo et al. (1992)
Vibrio aquamarinus	Wastewater	Sazykin et al. (2016)

this technique with microbial ecotoxicity tests, it will be easier to assess the toxicity level for the whole community. Therefore, one of the challenges in microbial ecotoxicology may be to incorporate and establish metagenomics as an aid in the collection of reliable toxicity levels.

Moreover, there is a need for bioremediation studies to conduct post-treatment toxicity tests in order to confirm the lack of toxicity at the site. Some bioremediation studies aim to reduce the contaminant below the local Environmental Protection Agency maximum limit; however, even if the contaminant concentration (or that of the intermediates) is within the desired limits it may still be toxic to several organisms.

Furthermore, apart from the classical toxicity indicators such as EC_{50}, LC_{50} and LD_{50}, it is important for microbial toxicological tests such as bioreporters to be accepted internationally in order to have a consensus on maximum toxicity levels. The establishment of microbial ecotoxicology as a new discipline should aid in such a task.

2.7 Conclusions

Molecular techniques such as clone libraries, DGGE, TRFLP and next generation sequencing combined with analytical methods (HPLC, GC-MS, ICPMS, etc.) have been extremely useful for microbial ecologists to identify the microorganisms responsible for a specific function such as the degradation of the contaminants or the production of a specific enzyme, while the analytical methods have helped to reliably measure chemical compounds of interest within a process.

Although microbial toxicology studies have also been performed previously, the results have generally been reported elsewhere in specialized toxicology journals.

Thus, one of the challenges of the recently established microbial ecotoxicology will be to bring together the results from the analytical methods, microbial toxicology and microbial ecology.

There is no doubt that the development of microbial ecotoxicology as a new discipline was both timely and crucial and its establishment will be fundamental to scientists assessing not only the impact of environmental contamination but also that of natural changes in the environment as well as the success of any remediation intervention.

References

Abd-El-Haleem D, Ripp S, Scott C, Sayler SG (2002) A luxCDABE-based bioluminescent bioreporter for the detection of phenol. J Ind Microbiol Biotechnol 29(5):233–237. doi:10.1038/sj.jim.7000309

Aburto-Medina A, Adetutu E, Aleer S, Weber J, Patil S, Sheppard P, Ball A, Juhasz A (2012) Comparison of indigenous and exogenous microbial populations during slurry phase biodegradation of long-term hydrocarbon-contaminated soil. Biodegradation 23(6):813–822. doi:10.1007/s10532-012-9563-8

Aburto A, Ball AS (2009) Bacterial population dynamics and separation of active degraders by stable isotope probing during benzene degradation in a BTEX-impacted aquifer. Rev Int Contaminacion Ambiental 25(3)

Aburto A, Fahy A, Coulon F, Lethbridge G, Timmis KN, Ball AS, McGenity TJ (2009) Mixed aerobic and anaerobic microbial communities in benzene-contaminated groundwater. J Appl Microbiol 106(1):317–328

Adetutu E, Weber J, Aleer S, Dandie CE, Aburto-Medina A, Ball AS, Juhasz AL (2013) Assessing impediments to hydrocarbon biodegradation in weathered contaminated soils. J Hazard Mater 261(0):847–853. doi:10.1016/j.jhazmat.2013.01.052

Adetutu EM, Gundry TD, Patil SS, Golneshin A, Adigun J, Bhaskarla V, Aleer S, Shahsavari E, Ross E, Ball AS (2015) Exploiting the intrinsic microbial degradative potential for field-based in situ dechlorination of trichloroethene contaminated groundwater. J Hazard Mater 300:48–57. doi:10.1016/j.jhazmat.2015.06.055

Ahtiainen J, Valo R, Järvinen M, Joutti A (2002) Microbial toxicity tests and chemical analysis as monitoring parameters at composting of creosote-contaminated soil. Ecotoxicol Environ Saf 53(2):323–329. doi:10.1006/eesa.2002.2225

Aleer S, Adetutu EM, Makadia TH, Patil S, Ball AS (2011) Harnessing the hydrocarbon-degrading potential of contaminated soils for the bioremediation of waste engine oil. Water Air Soil Pollut 218(1–4):121–130

Amann RI, Ludwig W, Schleifer K-H (1995) Phylogenetic identification and in situ detection of individual microbial cells without cultivation. Microbiol Rev 59(1):143–169

Arias ME, Gonzalez-Perez JA, Gonzalez-Vila FJ, Ball AS (2005) Soil health—a new challenge for microbiologists and chemists. Int Microbiol 8(1):13–21

Bachmann T (2003) Transforming cyanobacteria into bioreporters of biological relevance. Trends Biotechnol 21(6):247–249. doi:10.1016/S0167-7799(03)00114-8

Bastiaens L, Springael D, Dejonghe W, Wattiau P, Verachtert H, Diels L (2001) A transcriptional luxAB reporter fusion responding to fluorene in *Sphingomonas* sp. LB126 and its initial characterisation for whole-cell bioreporter purposes. Res Microbiol 152(10):849–859. doi:10.1016/S0923-2508(01)01268-2

Bechor O, Smulski DR, Van Dyk TK, LaRossa RA, Belkin S (2002) Recombinant microorganisms as environmental biosensors: pollutants detection by *Escherichia coli* bearing fabA′:lux fusions. J Biotechnol 94(1):125–132. doi:10.1016/S0168-1656(01)00423-0

Benitez L, Martin-Gonzalez A, Gilardi P, Soto T, de Lecea JR, Gutiérrez JC (1994) The ciliated protozoa *Tetrahymena thermophila* as a bionsensor to detect mycotoxins. Lett Appl Microbiol 19(6):489–491. doi:10.1111/j.1472-765X.1994.tb00989.x

Boehm MJ, Wu T, Stone AG, Kraakman B, Iannotti DA, Wilson GE, Madden LV, Hoitink HAJ (1997) Cross-polarized magic-angle spinning 13C nuclear magnetic resonance spectroscopic characterization of soil organic matter relative to culturable bacterial species composition and sustained biological control of pythium root rot. Appl Environ Microbiol 63(1):162–168

Bonnet JL, Bonnemoy F, Dusser M, Bohatier J (2007) Assessment of the potential toxicity of herbicides and their degradation products to nontarget cells using two microorganisms, the bacteria *Vibrio fischeri* and the ciliate *Tetrahymena pyriformis*. Environ Toxicol 22(1):78–91. doi:10.1002/tox.20237

Bonnet JL, Bonnemoy F, Dusser M, Bohatier J (2008) Toxicity assessment of the herbicides sulcotrione and mesotrione toward two reference environmental microorganisms: *Tetrahymena pyriformis* and *Vibrio fischeri*. Arch Environ Contam Toxicol 55(4):576–583. doi:10.1007/s00244-008-9145-2

Broos K, Warne MSJ, Heemsbergen DA, Stevens D, Barnes MB, Correll RL, McLaughlin MJ (2007) Soil factors controlling the toxicity of copper and zinc to microbial processes in Australian soils. Environ Toxicol Chem 26(4):583–590

Brussaard CPD, Peperzak L, Beggah S, Wick LY, Wuerz B, Weber J, Samuel Arey J, van der Burg B, Jonas A, Huisman J, van der Meer JR (2016) Immediate ecotoxicological effects of short-lived oil spills on marine biota. Nat Commun 7. doi:10.1038/ncomms11206

Bundy JG, Paton GI, Campbell CD (2004) Combined microbial community level and single species biosensor responses to monitor recovery of oil polluted soil. Soil Biol Biochem 36(7):1149–1159. doi:10.1016/j.soilbio.2004.02.025

Burns A, Ryder DS (2001) Potential for biofilms as biological indicators in Australian riverine systems. Ecol Manage Restor 2(1):53–64. doi:10.1046/j.1442-8903.2001.00069.x

Campanella L, Cubadda F, Sammartino MP, Saoncella A (2001) An algal biosensor for the monitoring of water toxicity in estuarine environments. Water Res 35(1):69–76. doi:10.1016/S0043-1354(00)00223-2

Cho J-C, Park K-J, Ihm H-S, Park J-E, Kim S-Y, Kang I, Lee K-H, Jahng D, Lee D-H, Kim S-J (2004) A novel continuous toxicity test system using a luminously modified freshwater bacterium. Biosens Bioelectron 20(2):338–344. doi:10.1016/j.bios.2004.02.001

Cornelissen G, Rigterink H, Ferdinandy MMA, van Noort PCM (1998) Rapidly desorbing fractions of PAHs in contaminated sediments as a predictor of the extent of bioremediation. Environ Sci Technol 32(7):966–970. doi:10.1021/es9704038

Costa PS, Reis MP, Ávila MP, Leite LR, de Araújo FMG, Salim ACM, Oliveira G, Barbosa F, Chartone-Souza E, Nascimento AMA (2015) Metagenome of a microbial community inhabiting a metal-rich tropical stream sediment. PLoS ONE 10(3):e0119465. doi:10.1371/journal.pone.0119465

Cupples AM (2011) The use of nucleic acid based stable isotope probing to identify the microorganisms responsible for anaerobic benzene and toluene biodegradation. J Microbiol Methods 85(2):83–91. doi:10.1016/j.mimet.2011.02.011

Curtis TP, Sloan WT, Scannell JW (2002) Estimating prokaryotic diversity and its limits. Proc Natl Acad Sci 99(16):10494–10499

D'Souza SF (2001) Microbial biosensors. Biosens Bioelectron 16(6):337–353. doi:10.1016/S0956-5663(01)00125-7

Durrieu C, Tran-Minh C (2002) Optical algal biosensor using alkaline phosphatase for determination of heavy metals. Ecotoxicol Environ Saf 51(3):206–209. doi:10.1006/eesa.2001.2140

Dutka BJ, Kwan KK (1981) Comparison of three microbial toxicity screening tests with the microtox test. Bull Environ Contam Toxicol 27(1):753–757. doi:10.1007/BF01611091

Dykhuizen DE (1998) Santa rosalia revisited: why are there so many species of bacteria? Antonie Van Leeuwenhoek 73(1):25–33. doi:10.1023/a:1000665216662

Fahy A, McGenity TJ, Timmis KN, Ball AS (2006) Heterogeneous aerobic benzene-degrading communities in oxygen-depleted groundwaters. FEMS Microbiol Ecol 58(2):260–270

Fang H, Cai L, Yang Y, Ju F, Li X, Yu Y, Zhang T (2014) Metagenomic analysis reveals potential biodegradation pathways of persistent pesticides in freshwater and marine sediments. Sci Total Environ 470–471:983–992. doi:10.1016/j.scitotenv.2013.10.076

Fang H, Cai L, Yu Y, Zhang T (2013) Metagenomic analysis reveals the prevalence of biodegradation genes for organic pollutants in activated sludge. Biores Technol 129:209–218. doi:10.1016/j.biortech.2012.11.054

Farré M, Barceló D (2009) Biosensors for aquatic toxicology evaluation. In: Barceló D, Hansen P-D (eds) Biosensors for environmental monitoring of aquatic systems: bioanalytical and chemical methods for endocrine disruptors. Springer, Berlin, pp 115–160. doi:10.1007/978-3-540-36253-1_5

Farré M, Fernandez J, Paez M, Granada L, Barba L, Gutierrez H, Pulgarin C, Barceló D (2014) Analysis and toxicity of methomyl and ametryn after biodegradation. Anal Bioanal Chem 373 (8):704–709. doi:10.1007/s00216-002-1413-9

Ghiglione J-F, Martin-Laurent F, Pesce S (2016) Microbial ecotoxicology: an emerging discipline facing contemporary environmental threats. Environ Sci Pollut Res 23(5):3981–3983. doi:10.1007/s11356-015-5763-1

Ghiglione J-F, Martin-Laurent F, Stachowski-Haberkorn S, Pesce S, Vuilleumier S (2014) The coming of age of microbial ecotoxicology: report on the first two meetings in France. Environ Sci Pollut Res 21(24):14241–14245. doi:10.1007/s11356-014-3390-x

Gołębiewski M, Deja-Sikora E, Cichosz M, Tretyn A, Wróbel B (2014) 16S rDNA pyrosequencing analysis of bacterial community in heavy metals polluted soils. Microb Ecol 67(3):635–647. doi:10.1007/s00248-013-0344-7

Gong P, Hawari J, Thiboutot S, Ampleman G, Sunahara GI (2001) Ecotoxicological effects of hexahydro-1,3,5-trinitro-1,3,5-triazine on soil microbial activities. Environ Toxicol Chem 20 (5):947–951. doi:10.1002/etc.5620200502

González-Barreiro O, Rioboo C, Herrero C, Cid A (2006) Removal of triazine herbicides from freshwater systems using photosynthetic microorganisms. Environ Pollut 144(1):266–271. doi:10.1016/j.envpol.2005.12.014

Green SJ, Leigh MB, Neufeld JD (2010) Denaturing Gradient Gel Electrophoresis (DGGE) for microbial community analysis. In: Timmis K (ed) Handbook of hydrocarbon and lipid microbiology. Springer, Berlin, pp 4137–4158. doi:10.1007/978-3-540-77587-4_323

Gu J-D, Wang Y (2014) Geomicrobial ecotoxicology as a new subject in environmental sciences is proposed. Ecotoxicology 23(10):1823–1825. doi:10.1007/s10646-014-1359-7

Gu MB, Chang ST (2001) Soil biosensor for the detection of PAH toxicity using an immobilized recombinant bacterium and a biosurfactant. Biosens Bioelectron 16(9–12):667–674. doi:10.1016/S0956-5663(01)00230-5

Gutiérrez JC, Martín-González A, Díaz S, Ortega R (2003) Ciliates as a potential source of cellular and molecular biomarkers/biosensors for heavy metal pollution. Eur J Protistol 39(4):461–467. doi:10.1078/0932-4739-00021

Guzzo J, Guzzo A, DuBow MS (1992) Characterization of the effects of aluminum on luciferase biosensors for the detection of ecotoxicity. Toxicol Lett 64:687–693. doi:10.1016/0378-4274(92)90248-I

Hansen P-D, Usedom A (1997) New biosensors for environmental analysis. In: Scheller FW, Schubert F, Fedrowitz J (eds) Frontiers in biosensorics II: practical applications. Birkhäuser Basel, Basel, pp 109–120. doi:10.1007/978-3-0348-9045-8_8

Hansen PD (2008) Biosensors and eco-toxicology. Eng Life Sci 8(1):26–31. doi:10.1002/elsc.200720228

Herrmann S, Kleinsteuber S, Chatzinotas A, Kuppardt S, Lueders T, Richnow H-H, Vogt C (2010) Functional characterization of an anaerobic benzene-degrading enrichment culture by DNA

stable isotope probing. Environ Microbiol 12(2):401–411. doi:10.1111/j.1462-2920.2009.02077.x
Hill GT, Mitkowski NA, Aldrich-Wolfe L, Emele LR, Jurkonie DD, Ficke A, Maldonado-Ramirez S, Lynch ST, Nelson EB (2000) Methods for assessing the composition and diversity of soil microbial communities. Appl Soil Ecol 15(1):25–36. doi:10.1016/S0929-1393(00)00069-X
Hirsch PR, Mauchline TH, Clark IM (2010) Culture-independent molecular techniques for soil microbial ecology. Soil Biol Biochem 42(6):878–887
Ingham ER, Cambardella C, Coleman DC (1986) Manipulation of bacteria, fungi and protozoa by biocides in lodgepole pine forest soil microcosms: effects on organism interactions and nitrogen mineralization. Can J Soil Sci 66(2):261–272. doi:10.4141/cjss86-028
Jones RJ, Muller J, Haynes D, Schreiber U (2003) Effects of herbicides diuron and atrazine on corals of the great barrier reef, Australia. Mar Ecol Prog Ser 251:153–167
Juneau P, Popovic R (1999) Evidence for the rapid phytotoxicity and environmental stress evaluation using the PAM fluorometric method: importance and future application. Ecotoxicology 8(6):449–455. doi:10.1023/a:1008955819527
Kahru A, Ivask A, Kasemets K, Pollumaa L, Kurvet I, François M, Dubourguier H-C (2005) Biotests and biosensors in ecotoxicological risk assessment of field soils polluted with zinc, lead, and cadmium. Environ Toxicol Chem 24(11):2973–2982. doi:10.1897/05-002R1.1
Kallmeyer J, Pockalny R, Adhikari RR, Smith DC, D'Hondt S (2012) Global distribution of microbial abundance and biomass in subseafloor sediment. Proc Natl Acad Sci 109(40): 16213–16216
Kamlet MJ, Doherty RM, Veith GD, Taft RW, Abraham MH (1986) Solubility properties in polymers and biological media. 7. An analysis of toxicant properties that influence inhibition of bioluminescence in *Photobacterium phosphoreum* (the Microtox test). Environ Sci Technol 20 (7):690–695. doi:10.1021/es00149a007
Kao C-M, Liao H-Y, Chien C-C, Tseng Y-K, Tang P, Lin C-E, Chen S-C (2016) The change of microbial community from chlorinated solvent-contaminated groundwater after biostimulation using the metagenome analysis. J Hazard Mater 302:144–150. doi:10.1016/j.jhazmat.2015.09.047
Kasai Y (2011) Molecular Technologies for Analysis of Petroleum Bioremediation. In: Koukkou A-I (ed) Microbial bioremediation of non-metals: current research. Caister Academic Press, Norfolk, UK, pp 233–252
Khudur L, Shahsavari E, Miranda A, Morrison P, Nugegoda D, Ball A (2015) Evaluating the efficacy of bioremediating a diesel-contaminated soil using ecotoxicological and bacterial community indices. Environ Sci Pollut Res 1–11. doi:10.1007/s11356-015-4624-2
King J, DiGrazia P, Applegate B, Burlage R, Sanseverino J, Dunbar P, Larimer F, Sayler Ga (1990) Rapid, sensitive bioluminescent reporter technology for naphthalene exposure and biodegradation. Science(Washington) 249(4970):778–781
Kováts N, Horváth E (2016) Bioluminescence-based assays for assessing eco-and genotoxicity of airborne emissions. Luminescence
Kumpiene J, Guerri G, Landi L, Pietramellara G, Nannipieri P, Renella G (2009) Microbial biomass, respiration and enzyme activities after in situ aided phytostabilization of a Pb- and Cu-contaminated soil. Ecotoxicol Environ Saf 72(1):115–119. doi:10.1016/j.ecoenv.2008.07.002
Kunapuli U, Lueders T, Meckenstock RU (2007) The use of stable isotope probing to identify key iron-reducing microorganisms involved in anaerobic benzene degradation. ISME J\ 1\ (7\):643\-653\
Lanno R, Wells J, Conder J, Bradham K, Basta N (2004) The bioavailability of chemicals in soil for earthworms. Ecotoxicol Environ Saf 57(1):39–47. doi:10.1016/j.ecoenv.2003.08.014
Lehman RM, Colwell FS, Ringelberg DB, White DC (1995) Combined microbial community-level analyses for quality assurance of terrestrial subsurface cores. J Microbiol Methods 22(3):263–281. doi:10.1016/0167-7012(95)00012-A

Liu F, Ying G-G, Yang L-H, Zhou Q-X (2009) Terrestrial ecotoxicological effects of the antimicrobial agent triclosan. Ecotoxicol Environ Saf 72(1):86–92. doi:10.1016/j.ecoenv.2008.06.009

Lors C, Damidot D, Ponge J-F, Périé F (2012) Comparison of a bioremediation process of PAHs in a PAH-contaminated soil at field and laboratory scales. Environ Pollut 165:11–17. doi:10.1016/j.envpol.2012.02.004

Maderova L, Paton GI (2013) Deployment of microbial sensors to assess zinc bioavailability and toxicity in soils. Soil Biol Biochem 66:222–228. doi:10.1016/j.soilbio.2013.07.017

Maderova L, Watson M, Paton GI (2011) Bioavailability and toxicity of copper in soils: integrating chemical approaches with responses of microbial biosensors. Soil Biol Biochem 43(6):1162–1168. doi:10.1016/j.soilbio.2011.02.004

Malik S, Beer M, Megharaj M, Naidu R (2008) The use of molecular techniques to characterize the microbial communities in contaminated soil and water. Environ Int 34(2):265–276

Manefield M, Whiteley AS, Griffiths RI, Bailey MJ (2002) RNA stable isotope probing, a novel means of linking microbial community function to phylogeny. Appl Environ Microbiol 68 (11):5367–5373

Margesin R, Zimmerbauer A, Schinner F (2000) Monitoring of bioremediation by soil biological activities. Chemosphere 40(4):339–346. doi:10.1016/S0045-6535(99)00218-0

Marty JL, Olive D, Asano Y (1997) Measurement of BOD: correlation between 5-day BOD and commercial BOD biosensor values. Environ Technol 18(3):333–337. doi:10.1080/09593331808616544

Mateo P, Leganés F, Perona E, Loza V, Fernández-Piñas F (2015) Cyanobacteria as bioindicators and bioreporters of environmental analysis in aquatic ecosystems. Biodivers Conserv 24 (4):909–948. doi:10.1007/s10531-015-0903-y

Mejáre M, Bülow L (2001) Metal-binding proteins and peptides in bioremediation and phytoremediation of heavy metals. Trends Biotechnol 19(2):67–73. doi:10.1016/S0167-7799 (00)01534-1

Mekki A, Dhouib A, Feki F, Sayadi S (2008) Assessment of toxicity of the untreated and treated olive mill wastewaters and soil irrigated by using microbiotests. Ecotoxicol Environ Saf 69 (3):488–495. doi:10.1016/j.ecoenv.2007.04.008

Metzker ML (2010) Sequencing technologies [mdash] the next generation. Nat Rev Genet 11(1):31–46

Min J, Pham CH, Gu MB (2003) Specific responses of bacterial cells to dioxins. Environ Toxicol Chem 22(2):233–238. doi:10.1002/etc.5620220201

Molina-Barahona L, Vega-Loyo L, Guerrero M, Ramírez S, Romero I, Vega-Jarquín C, Albores A (2005) Ecotoxicological evaluation of diesel-contaminated soil before and after a bioremediation process. Environ Toxicol 20(1):100–109. doi:10.1002/tox.20083

Mora APd, Ortega-Calvo JJ, Cabrera F, Madejón E (2005) Changes in enzyme activities and microbial biomass after "in situ" remediation of a heavy metal-contaminated soil. Appl Soil Ecol 28(2):125–137. doi:10.1016/j.apsoil.2004.07.006

Mueller DC, Bonner JS, McDonald SJ, Autenrieth RL, Donnelly KC, Lee K, Doe K, Anderson J (2003) The use of toxicity bioassays to monitor the recovery of oiled wetland sediments. Environ Toxicol Chem 22(9):1945–1955. doi:10.1897/02-325

Muñoz-Leoz B, Ruiz-Romera E, Antigüedad I, Garbisu C (2011) Tebuconazole application decreases soil microbial biomass and activity. Soil Biol Biochem 43(10):2176–2183. doi:10.1016/j.soilbio.2011.07.001

Muyzer G, de Waal EC, Uitterlinden AG (1993) Profiling of complex microbial populations by denaturing gradient gel electrophoresis analysis of polymerase chain reaction-amplified genes coding for 16S rRNA. Appl Environ Microbiol 59(3):695–700

Naessens M, Leclerc JC, Tran-Minh C (2000) Fiber optic biosensor using chlorella vulgaris for determination of toxic compounds. Ecotoxicol Environ Saf 46(2):181–185. doi:10.1006/eesa.1999.1904

Naidu R (2011) Chemical bioavailability in terrestrial environments. Elsevier Science

Nakatsu C (2007) Soil microbial community analysis using denaturing gradient gel electrophoresis. Soil Sci Soc Am J 71(2):562–571

Nomura Y, Ikebukuro K, Yokoyama K, Takeuchi T, Arikawa Y, Ohno S, Karube I (1998) Application of a linear alkylbenzene sulfonate biosensor to river water monitoring1. Biosens Bioelectron 13(9):1047–1053. doi:10.1016/S0956-5663(97)00077-8

Ogawa T, Shibata M, Yatome C, Idaka E (1988) Growth inhibition ofBacillus subtilis by basic dyes. Bull Environ Contam Toxicol 40(4):545–552. doi:10.1007/bf01688379

Oka AR, Phelps CD, McGuinness LM, Mumford A, Young LY, Kerkhof LJ (2008) Identification of critical members in a sulfidogenic benzene-degrading consortium by DNA stable isotope probing. Appl Environ Microbiol 74(20):6476–6480. doi:10.1128/aem.01082-08

Oulas A, Pavloudi C, Polymenakou P, Pavlopoulos GA, Papanikolaou N, Kotoulas G, Arvanitidis C, Iliopoulos I (2015) Metagenomics: tools and insights for analyzing next-generation sequencing data derived from biodiversity studies. Bioinform Biol Insights 9:75

Paszczynski A, Crawford RL (1995) Potential for bioremediation of xenobiotic compounds by the white-rot fungus *Phanerochaete chrysosporium*. Biotechnol Prog 11(4):368–379. doi:10.1021/bp00034a002

Patil SS, Adetutu EM, Aburto-Medina A, Menz IR, Ball AS (2013) Biostimulation of indigenous communities for the successful dechlorination of tetrachloroethene (perchloroethylene)-contaminated groundwater. Biotech Lett 36(1):75–83. doi:10.1007/s10529-013-1369-1

Pimentel D, Whitecraft M, Scott ZR, Zhao L, Satkiewicz P, Scott TJ, Phillips J, Szimak D, Singh G, Gonzalez DO (2010) Will limited land, water, and energy control human population numbers in the future? Hum Ecol 38(5):599–611

Płaza GA, Nałęcz-Jawecki G, Pinyakong O, Illmer P, Margesin R (2009) Ecotoxicological and microbiological characterization of soils from heavy-metal- and hydrocarbon-contaminated sites. Environ Monit Assess 163(1):477–488. doi:10.1007/s10661-009-0851-7

Preston S, Coad N, Townend J, Killham K, Paton GI (2000) Biosensing the acute toxicity of metal interactions: are they additive, synergistic, or antagonistic? Environ Toxicol Chem 19(3): 775–780. doi:10.1002/etc.5620190332

Radajewski S, Ineson P, Parekh NR, Murrell JC (2000) Stable-isotope probing as a tool in microbial ecology. Nature 403(6770):646–649

Reid BJ, Jones KC, Semple KT (2000) Bioavailability of persistent organic pollutants in soils and sediments—a perspective on mechanisms, consequences and assessment. Environ Pollut 108 (1):103–112. doi:10.1016/S0269-7491(99)00206-7

Reith F, Zammit CM, Shar SS, Etschmann B, Bottrill R, Southam G, Ta C, Kilburn M, Oberthür T, Ball AS (2016) Biological role in the transformation of platinum-group mineral grains. Nat Geosci

Renella G, Landi L, Ascher J, Ceccherini MT, Pietramellara G, Mench M, Nannipieri P (2008) Long-term effects of aided phytostabilisation of trace elements on microbial biomass and activity, enzyme activities, and composition of microbial community in the Jales contaminated mine spoils. Environ Pollut 152(3):702–712. doi:10.1016/j.envpol.2007.06.053

Robbens J, Dardenne F, Devriese L, Coen W, Blust R (2010) Escherichia coli as a bioreporter in ecotoxicology. Appl Microbiol Biotechnol 88(5):1007–1025. doi:10.1007/s00253-010-2826-6

Rodriguez-Mozaz S, Marco M-P, de Alda Lopez, Maria J, Barceló D (2003) Biosensors for environmental monitoring of endocrine disruptors: a review article. Anal Bioanal Chem 378 (3):588–598. doi:10.1007/s00216-003-2385-0

Sabater S, Guasch H, Ricart M, Romaní A, Vidal G, Klünder C, Schmitt-Jansen M (2007) Monitoring the effect of chemicals on biological communities. The biofilm as an interface. Anal Bioanal Chem 387(4):1425–1434. doi:10.1007/s00216-006-1051-8

Sazykin IS, Sazykina MA, Khmelevtsova LE, Mirina EA, Kudeevskaya EM, Rogulin EA, Rakin AV (2016) Biosensor-based comparison of the ecotoxicological contamination of the wastewaters of Southern Russia and Southern Germany. Int J Environ Sci Technol 13(3): 945–954. doi:10.1007/s13762-016-0936-0

Schuster SC (2007) Next-generation sequencing transforms today's biology. Nature 200(8):16–18

Scott C, Pandey G, Hartley CJ, Jackson CJ, Cheesman MJ, Taylor MC, Pandey R, Khurana JL, Teese M, Coppin CW, Weir KM, Jain RK, Lal R, Russell RJ, Oakeshott JG (2008) The enzymatic basis for pesticide bioremediation. Indian J Microbiol 48(1):65–79. doi:10.1007/s12088-008-0007-4

Senan RC, Abraham TE (2004) Bioremediation of textile azo dyes by aerobic bacterial consortium aerobic degradation of selected azo dyes by bacterial consortium. Biodegradation 15(4): 275–280. doi:10.1023/B:BIOD.0000043000.18427.0a

Shahi A, Aydin S, Ince B, Ince O (2016) Evaluation of microbial population and functional genes during the bioremediation of petroleum-contaminated soil as an effective monitoring approach. Ecotoxicol Environ Saf 125:153–160. doi:10.1016/j.ecoenv.2015.11.029

Shahsavari E, Aburto-Medina A, Taha M, Ball AS (2016) A quantitative PCR approach for quantification of functional genes involved in the degradation of polycyclic aromatic hydrocarbons in contaminated soils. MethodsX 3:205–211. doi:10.1016/j.mex.2016.02.005

Shahsavari E, Adetutu E, Ball A (2015a) Phytoremediation and necrophytoremediation of petrogenic hydrocarbon-contaminated soils. In: Ansari AA, Gill SS, Gill R, Lanza GR, Newman L (eds) Phytoremediation. Springer International Publishing. pp 321–334. doi:10.1007/978-3-319-10969-5_26

Shahsavari E, Adetutu EM, Anderson PA, Ball AS (2013) Plant residues—A low cost, effective bioremediation treatment for petrogenic hydrocarbon-contaminated soil. Sci Total Environ 443:766–774. doi:10.1016/j.scitotenv.2012.11.029

Shahsavari E, Adetutu EM, Taha M, Ball AS (2015b) Rhizoremediation of phenanthrene and pyrene contaminated soil using wheat. J Environ Manage 155:171–176. doi:10.1016/j.jenvman.2015.03.027

Shao CY, Howe CJ, Porter AJR, Glover LA (2002) Novel cyanobacterial biosensor for detection of herbicides. Appl Environ Microbiol 68(10):5026–5033. doi:10.1128/aem.68.10.5026-5033.2002

Simons KL, Sheppard PJ, Adetutu EM, Kadali K, Juhasz AL, Manefield M, Sarma PM, Lal B, Ball AS (2013) Carrier mounted bacterial consortium facilitates oil remediation in the marine environment. Biores Technol 134:107–116. doi:10.1016/j.biortech.2013.01.152

Smolders E, Brans K, Coppens F, Merckx R (2001) Potential nitrification rate as a tool for screening toxicity in metal-contaminated soils. Environ Toxicol Chem 20(11):2469–2474

Su L, Jia W, Hou C, Lei Y (2011) Microbial biosensors: A review. Biosens Bioelectron 26(5):1788–1799. doi:10.1016/j.bios.2010.09.005

Sütterlin H, Alexy R, Kümmerer K (2008) The toxicity of the quaternary ammonium compound benzalkonium chloride alone and in mixtures with other anionic compounds to bacteria in test systems with Vibrio fischeri and Pseudomonas putida. Ecotoxicol Environ Saf 71(2):498–505. doi:10.1016/j.ecoenv.2007.12.015

Sverdrup LE, Ekelund F, Krogh PH, Nielsen T, Johnsen K (2002) Soil microbial toxicity of eight polycyclic aromatic compounds: effects on nitrification, the genetic diversity of bacteria, and the total number of protozoans. Environ Toxicol Chem 21(8):1644–1650. doi:10.1002/etc.5620210815

Taranova L, Semenchuk I, Manolov T, Iliasov P, Reshetilov A (2002) Bacteria-degraders as the base of an amperometric biosensor for detection of anionic surfactants. Biosens Bioelectron 17 (8):635–640. doi:10.1016/S0956-5663(01)00307-4

Thomas DJL, Tyrrel SF, Smith R, Farrow S (2009) Bioassays for the evaluation of landfill leachate toxicity. J Toxicol Environ Health Part B 12(1):83–105. doi:10.1080/10937400802545292

Thomas T, Gilbert J, Meyer F (2012) Metagenomics—a guide from sampling to data analysis. Microb Informatics Exp 2(1):1–12. doi:10.1186/2042-5783-2-3

Tscherko D, Kandeler E (1997) Ecotoxicological effects of fluorine deposits on microbial biomass and enzyme activity in grassland. Eur J Soil Sci 48(2):329–335. doi:10.1111/j.1365-2389.1997.tb00553.x

Ulitzur S, Lahav T, Ulitzur N (2002) A novel and sensitive test for rapid determination of water toxicity. Environ Toxicol 17(3):291–296. doi:10.1002/tox.10060 van Beelen P, Doelman P (1997) Significance and application of microbial toxicity tests in assessing ecotoxicological risks of contaminants in soil and sediment. Chemosphere 34(3):455–499. doi:10.1016/S0045-6535(96)00388-8

van der Meer JR, Belkin S (2010) Where microbiology meets microengineering: design and applications of reporter bacteria. Nat Rev Micro 8(7):511–522. doi:10.1038/nrmicro2392

van Dijk EL, Auger H, Jaszczyszyn Y, Thermes C (2014) Ten years of next-generation sequencing technology. Trends Genet 30(9):418–426. doi:10.1016/j.tig.2014.07.001

Vollmer AC, Belkin S, Smulski DR, Van Dyk TK, LaRossa RA (1997) Detection of DNA damage by use of Escherichia coli carrying recA':lux, uvrA':lux, or alkA':lux reporter plasmids. Appl Environ Microbiol 63(7):2566–2571

Watanabe K (2001) Microorganisms relevant to bioremediation. Curr Opin Biotechnol 12(3): 237–241

Weitz HJ, Ritchie JM, Bailey DA, Horsburgh AM, Killham K, Glover LA (2001) Construction of a modified mini-Tn5 luxCDABE transposon for the development of bacterial biosensors for ecotoxicity testing. FEMS Microbiol Lett 197(2):159–165. doi:10.1111/j.1574-6968.2001.tb10598.x

Whyte LG, Greer CW (2005) Molecular techniques for monitoring and assessing soil bioremediation. In: Margesin R, Schinner F (eds) Monitoring and assessing soil bioremediation, vol 5. Springer, Berlin Heidelberg, pp 201–231

Widenfalk A, Bertilsson S, Sundh I, Goedkoop W (2008) Effects of pesticides on community composition and activity of sediment microbes—responses at various levels of microbial community organization. Environ Pollut 152(3):576–584. doi:10.1016/j.envpol.2007.07.003

Woese CR, Kandler O, Wheelis ML (1990) Towards a natural system of organisms: proposal for the domains Archaea, Bacteria, and Eucarya. Proc Natl Acad Sci USA 87(12):4576–4579. doi:10.1073/pnas.87.12.4576

Wu X, McLaren J, Madl R, Wang D (2010) Biofuels from lignocellulosic biomass. In: Sustainable biotechnology. Springer, Berlin, pp 19–41

Xu X, Oliff K, Xu T, Ripp S, Sayler G, Zhuang J (2015) Microbial availability of mercury: effective detection and organic ligand effect using a whole-cell bioluminescent bioreporter. Ecotoxicology 24(10):2200–2206. doi:10.1007/s10646-015-1553-2

Yergeau E, Sanschagrin S, Beaumier D, Greer CW (2012) Metagenomic analysis of the bioremediation of diesel-contaminated canadian high arctic soils. PLoS ONE 7(1):e30058. doi:10.1371/journal.pone.0030058

Yoon Y, Kang Y, Chae Y, Kim S, Lee Y, Jeong S-W, An Y-J (2015) Arsenic bioavailability in soils before and after soil washing: the use of Escherichia coli whole-cell bioreporters. Environ Sci Pollut Res 23(3):2353–2361. doi:10.1007/s11356-015-5457-8

Zhang DC, Mörtelmaier C, Margesin R (2012) Characterization of the bacterial archaeal diversity in hydrocarbon-contaminated soil. Sci Total Environ 421–422:184–196

Part II
Concepts and Case Studies

Chapter 3
Microbiology and Chemistry in the Environment: Two Symbiotic Species in the Same Biotope

Philippe Garrigues

Abstract Apparently the global definition of each specific research niche, i.e. microbiology versus chemistry seems to be very exclusive. Environmental Microbiology is defined as the study of microbial processes in the environment, microbial communities and microbial interactions. Environmental Chemistry is often described as the study of the fate and effects of chemicals in the environment. The intersections and the interactions of these 2 disciplines appear a priori very weak looking at their respective definition, but this is not the case. The Environment appears as a common playground for both disciplines. This chapter will briefly demonstrate such an interdisciplinary approach.

Keywords Abiotic processes · Biotic processes · Bioavailability · Bioaccessibility · Biodegradation · Chemotaxis · Interactions

Environmental chemists were used to study the behaviour of chemicals in the environment, mainly based by the observations of processes strongly supported by the (abiotic) physical-chemical properties of the chemicals of interest. But some discrepancies with chemical rules were rapidly observed. It led to the necessary need to introduce biotic processes in the global processes of chemicals in the environment. Indeed the biotic processes are taking a major role and microbes are one of the essential parts of the food chains. Bacteria may act apparently against some chemical rules. For instance stable compounds according to the thermodynamics maybe less present than expected. This is particularly the case when preferential biodegradation of chemicals is favoured based on the molecular shape of the chemicals. Some substituent group (alkyl, n-alkyl chain) in specific position more accessible to bacteria and then more sensitive to biodegradation (Malcom and Jones 2000).

P. Garrigues (✉)
Institut des Sciences Moléculaires, UMR 5255 CNRS/Université de Bordeaux, 33405 Talence, France
e-mail: p.garrigues@ism.u-bordeaux1.fr

C. Cravo-Laureau et al. (eds.), *Microbial Ecotoxicology*,
DOI 10.1007/978-3-319-61795-4_3

Environmental chemists and microbiologists are struggling since several decades, with the concept of (bio)availability and/or (bio)accessibility. Indeed, microbiologists are interested in predicting the biodegradation progress and chemists in predicting the concentrations in the environment. There are a lot of different definitions regard availability/accessibility but Semple et al. (2004) have given by far the ultimate definitions. Bioavailability is "that which is freely available to cross an organism's (cellular) membrane from the medium the organism inhabits at a given point in time". Bioaccessibility is defined as "that which is available to cross an organisms' (cellular) membrane from the environment it inhabits, if the organism had access to it; however, it may be either physically removed from the organism or only bioavailable after a period of time". The important aspect in these definitions is the introduction of a time period. Then chemicals may belong to rapidly, slowly, very slowly desorbing fraction or even to a non-desorbable from the considered environmental matrices. Bioaccessibility maybe mimic by chemical amount available in a rapid/slow desorbing fraction obtained by chemical extractions techniques.

Environmental analytical chemistry has been of major importance in the identification of degraded forms of chemicals both under biotic or abiotic stresses. Such identification provides information on degradation pathways of molecules particularly in the case of microbial degradation. The detection of degraded molecular forms may provide information on the specific enzymes or genes involved in the degradation processes. Particularly a lot of studies have been conducted in the last 30 years on microbial biodegradation of aromatic compounds leading to a better knowledge of microbial enzymes involved in ring cleavage or ring hydroxylation or oxidation (for a review, see Peng et al. 2008).

Chemotaxis is the movement of an organism in response to a chemical stimulus. This is a physiological adaptation which include the evolutionary acquisition of uptake systems for specific contaminant, which enable bacteria to drive diffusive transfer of a contaminant faster among numerous other chemicals. It has been observed a direct relationship between the specific affinity and the diffusive transfer to a microbe in an environment of limited substrate availability (Harms and Wick 2006). Solubilising agents such as extracellular polymeric substances (EPS) play an important role in chemical accessibility. Chemotaxis has demonstrated in various publications the involvement of mathematicians, physicists and chemists to integrate diffusion mechanisms of chemicals in the chemotaxis phenomenon (Chalub et al. 2002).

Extracellular Polymeric Substances (EPS) are the major components of biofilms composed of polysaccharides, proteins, glycoproteins, glycolipids, nucleic acids, and amphiphilic compounds (Jamal et al. 2015). Biofilms themselves are composed of microbial cells, DNA/RNA, polysaccharides, proteins and a majority of water (up to 97%). One benefit of the biofilm is increased resistance since the dense matrix and the outer layer of cells protect the microbial community. Gene transfer is greatly facilitated in such environments and leads to a more stable biofilm structure. Various separation techniques and spectroscopic measurements (mass spectrometry, lase microscopy) have led to the chemical identifications of EPS (as example, see Zippel and Neu 2011).

Methods for bacterial identification can be split into genotypic techniques based on getting a pattern of an organism's genetic material (DNA) and phenotypic techniques based on metabolomic profile and/or its chemical composition. New proteomics tools based mass spectrometry (matrix-assisted laser desorption/ionization time-of-flight mass spectrometry (MALDI-TOF-MS); electro spray ionization mass spectrometry (ESI-MS)) allow detection of biomolecules produced by bacteria and complement to classical microbiological techniques for bacterial classification and phenotypic characterization Emerson et al. (2008).

Bacteria play an important role in recycling substances and degrading xenobiotics, since they are able to metabolize complex organic matter. Actinobacteria is a microbiological group capable to remove chemicals as pesticides, and heavy metals, among others substances (Alvarez et al. 2017). Actinobacteria have demonstrated their potential as tools for bioremediation of several contaminants (oil, rubber, plastics, pesticides, and heavy metals). Strategies such as biostimulation, cell immobilization, phytoremediation, production of biosurfactants have been developed to enhance the capabilities of Actinobacteria in bioremediation.

The very short examples presented above show the strong interactions of environmental microbiology and chemistry. All the joined studies have demonstrated the reciprocal benefit of such approaches for a better knowledge for both chemical and bacterial environmental processes. Such knowledge is of paramount importance for microbial ecotoxicology.

References

Alvarez A, Saez JM, Davila Costa JS, Colin VL, Fuentes MS, Cuozzo SA, Benimelli CS, Polti MA, Amoros MJ (2017) Actinobacteria: current research and pespective for bioremediation of pesticides and heavy metals. Chemosphere 166(41–62):2017

Chaluby FACC, Markowichz P, perthamex B, Schmeiser C (2002) Kinetic models for chemotaxis and their drift-diffusion limits. Nonlinear Differ Equ Models, 123–141

Emerson D, Agulto L, Liu H, Liu L (2008) Identifyng and characterizing bacteria in an era of genomics and proteomics. BioScience 58(10):925–936

Harms H, Wick LY (2006) Dispersing pollutant-degrading bacteria in contaminated soil without touching it. Eng Life Sci 6(3):252–260

Jamal M, Tasmeem U, Hussain T, Andleeb S (2015) Bacterial biofilm: its composition, formation and role in human infections Research and Reviews. J Microbiol Biotechnol 4(3):1–13

Peng RH, Hiong AS, Xue Y, Fu XY, Gao F, Zhao W, Tian YS, Yao QH (2008) Microbial biodegradation of polyaromatic hydrocarbons. FEMS Microbiol Rev 32:927–955

Scott MJ, Jones MM (2000) The biodegradation of surfactants in the environment. Biochimica et Biophysica Acta 1058(1–2):235–251

Semple KT, Doick KJ, Jones KC, Burauel P, Craven A, Harms H (2004) Defining bioavailability and bioaccessibility of contaminated soil and sediment is complicated. Env Sci Tech 38 (12):228A–231A

Zippel B, Neu TR (2011) Characterization of glycoconjugates of extracellular polymeric substances in tufa-associated biofilms by using fluorescence lectin-binding analysis. Appl Environ Microbiol 77:505–516

Chapter 4
Microbial Responses to Pollution—Ecotoxicology: Introducing the Different Biological Levels

Cristiana Cravo-Laureau, Béatrice Lauga, Christine Cagnon and Robert Duran

Abstract The environmental pollutions generated by human activities are important concerns that environmental risk assessment procedures have the purpose to evaluate and mitigate the effects. Microorganisms are among the first impacted by human generated pollutions. Furthermore, because they are essential actors in ecosystem functioning the evaluation of the pollution effects on microorganisms is of paramount importance. Their response may serve as proxy to report the effects on, and the recovering capacities of, the ecosystem. The behaviour of microorganisms in response to chemical pollution has been largely studied. In this chapter, we introduce the mechanisms underlying the microbial adaptation capacities involved in response to pollutants. We also discuss the basic knowledge inspiring microbial ecotoxicological tools reporting the pollutant effects that have been developed at the different biological organization levels, from genes and cellular processes to population and microbial community responses.

Keywords Microbial physiology · Microbial metabolism · Community ecology · Microbial adaptation · Metabolic versatility

---

C. Cravo-Laureau (✉) · B. Lauga · C. Cagnon · R. Duran
Equipe Environnement et Microbiologie, IPREM UMR 5254,
MELODY Group, Université de Pau et Des Pays de L'Adour,
BP 1155, 64013 Pau, France
e-mail: cristiana.cravo-laureau@univ-pau.fr

B. Lauga
e-mail: beatrice.lauga@univ-pau.fr

C. Cagnon
e-mail: christine.cagnon@univ-pau.fr

R. Duran
e-mail: robert.duran@univ-pau.fr

C. Cravo-Laureau et al. (eds.), *Microbial Ecotoxicology*,
DOI 10.1007/978-3-319-61795-4_4

4.1 Introduction

Modifications in microbial community structure and composition have been largely reported after the addition of pollutant, whether organic (Bordenave et al. 2004, 2007; Hjorth et al. 2007; Vercraene-Eairmal et al. 2010; Chronopoulou et al. 2013; Stauffert et al. 2013, 2014, Stauffert et al. 2015a, b; Cravo-Laureau and Duran 2014; Ben Said et al. 2015) or metallic/metalloid (Duran et al. 2003, 2008; Viret et al. 2006; Dominique et al. 2007; Ramond et al. 2009). In agreement with these studies, specific microbial communities have been described in polluted environments according to the nature of the pollutant (Bruneel et al. 2008; Paissé et al. 2008; Volant et al. 2014; Bargiela et al. 2015; Duran et al. 2015; Rodriguez-R et al. 2015; Misson et al. 2016). Such observations made at the microbial community level are of ecological relevance providing information on the behaviour of microbial communities in response to pollutants. The information indicating whether microbial community composition is resistant, resilient, or functionally redundant is of primary importance at the ecotoxicological point of view. It has been proposed to include the modification of microbial community composition into ecosystem process models in order to predict the response of ecosystem to disturbances (Allison and Martiny 2008). Actually, the modification of microbial community structures translates the metabolic versatility of microorganisms that is expressed at the cellular and population levels by the capacities to resist, transform and/or degrade the different classes of pollutants (Parales and Haddock 2004). However, it is noteworthy that the degradation and transformation capacities also depend on the interactions between microorganisms resulting in microbial networks performing complex task as demonstrated for the degradation of hydrocarbon compounds in marine environments (for review, see Head et al. 2006; McGenity et al. 2012). Several studies have demonstrated that assessing microbial activities at work during a pollution event provides relevant knowledge on the metabolic capacities affected by a pollutant and on the potential for ecosystem recovery in presence of metals (Bruins et al. 2000), pharmaceuticals (Barra Caracciolo et al. 2015) and other organic pollutants (Diepens et al. 2014). The metabolic versatility of microorganisms is related to genetic adaptation mechanisms that include mutations and horizontal gene transfer (Pieper et al. 2004; Stokes and Gillings 2011; Guieysse and Wuertz 2012; Puglisi et al. 2012). A large number of genotoxicity tests have been developed using microorganisms (Kokkali and van Delft 2014; Ma et al. 2014) for toxicity evaluation of polluted environmental sites.

As outlined in Fig. 4.1, the microbial processes involved in the response to pollutants provide the opportunity to develop microbial ecotoxicological tools at the different biological organization levels from gene to ecosystem, which at the academic point of view are relevant to assess microbial mechanisms and address ecological considerations respectively. In this chapter we summarize the microbial responses at different biological levels, which provide the basic knowledge of tools now available allowing ecotoxicological observations from genotoxicity tests to the development of ecosystem process models.

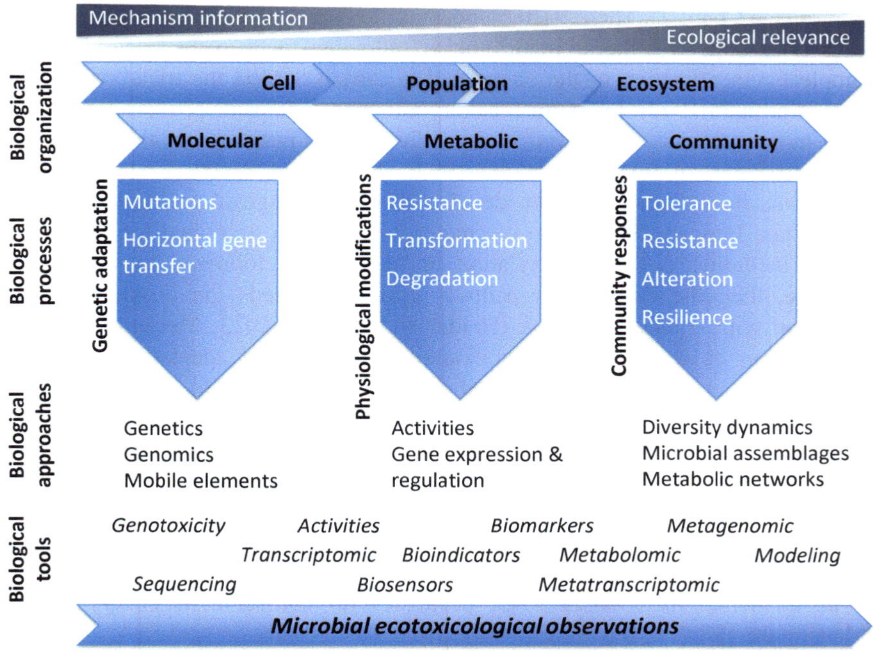

Fig. 4.1 Microbial ecotoxicological observations and biological organisation levels. The microbial ecotoxicological tools are based on microbial processes including genetic adaptation, physiological modifications and community responses that translate mechanisms operating at cell, population and ecosystem levels respectively

4.2 Microbial Physiology and Metabolism

The physiological responses of microorganisms to pollutants have been initially mainly addressed using culture-based approaches. Many pollutants could be degraded or transformed by microbial action, microorganisms being adapted and selected to xenobiotic compounds introduced into the environment. In most cases, biodegradation capacities or metabolic pathways have been described in model strains. Thereby, physiological, morphological, taxonomic, and metabolic characteristics have been studied to better understand potential capacities and behaviour of microorganisms face to pollutants.

Microbial pathways involved in the degradation of organic pollutants have been intensively studied for decades. Microorganisms have developed diverse strategies to degrade organic pollutants in presence and in absence of molecular oxygen. Under aerobic conditions the oxygen could be not only the final electron acceptor but also a co-substrate for some catabolic processes, as described for some aromatic compounds (Fuchs et al. 2011; Diaz et al. 2013). For example, bacteria have the capacity to use polycyclic aromatic hydrocarbons (PAHs) as carbon and energy sources (see Chap. 7). These bacteria possess dioxygenases (RHD) which introduce hydroxyl groups into the aromatic nucleus allowing to open the cycle, and then complete mineralization of the compound is carried out via the tricarboxylic acid cycle (Cerniglia 1992). In contrast, eukaryotic microorganisms possess cytochrome P450 monooxygenases involved in detoxification pathways rather than in assimilation processes of PAHs (Cerniglia 1992; Doyle et al. 2008). Under anaerobic conditions, microorganisms have developed a wide range of catabolic strategies (Zhang and Bennett 2005). Two mechanisms have been described for PAHs activation, including direct carboxylation or methylation followed by addition to fumarate, then the degradation pathway further proceeds via β-oxidation after activation with coenzyme (Co)A (Heider and Schühle 2013; Meckenstock et al. 2016; Rabus et al. 2016). Further examples of microbial adaptation to organic pollutants are provided by the degradation capacities for polychlorinated biphenyls (PCBs) and pesticides. PCBs degradation includes anaerobic reductive dechlorination, an energy-yielding process where PCBs serve as electron acceptor, and aerobic breakdown of the biphenyl structure through an oxidation reaction (Field and Sierra-Alvarez 2008; Sowers and May 2013; Passatore et al. 2014). Regarding pesticides, microbial metabolic transformation could be classified as a catabolic response where pesticides serve as energy sources, as a detoxification metabolism or as incidental metabolism when pesticides do not serve as energy sources (Matsumura 1989).

Coping with organic compounds involves a metabolic response, connecting the specific catabolic pathway with the energetic/biosynthetic metabolism of the cell, and a stress response for protection from the toxic effect of organic pollutants and adaptation to suboptimal growth conditions. Although extended research has been carried out focusing on degradation, other physiological responses may constitute important events preceding catabolism of organic pollutants, as bioavailability, chemotaxis, intracellular accumulation, tolerance mediated by physical and biochemical barriers (Sardessai and Bhosle 2002; Jain et al. 2005; Zhang and Bennett 2005; Chavez et al. 2006; Murinova and Dercova 2014; Parales et al. 2015; Duran and Cravo-Laureau 2016). Knowledge on metabolism of organic pollutants is still rather fragmentary and the diversity of bacterial strategies is highly underestimated.

Regarding metals, transformations could be related to energetic metabolisms, when used as electron donor or acceptor. Resistance and detoxification mechanisms have been also developed by microorganisms to cope with toxic metals. In some cases, oxidases or reductases are synthesized transforming metals into a volatile compound (e.g. mercury, Barkay and Wagner-Dobler 2005) or in less toxic compounds (e.g. arsenic, Cervantes et al. 1994). Incidental or indirect mechanisms (biomethylation, indirect reduction), as well as the presence of metal carriers, or even the formation of structures able to immobilize toxic compounds have been described (Prabhakaran et al. 2016). Research on the fate and ecological effects of some *emerging* pollutants, as nanoparticles, has become a focus of attention only recently (Concha-Guerrero et al. 2014; Cervantes-Avilés et al. 2016; Simonin et al. 2017). The interaction of nanoparticles with microorganisms is addressed in Chap. 5.

The impact of pollutants and degradation capacities reported in laboratory studies reflect only potential degradation that may occur in the natural environment. Environmental parameters, as temperature, salinity and pH, physical-chemical properties of pollutants, their concentration, as well as concentration and diversity of microorganisms, are all factors that play an important role in the biodegradation process (Pieper 2005; Shahgholi and Gholamalizadeh Ahangar 2014; Duran and Cravo-Laureau 2016). Therefore it is essential to consider all these parameters to characterize in-depth physiological and metabolic response of microorganisms to pollutants. Nowadays the integration of all these parameters via in situ studies is still difficult. Thus, experimental ecology approaches have been developed; mimicking as close as possible the environmental conditions. These approaches combine the advantages of lab-controlled systems with the possibility of extrapolation to the real situation found in complex ecosystems (Cravo-Laureau and Duran 2014).

The physiological and metabolic versatility of microorganisms is a key advantage in the response and in the adaptation to pollutants. Although culture-dependent methods generally recover a small portion of the diversity from environments, they are still a critical component of research and bioremediation development (Watanabe 2001). However, proteogenomic, metabolomic, transcriptomic and metagenomic studies revealed novel degradation pathways, allowing to consider metabolism of viable but non-cultivable microorganisms. The use of integrative culture-dependant and culture-independent methods, including *omics* approaches, has enabled an unprecedented view of metabolic pathways and clues to the evolution of degradation pathways and physiological and metabolic adaptation strategies to changing environmental conditions (Cravo-Laureau and Duran 2014; Ufarte et al. 2015).

4.3 Microbial Community Responses

Recent cultivation-independent genome approaches and sampling of previously unexamined environments have revealed the unsuspected huge diversity of microorganisms, both eukaryotic and prokaryotic (Hug et al. 2016; Lennon and Locey 2016). This considerable taxonomic diversity strongly echoes their capabilities to thrive in a large range compartments on earth where they ensure through their activities the sustainability and functioning of the ecosystems (Azam 1998; Field et al. 1998; Guerrero and Berlanga 2006; Pomeroy et al. 2007; Falkowski et al. 2008; Van Der Heijden et al. 2008; Bardgett and Van Der Putten 2014). In the environment microorganisms are assembled in communities of various degrees of complexity. These assemblages result of complex interactions that maintain the cohesion of the communities. Interactions are of two orders:

- biotic interactions, either trophic or non-trophic, among microorganisms of the community and among microorganisms and surrounding or host macroorganisms: plants and animals.
- abiotic interactions defined by physical-chemical conditions relevant to the environment in which the communities thrive.

Determining and predicting the effect of contaminants in natural environments, that are the ultimate goals of ecotoxicology, involves to address complex biological organization, communities, ecosystems, or landscapes and necessitate to cover large spatial scales (Beketov and Liess 2012; Newman 2015). Also a critical issue in ecotoxicology studies consists in disentangling the part of the response that is due to the contaminant (either chemical or physical properties), that can be termed as a direct effect, from that due to biological interactions among organisms, thus an indirect effect (see Chap. 14 for more development). Investigating the impact of pollutants on microbial communities are often addressed by two main approaches, namely in situ studies where polluted sites are compared to reference sites or investigated along gradients of contamination (Païssé et al. 2008; Volant et al. 2014), and microcosm or mesocosm studies that aimed at approaching environmental conditions similar to those of natural ecosystems while keeping under control their fluctuations (Vercraene-Eairmal et al. 2010; Paule et al. 2013; Stauffert et al. 2013; Bour et al. 2015).

Although microorganisms have developed tremendous ranges of metabolic capacities and stress-related pathways and strategies (see previous section and Chap. 5), environmental pollutants such as toxic metals and hazardous organic compounds constitute nevertheless important environmental pressures that may have adverse effects on the metabolism and the survival of several taxa. Indeed taxon owns only a limited fraction of whole microorganism metabolic repertoire even though some taxa may exhibit larger metabolic capacities. In this case such ecological versatility frequently allows the strains to cope with several pollutants and face large ranges of environmental conditions (Brazilian National Genome Project Consortium 2003; Nelson and Fraser 2004; Mongodin et al. 2006). Impact of pollutants on microbial communities depends greatly on their chemical properties, bioavailability and persistence (Calvet 1989; Bonnet et al. 2007; Spagnuolo et al. 2010; Xiao et al. 2013), as well as their physical properties that, in turn, can modify the properties of the milieu (e.g. crude oil pollution, Dachs et al. 2000).

On the other side physical and chemical properties of the milieu influence also the time of residence and the bioavailability of the pollutant in the environment (Barriuso et al. 1996), for instance several authors demonstrated a clear relationship between organic matter content in soils and the sequestration of pesticides (Chung and Alexander 2002; Bogan and Sullivan 2003; Moreno-Jiménez et al. 2013; Woignier et al. 2013). Pollution history at site is also determinant because microbial community previously exposed to pollutant may be promptly mobilised in subsequent exposure, and, when effective, biodegradation of the pollutant can be enhanced and can occur faster compared to an environment exposed for the first time (Walker 1987; Head et al. 2006; Baxter and Cummings 2008; Lauga et al. 2013).

This memory effect may result of an increased tolerance of the community to the pollutant as a consequence of physiological adaptations or community shifts (Widenfalk et al. 2008; Azarbad et al. 2015; Mauffret et al. 2017). As a consequence of differences in pollutants sensitivity among microbial species disruption on microbial communities was evidenced. Studies conducted to investigate the impact of pollutants on microbial communities demonstrated that both diversity (richness and evenness) and function structures host by microorganisms might be impaired under pollutants pressure. Several studies have reported shifts in the community structure, decrease of richness or changes in abundance of some taxa either at environmentally relevant or high concentrations of pollutants (Li et al. 2006; Foley et al. 2007; Johnston and Roberts 2009; Lubarsky et al. 2012; Pascault et al. 2014; Ibekwe et al. 2016; Jiao et al. 2016; Misson et al. 2016; Mustafa et al. 2016; Wang et al. 2016). Interestingly in a study that aimed at investigate the impact of diuron (an herbicide) on river epilithic biofilms, (Vercraene-Eairmal et al. 2010) demonstrated that bacterial communities at the most contaminated site were less affected under realistic exposure to diuron than communities developing under lower exposure in their native site. This result suggests that adaptation and resistance may have emerged in the former community under selection pressure and then spread in the communities or, alternatively, that the community was shaped and stabilized in such contamination background allowing for resistance to subsequent exposure. Cases of resistance and adaptation at the community level have also been reported in different studies among which Acosta-González and Marqués (2016) in oil-polluted marine coastal sediments or Mukherjee et al. (2014) in a creosote-contaminated site. Resistance and adaptation may operate through over-expression or higher frequency of genes conferring pollutant tolerance, by selective growth of metal-tolerant microorganisms or via acquisition of new genetic tolerance-related capabilities through mutation or horizontal gene transfer, this latter aspect is presented in the following section and Chap. 6. In contrast some authors did not observe any or poor impact on community structure. Albeit such result could indeed correspond to a real situation revealing non-toxic effect of the pollutant on microbial communities, it can also indicate the onset of resistance phenomenon at the whole community level that may have occurred in the past. Additionally it cannot be exclude that methods of investigation or data analysis may have fail to detect a marginal effect of the pollutant (Wu et al. 2016).

Several studies indicated that the incidence of the pollutant on the microbial community may be transient, i.e. that once the contaminant was removed, the recovery of at least functional or even taxonomic diversity was observed (Tobor-Kapłon et al. 2005; Boivin et al. 2006; Bordenave et al. 2007; Mertens et al. 2007; Ma et al. 2015). This resilience may by explain by manifold reasons among which microorganism colonization from undisturbed site arising from the vicinity, population dormancy, growth rate (r *vs* K strategies), short time of exposure and functional redundancy.

As a consequence of shift in microbial diversity, it is important to research impact of pollutants on functional traits in the microbial community since their alteration may also seriously jeopardize ecosystem functioning or because they sustain community resistance, tolerance or resilience when traits are related to biodegradation. Hence several authors showed that pollutants might impair biomass, carbon mineralization, microbial respiration, nitrification and denitrification (Monard et al. 2011; Kumar et al. 2012; Singh et al. 2014; Delgado-Baquerizo et al. 2016; Simonin et al. 2016; Wu et al. 2016) or biodegradation (Caliz et al. 2011; Delgado-Baquerizo et al. 2016). However shifts in community structure could also be neutral on ecosystems functioning demonstrating that functional redundancy hold true in certain situation. This is conceivable if the loss of functionality bear by sensitive microorganisms does not affect trophic structure and is compensated by the functions of tolerant microorganisms still present in the community (Widenfalk et al. 2008; Azarbad et al. 2015). As suggested under the insurance hypothesis, this scheme is all the more likely if diversity is high in the ecosystem (Yachi and Loreau 1999).

The rise of high-throughput sequencing techniques contributes nowadays to uncover the huge diversity of the microbial realm. Our knowledge on bacterial community composition, functions and dynamics know since a decade unprecedented advances. Given this new development alongside other analytical tool important research topics are open. Hence, in 2006, this technologies had shed light on vast pool of low-abundance populations, the rare biosphere, that account for most of the observed phylogenetic diversity in every environment (Sogin et al. 2006). Importantly, microorganisms that constitute this pool, although still neglected in our investigation, may harbour ecologically critical functions in the ecosystem as demonstrated by Pester et al. (2010). Also in the context of future development in microbial ecotoxicology it would be worth asking what could be the role of this rare populations in microbial communities that had to face toxic agents. Additionally it seems necessary to develop new bioinformatics and statistical tools to extract meaningful information and decipher the ecologically relevant information from high-throughput sequencing data and more generally 'omics' technologies. Alongside to data collected at site, integrated approaches should allow to gain important knowledge on ecosystem functioning and has to ultimately lead to a better risk assessment and management of pollution at local but also importantly global scales.

4.4 Microbial Genetic Adaptation

At genome scale, the adaptation is the consequence of genetic variability and evolvability. Indeed, except the core genome involved in essential functions, a part of the genome is suitable to strong variations (Baquero 2009). Three major ways generate genome variations (Arber 2000). First, spontaneous mutations occur in a regular manner at each generation allowing local genomic changes (Feldgarden et al. 2003). Second, the rearrangement of segments of genomic sequences can be mediated by mechanisms such as homologous recombination or transposition (Thomas and Smalla 2000; Sota et al. 2006). Third, the fastest and powerful way to acquire new functions in bacteria and archaea is the acquisition of sequences from other organisms by horizontal gene transfer (HGT) (Garcia-Vallve et al. 2000; Ochman et al. 2000; Springael and Top 2004; Brochier-Armanet et al. 2011). At least a part of the genes acquired by HGT have a role in adaptation (Lawrence 1999; Marri et al. 2006). The mechanisms that allow the entrance and the establishment of foreign DNA in a genome are well-known (see Chap. 6). The establishment of the new genetic material will be possible if it is autonomous for replication in the recipient cell, or if it is capable of insertion in the chromosomal DNA (without major damage for the integrity of the host genome). Then, numerous genetic elements are involved in HGT, most of them are mobile genetic elements (Smalla and Sobecky 2002; Koonin and Wolf 2008; Boyd et al. 2009; Sentchilo et al. 2013; Darmon and Leach 2014). Moreover, mobile genetic elements are often more abundant in bacterial genomes in extreme environments (Bickhart et al. 2009; Lin et al. 2011), suggesting their role in the adaptation to unfavorable habitats. Thus mobile genetic elements play a major role in the spread and even *de novo* construction of new functions (Top and Springael 2003), and are thus central vectors for diversification and adaptation (Frost et al. 2005).

It is well known that stress conditions enhance the processes of genetic adaptation (Matic et al. 1995; Beaber et al. 2004; Ubeda et al. 2005; Galhardo et al. 2007; Baquero 2009). Therefore, contamination by xenobiotics is one factor that can stimulate microbial genomic adaptation (Top and Springael 2003; Springael and Top 2004; Marri et al. 2006; Heuer et al. 2008; Monard et al. 2011). Numerous characterized genetic elements, such as plasmids, transposons, genomic islands and integrons, carry adaptive genes involved in the resistance of antibiotics (Stokes and Hall 1991; Hansson et al. 2002; Del Grosso et al. 2007; Barraud et al. 2013; Giakkoupi et al. 2015; Korona-Glowniak et al. 2015), metals (Ji and Silver 1992; Liebert et al. 1999; Tuffin et al. 2005; Novais et al. 2010) and the degradation of organic pollutants (Nakatsu et al. 1991; van der Meer et al. 1991; Romine et al. 1999; Fong et al. 2000; Top and Springael 2003; Chae et al. 2007; Yano et al. 2007; Koenig et al. 2009; Ilori et al. 2015). Even some (e.g. integrons) are also able to exchange these genes in accordance with the contamination pressure imposed in the habitat (Stalder et al. 2012; Abella et al. 2015a). These observations support that all these genetic elements are important actors in adaptive responses to chemical contaminations.

The adaptation acquired by one organism is not only beneficial to the concerned individual, but also advantages the entire community, as for example the acquisition of a degradation function involved in the pollutant removal (Sentchilo et al. 2013). Also, the adaptive function acquired by HGT can be transmitted again to other members of the community. In this way, although the acquisition of new functions can be to the detriment of other functions (Ferenci 2016), it is doubtless an asset for the community. Our current knowledge on the adaptive genetic elements results essentially from analyses of isolated bacterial strains. The analysis of microbial complete-sequenced genomes allowed to estimate the part of sequences acquired by HGT in a given genome (Ochman et al. 2000; Brochier-Armanet et al. 2011). Nevertheless, we know that, within a community, microorganisms are organized in networks, sometimes showing strong interactions. Then the adaptation must be also studied at the community level. The new techniques of high throughput sequencing should give new information. On one hand, targeting directly the genetic elements involved in HGT (Zaneveld et al. 2008, 2011; Jacquiod et al. 2014), they enable the characterization of new adaptive genes acquired within the community (Huang et al. 2009) as well as the hierarchization of the involvement of the genetic elements in the adaptation mechanisms. On the other hand, metagenomic and metatranscriptomic studies may also contribute to better understand the mechanisms of genetic adaptation within communities in response to chemical pollutants. In particular, the pollution history influences the spread of adaptive genes, which are easily spread within a community subjected to an already experimented pollutant, while it is slower when submitted to a recent pollution or to a new pollutant (Abella et al. 2015a, b; Chessa et al. 2016). Research efforts must be undertaken in this sense in order to complete our knowledge on the genetic mechanisms involved in the adaptation of microbial communities.

4.5 Overview—Concluding Remarks

The metabolic versatility and genetic flexibility, together with community strategies are crucial assets allowing microorganisms to withstand the presence of pollutants. The microbial mechanisms discussed in this chapter provide the basic knowledge for the development of ecotoxicological tools reporting environmental quality. In contrast to chemical analysis methods, microbial ecotoxicological tools enable not only to determine pollutant concentration (biosensors, Chaps. 12 and 13) but also to assess the toxic effect at different biological levels including the genetic/genomic levels (Chaps. 6 and 8), the metabolic level (biomarkers, Chap. 11; bioindicators, Chap. 10) and the community level (Chaps. 8 and 9). Furthermore, microbial ecotoxicological tools allow to determine the microbial capacities to remove pollutants and represent thus useful tools for the implementation and the follow up of bioremediation processes. Because microorganisms are ubiquitous, microbial ecotoxicological tools can be potentially developed and exploited for every ecosystem and for any pollutant. However, the future challenges for the microbial ecotoxicology will be to propose integrated approaches to evaluate the impact of multi-contamination, including emergent contaminants. For this purpose basic knowledge on microbial ecology with a holistic point of view is of paramount importance and therefore such basic research should be encouraged.

Acknowledgements We would like to thank all partners of the MELODY group for their useful discussions.

References

Abella J, Bielen A, Huang L, Delmont TO, Vujaklija D, Duran R, Cagnon C (2015a) Integron diversity in marine environments. Environ Sci Pollut Res 22:15360–15369

Abella J, Fahy A, Duran R, Cagnon C (2015b) Integron diversity in bacterial communities of freshwater sediments at different contamination levels. FEMS Microbiol Ecol 91:fiv140

Acosta-González A, Marqués S (2016) Bacterial diversity in oil-polluted marine coastal sediments. Curr Opin Biotechnol 38:24–32

Allison SD, Martiny JBH (2008) Resistance, resilience, and redundancy in microbial communities. Proc Natl Acad Sci USA 105:11512–11519

Arber W (2000) Genetic variation: molecular mechanisms and impact on microbial evolution. FEMS Microbiol Rev 24:1–7

Azam F (1998) Microbial control of oceanic carbon flux: the plot thickens. Science 280:694–696

Azarbad H, Niklińska M, Nikiel K, van Straalen NM, Röling WFM (2015) Functional and compositional responses in soil microbial communities along two metal pollution gradients: does the level of historical pollution affect resistance against secondary stress? Biol Fertil Soils 51:879–890

Baquero F (2009) Environmental stress and evolvability in microbial systems. Clin Microbiol Infect 15:5–10

Bardgett RD, Van Der Putten WH (2014) Belowground biodiversity and ecosystem functioning. Nature 515:505–511

Bargiela R, Mapelli F, Rojo D et al (2015) Bacterial population and biodegradation potential in chronically crude oil-contaminated marine sediments are strongly linked to temperature. Scientific Reports 5

Barkay T, Wagner-Dobler I (2005) Microbial transformations of mercury: potentials, challenges, and achievements in controlling mercury toxicity in the environment. In: Laskin AI, Bennett JW, Gadd GM (eds) Advances in applied microbiology, Vol 57. pp 1–52

Barra Caracciolo A, Topp E, Grenni P (2015) Pharmaceuticals in the environment: Biodegradation and effects on natural microbial communities. A review. J Pharm Biomed Anal 106:25–36

Barraud O, Casellas M, Dagot C, Ploy MC (2013) An antibiotic-resistant class 3 integron in an Enterobacter cloacae isolate from hospital effluent. Clin Microbiol Infect 19:E306–E308

Barriuso E, Calvet R, Schiavon M, Soulas G (1996) Les pesticides et les polluants organiques des sols. Etude et gestion des sols 3:279–295

Baxter J, Cummings SP (2008) The degradation of the herbicide bromoxynil and its impact on bacterial diversity in a top soil. J Appl Microbiol 104:1605–1616

Beaber JW, Hochhut B, Waldor MK (2004) SOS response promotes horizontal dissemination of antibiotic resistance genes. Nature 427:72–74

Beketov MA, Liess M (2012) Ecotoxicology and macroecology—Time for integration. Environ Pollut 162:247–254

Ben Said O, Louati H, Soltani A et al. (2015) Changes of benthic bacteria and meiofauna assemblages during bio-treatments of anthracene-contaminated sediments from Bizerta lagoon (Tunisia). Environ Sci Pollut Res

Bickhart DM, Gogarten JP, Lapierre P, Tisa LS, Normand P, Benson DR (2009) Insertion sequence content reflects genome plasticity in strains of the root nodule actinobacterium Frankia. Bmc Genomics 10

Bogan BW, Sullivan WR (2003) Physicochemical soil parameters affecting sequestration and mycobacterial biodegradation of polycyclic aromatic hydrocarbons in soil. Chemosphere 52:1717–1726

Boivin MEY, Massieux B, Breure AM, Greve GD, Rutgers M, Admiraal W (2006) Functional recovery of biofilm bacterial communities after copper exposure. Environ Pollut 140:239–246

Bonnet JL, Bonnemoy F, Dusser M, Bohatier J (2007) Assessment of the potential toxicity of herbicides and their degradation products to nontarget cells using two microorganisms, the bacteria *Vibrio fischeri* and the ciliate *Tetrahymena pyriformis*. Environ Toxicol 22:78–91

Bordenave S, Goñi-Urriza MS, Caumette P, Duran R (2007) Effects of heavy fuel oil on the bacterial community structure of a pristine microbial mat. Appl Environ Microbiol 73: 6089–6097

Bordenave S, Fourçans A, Blanchard S, Goñi MS, Caumette P, Duran R (2004) Structure and functional analyses of bacterial communities changes in microbial mats following petroleum exposure. Ophelia 58:195–203

Bour A, Mouchet F, Silvestre J, Gauthier L, Pinelli E (2015) Environmentally relevant approaches to assess nanoparticles ecotoxicity: a review. J Hazard Mater 283:764–777

Boyd EF, Almagro-Moreno S, Parent MA (2009) Genomic islands are dynamic, ancient integrative elements in bacterial evolution. Trends Microbiol 17:47–53

Brazilian National Genome Project Consortium (2003) The complete genome sequence of *Chromobacterium violaceum* reveals remarkable and exploitable bacterial adaptability. Proc Natl Acad Sci USA 100:11660–11665

Brochier-Armanet C, Deschamps P, Lopez-Garcia P, Zivanovic Y, Rodriguez-Valera F, Moreira D (2011) Complete-fosmid and fosmid-end sequences reveal frequent horizontal gene transfers in marine uncultured planktonic archaea. ISME J 5:1291–1302

Bruins MR, Kapil S, Oehme FW (2000) Microbial resistance to metals in the environment. Ecotoxicol Environ Saf 45:198–207

Bruneel O, Pascault N, Egal M et al (2008) Archaeal diversity in a Fe-As rich acid mine drainage at Carnoulès (France). Extremophiles 12:563–571

Caliz J, Vila X, Marti E, Sierra J, Cruanas R, Garau MA, Montserrat G (2011) Impact of chlorophenols on microbiota of an unpolluted acidic soil: microbial resistance and biodegradation. FEMS Microbiol Ecol 78:150–164

Calvet R (1989) Adsorption of organic chemicals in soils. Environ Health Perspect 83:145–177

Cerniglia C (1992) Biodegradation of polycyclic aromatic hydrocarbons. Biodegradation 3: 351–368

Cervantes-Avilés P, Brito EMS, Duran R, Martínez AB, Cuevas-Rodríguez G (2016) Effect of ZnO nanoparticles in the oxygen uptake during aerobic wastewater treatment. J Nanopart Res 18

Cervantes C, Ji GY, Ramirez JL, Silver S (1994) Resistance to arsenic compounds in microorganisms. FEMS Microbiol Rev 15:355–367

Chae JC, Kim E, Bini E, Zystra GJ (2007) Comparative analysis of the catechol 2,3-dioxygenase gene locus in thermoacidophilic archaeon Sulfolobus solfataricus strain 98/2. Biochem Biophys Res Commun 357:815–819

Chavez FP, Gordillo F, Jerez CA (2006) Adaptive responses and cellular behaviour of biphenyl-degrading bacteria toward polychlorinated biphenyls. Biotechnol Adv 24:309–320

Chessa L, Jechalke S, Ding GC, Pusino A, Mangia NP, Smalla K (2016) The presence of tetracycline in cow manure changes the impact of repeated manure application on soil bacterial communities. Biol Fertil Soils 52:1121–1134

Chronopoulou P-M, Fahy A, Coulon F et al (2013) Impact of a simulated oil spill on benthic phototrophs and nitrogen-fixing bacteria in mudflat mesocosms. Environ Microbiol 15: 242–252

Chung N, Alexander M (2002) Effect of soil properties on bioavailability and extractability of phenanthrene and atrazine sequestered in soil. Chemosphere 48:109–115

Concha-Guerrero SI, Brito EMS, Piñón-Castillo HA et al. (2014) Effect of CuO nanoparticles over isolated bacterial strains from agricultural soil. J Nanomaterials

Cravo-Laureau C, Duran R (2014) Marine coastal sediments microbial hydrocarbon degradation processes: contribution of experimental ecology in the omics'era. Frontiers in Microbiol 5:39

Dachs J, Eisenreich SJ, Hoff RM (2000) Influence of eutrophication on air-water exchange, vertical fluxes, and phytoplankton concentrations of persistent organic pollutants. Environ Sci Technol 34:1095–1102

Darmon E, Leach DRF (2014) Bacterial genome instability. Microbiol Mol Biol Rev 78:1–39

Del Grosso M, Northwood JGE, Farrell DJ, Pantosti A (2007) The macrolide resistance genes erm (B) and mef(E) are carried by Tn2010 in dual-gene *Streptococcus pneumoniae* isolates belonging to clonal complex CC271. Antimicrob Agents Chemother 51:4184–4186

Delgado-Baquerizo M, Giaramida L, Reich PB et al (2016) Lack of functional redundancy in the relationship between microbial diversity and ecosystem functioning. J Ecol 104:936–946

Diaz E, Jimenez JI, Nogales J (2013) Aerobic degradation of aromatic compounds. Curr Opin Biotechnol 24:431–442

Diepens NJ, Arts GHP, Brock TCM, Smidt H, Van Den Brink PJ, Van Den Heuvel-Greve MJ, Koelmans AA (2014) Sediment toxicity testing of organic chemicals in the context of prospective risk assessment: A review. Crit Rev Environ Sci Technol 44:255–302

Dominique Y, Muresan B, Duran R, Richard S, Boudou A (2007) Simulation of the chemical fate and bioavailability of liquid elemental mercury drops from gold mining in Amazonian freshwater systems. Environ Sci Technol 41:7322–7329

Doyle E, Muckian L, Hickey AM, Clipson N (2008) Microbial PAH degradation. In: Laskin AI, Sariaslani S, Gadd GM (eds) Advances in applied microbiology, vol 65. pp 27–66

Duran R, Cravo-Laureau C (2016) Role of environmental factors and microorganisms in determining the fate of polycyclic aromatic hydrocarbons in the marine environment. FEMS Microbiol Rev 40:814–830

Duran R, Bielen A, Paradžik T et al (2015) Exploring Actinobacteria assemblages in coastal marine sediments under contrasted human influences in the West Istria Sea, Croatia. Environ Sci Pollut Res

Duran R, Menuet V, Monperrus M et al (2003) Monitoring bacterial communities adaptation to mercury contamination in estuarine sediments maintained in slurries. vol 107, pp 393–396

Duran R, Ranchou-Peyruse M, Menuet V et al (2008) Mercury methylation by a microbial community from sediments of the Adour Estuary (Bay of Biscay, France). Environ Pollut 156:951–958

Falkowski PG, Fenchel T, Delong EF (2008) The microbial engines that drive earth's biogeochemical cycles. Science 320:1034–1039

Feldgarden M, Byrd N, Cohan FM (2003) Gradual evolution in bacteria: evidence from Bacillus systematics. Microbiology-Sgm 149:3565–3573

Ferenci T (2016) Trade-off mechanisms shaping the diversity of bacteria. Trends Microbiol 24:209–223

Field CB, Behrenfeld MJ, Randerson JT, Falkowski P (1998) Primary production of the biosphere: Integrating terrestrial and oceanic components. Science 281:237–240

Field JA, Sierra-Alvarez R (2008) Microbial transformation and degradation of polychlorinated biphenyls. Environ Pollut 155:1–12

Foley ME, Sigler V, Gruden CL (2007) A multiphasic characterization of the impact of the herbicide acetochlor on freshwater bacterial communities. ISME J 2:56–66

Fong KPY, Goh CBH, Tan HM (2000) The genes for benzene catabolism in Pseudomonas putida ML2 are flanked by two copies of the insertion element IS1489, forming a class-I-type catabolic transposon, Tn5542. Plasmid 43:103–110

Frost LS, Leplae R, Summers AO, Toussaint A (2005) Mobile genetic elements: the agents of open source evolution. Nat Rev Microbiol 3:722–732

Fuchs G, Boll M, Heider J (2011) Microbial degradation of aromatic compounds—from one strategy to four. Nat Rev Microbiol 9:803–816

Galhardo RS, Hastings PJ, Rosenberg SM (2007) Mutation as a stress response and the regulation of evolvability. Crit Rev Biochem Mol Biol 42:399–435

Garcia-Vallve S, Romeu A, Palau J (2000) Horizontal gene transfer in bacterial and archaeal complete genomes. Genome Res 10:1719–1725

Giakkoupi P, Tryfinopoulou K, Polemis M, Pappa O, Miriagou V, Vatopoulos A (2015) Circulation of a multiresistant, conjugative, IncA/C plasmid within the nosocomial *Providencia stuartii* population in the Athens area. Diagn Microbiol Infect Dis 82:62–64

Guerrero R, Berlanga M (2006) Life's unity and flexibility: the ecological link. Int Microbiol 9:225–235

Guieysse B, Wuertz S (2012) Metabolically versatile large-genome prokaryotes. Curr Opin Biotechnol 23:467–473
Hansson K, Sundstrom L, Pelletier A, Roy PH (2002) IntI2 integron integrase in Tn7. J Bacteriol 184:1712–1721
Head IM, Jones DM, Roling WFM (2006) Marine microorganisms make a meal of oil. Nat Rev Micro 4:173–182
Heider J, Schühle K (2013) Anaerobic biodegradation of hydrocarbons including methane. In: Rosenberg E, DeLong EF, Lory S, Stackebrandt E, Thompson F (eds) The prokaryotes: prokaryotic physiology and biochemistry, pp 605–634,.Springer, Berlin, Heidelberg
Heuer H, Abdo Z, Smalla K (2008) Patchy distribution of flexible genetic elements in bacterial populations mediates robustness to environmental uncertainty. FEMS Microbiol Ecol 65: 361–371
Hjorth M, Vester J, Henriksen P, Forbes V, Dahllöf I (2007) Functional and structural responses of marine plankton food web to pyrene contamination. Mar Ecol Prog Ser 338:21–31
Huang L, Cagnon C, Caumette P, Duran R (2009) First gene cassettes of integrons as targets in finding adaptive genes in metagenomes. Appl Environ Microbiol 75:3823–3825
Hug LA, Baker BJ, Anantharaman K et al (2016) A new view of the tree of life. Nature Microbiology 1:16048
Ibekwe AM, Ma JC, Murinda SE (2016) Bacterial community composition and structure in an Urban River impacted by different pollutant sources. Sci Total Environ 566:1176–1185
Ilori MO, Picardal FW, Aramayo R, Adebusoye SA, Obayori OS, Benedik MJ (2015) Catabolic plasmid specifying polychlorinated biphenyl degradation in Cupriavidus sp strain SK-4: mobilization and expression in a pseudomonad. J Basic Microbiol 55:338–345
Jacquiod S, Demaneche S, Franqueville L et al (2014) Characterization of new bacterial catabolic genes and mobile genetic elements by high throughput genetic screening of a soil metagenomic library. J Biotechnol 190:18–29
Jain RK, Kapur M, Labana S, Lal B, Sarma PM, Bhattacharya D, Thakur IS (2005) Microbial diversity: application of microorganisms for the biodegradation of xenobiotics. Curr Sci 89:101–112
Ji GY, Silver S (1992) Regulation and expression of the arsenic resistance operon from Staphylococcus aureus plasmid PI258. J Bacteriol 174:3684–3694
Jiao S, Zhang Z, Yang F, Lin Y, Chen W, Wei G (2016) Temporal dynamics of microbial communities in microcosms in response to pollutants. Mol Ecol n/a-n/a
Johnston EL, Roberts DA (2009) Contaminants reduce the richness and evenness of marine communities: a review and meta-analysis. Environ Pollut 157:1745–1752
Koenig JE, Sharp C, Dlutek M, Curtis B, Joss M, Boucher Y, Doolittle WF (2009) Integron gene cassettes and degradation of compounds associated with industrial waste: the case of the sydney tar ponds. Plos One 4
Kokkali V, van Delft W (2014) Overview of commercially available bioassays for assessing chemical toxicity in aqueous samples. TrAC—Trends Anal Chem 61:133–155
Koonin EV, Wolf YI (2008) Genomics of bacteria and archaea: the emerging dynamic view of the prokaryotic world. Nucleic Acids Res 36:6688–6719
Korona-Glowniak I, Siwiec R, Malm A (2015) Resistance determinants and their association with different transposons in the antibiotic-resistant *Streptococcus pneumoniae*. Biomed Res Int Article ID 836496
Kumar N, Shah V, Walker VK (2012) Influence of a nanoparticle mixture on an arctic soil community. Environ Toxicol Chem 31:131–135
Lauga B, Girardin N, Karama S, Le Ménach K, Budzinski H, Duran R (2013) Removal of alachlor in anoxic soil slurries and related alteration of the active communities. Environ Sci Pollut Res 20:1089–1105
Lawrence JG (1999) Gene transfer, speciation, and the evolution of bacterial genomes. Curr Opin Microbiol 2:519–523
Lennon JT, Locey KJ (2016) The underestimation of global microbial diversity. mBio 7

Li ZJ, Xu JM, Tang CX, Wu JJ, Muhammad A, Wang HZ (2006) Application of 16S rDNA-PCR amplification and DGGE fingerprinting for detection of shift in microbial community diversity in Cu-, Zn-, and Cd-contaminated paddy soils. Chemosphere 62:1374–1380

Liebert CA, Hall RM, Summers AO (1999) Transposon Tn21, flagship of the floating genome. Microbiol Mol Biol Rev 63:507–522

Lin S, Haas S, Zemojtel T, Xiao P, Vingron M, Li RH (2011) Genome-wide comparison of cyanobacterial transposable elements, potential genetic diversity indicators. Gene 473:139–149

Lubarsky HV, Gerbersdorf SU, Hubas C, Behrens S, Ricciardi F, Paterson DM (2012) Impairment of the bacterial biofilm stability by Triclosan. Plos One 7

Ma J, Nossa CW, Alvarez PJJ (2015) Groundwater ecosystem resilience to organic contaminations: microbial and geochemical dynamics throughout the 5-year life cycle of a surrogate ethanol blend fuel plume. Water Res 80:119–129

Ma XY, Wang XC, Ngo HH, Guo W, Wu MN, Wang N (2014) Bioassay based luminescent bacteria: interferences, improvements, and applications. Sci Total Environ 468–469:1–11

Marri PR, Hao WL, Golding GB (2006) Gene gain and gene loss in Streptococcus: is it driven by habitat? Mol Biol Evol 23:2379–2391

Matic I, Rayssiguier C, Radman M (1995) Interspecies gene exchange in bacteria—the role of SOS and mismatch repair systems in evolution of species. Cell 80:507–515

Matsumura F (1989) Biotic degradation of pollutants. In: Bourdeau P, Haines JA, Klein W, Krishna Murti CR Ecotoxicology and Climate, pp 79–89, John Wiley

Mauffret A, Baran N, Joulian C (2017) Effect of pesticides and metabolites on groundwater bacterial community. Sci Total Environ 576:879–887

McGenity TJ, Folwell BD, McKew BA, Sanni GO (2012) Marine crude-oil biodegradation: a central role for interspecies interactions. Aquatic Biosystems 8

Meckenstock RU, Boll M, Mouttaki H et al (2016) Anaerobic degradation of benzene and polycyclic aromatic hydrocarbons. J Mol Microbiol Biotechnol 26:92–118

Mertens J, Ruyters S, Springael D, Smolders E (2007) Resistance and resilience of zinc tolerant nitrifying communities is unaffected in long-term zinc contaminated soils. Soil Biol Biochem 39:1828–1831

Misson B, Garnier C, Lauga B et al (2016) Chemical multi-contamination drives benthic prokaryotic diversity in the anthropized Toulon Bay. Sci Total Environ 556:319–329

Monard C, Vandenkoornhuyse P, Le Bot B, Binet F (2011) Relationship between bacterial diversity and function under biotic control: the soil pesticide degraders as a case study. ISME J 5:1048–1056

Mongodin EF, Shapir N, Daugherty SC et al (2006) Secrets of soil survival revealed by the genome sequence of *Arthrobacter aurescens* TC1. PLoS Genet 2:e214

Moreno-Jiménez E, Clemente R, Mestrot A, Meharg AA (2013) Arsenic and selenium mobilisation from organic matter treated mine spoil with and without inorganic fertilisation. Environ Pollut 173:238–244

Mukherjee S, Juottonen H, Siivonen P, Lloret Quesada C, Tuomi P, Pulkkinen P, Yrjala K (2014) Spatial patterns of microbial diversity and activity in an aged creosote-contaminated site. ISME J 8:2131–2142

Murinova S, Dercova K (2014) Response mechanisms of bacterial degraders to environmental contaminants on the level of cell walls and cytoplasmic membrane. Int J Microbiol 873081:26

Mustafa GA, Abd-Elgawad A, Ouf A, Siam R (2016) The Egyptian Red Sea coastal microbiome: a study revealing differential microbial responses to diverse anthropogenic pollutants. Environ Pollut 214:892–902

Nakatsu C, Ng J, Singh R, Straus N, Wyndham C (1991) Chlorobenzoate catabolic transposon TN5271 is a composite class-I element with flanking class-II insertion sequences. Proc Natl Acad Sci USA 88:8312–8316

Nelson KE, Fraser CM (2004) Champions of versatility. Trends Microbiol 12:111–112

Newman MC (2015) Fundamentals of ecotoxicology the science of pollution. CRC Press, Boca Raton (Fla.) London, New York

Novais A, Baquero F, Machado E, Canton R, Peixe L, Coque TM (2010) International spread and persistence of TEM-24 is caused by the confluence of highly penetrating Enterobacteriaceae clones and an IncA/C-2 plasmid containing Tn1696:Tn1 and IS5075-Tn21. Antimicrob Agents Chemother 54:825–834

Ochman H, Lawrence JG, Groisman EA (2000) Lateral gene transfer and the nature of bacterial innovation. Nature 405:299–304

Paissé S, Coulon F, Goñi-Urriza M, Peperzak L, McGenity TJ, Duran R (2008) Structure of bacterial communities along a hydrocarbon contamination gradient in a coastal sediment. FEMS Microbiol Ecol 66:295–305

Parales RE, Haddock JD (2004) Biocatalytic degradation of pollutants. Curr Opin Biotechnol 15:374–379

Parales RE, Luu RA, Hughes JG, Ditty JL (2015) Bacterial chemotaxis to xenobiotic chemicals and naturally-occurring analogs. Curr Opin Biotechnol 33:318–326

Pascault N, Roux S, Artigas J et al (2014) A high-throughput sequencing ecotoxicology study of freshwater bacterial communities and their responses to tebuconazole. FEMS Microbiol Ecol 90:563–574

Passatore L, Rossetti S, Juwarkar AA, Massacci A (2014) Phytoremediation and bioremediation of polychlorinated biphenyls (PCBs): state of knowledge and research perspectives. J Hazard Mater 278:189–202

Paule A, Roubeix V, Lauga B, Duran R, Delmas F, Paul E, Rols JL (2013) Changes in tolerance to herbicide toxicity throughout development stages of phototrophic biofilms. Aquat Toxicol 144–145:310–321

Pester M, Bittner N, Deevong P, Wagner M, Loy A (2010) A "rare biosphere" microorganism contributes to sulfate reduction in a peatland. ISME J 4:1591–1602

Pieper DH (2005) Aerobic degradation of polychlorinated biphenyls. Appl Microbiol Biotechnol 67:170–191

Pieper DH, Martins Dos Santos VAP, Golyshin PN (2004) Genomic and mechanistic insights into the biodegradation of organic pollutants. Curr Opin Biotechnol 15:215–224

Pomeroy LR, le Williams PJB, Azam F, Hobbie JE (2007) The microbial loop. Oceanography 20:28–33

Prabhakaran P, Ashraf MA, Aqma WS (2016) Microbial stress response to heavy metals in the environment. Rsc Advances 6:109862–109877

Puglisi E, Hamon R, Vasileiadis S, Coppolecchia D, Trevisan M (2012) Adaptation of soil microorganisms to trace element contamination: A review of mechanisms, methodologies, and consequences for risk assessment and remediation. Crit Rev Environ Sci Technol 42: 2435–2470

Rabus R, Boll M, Heider J et al (2016) Anaerobic microbial degradation of hydrocarbons: from enzymatic reactions to the environment. J Mol Microbiol Biotechnol 26:5–28

Ramond JB, Berthe T, Duran R, Petit F (2009) Comparative effects of mercury contamination and wastewater effluent input on Gram-negative merA gene abundance in mudflats of an anthropized estuary (Seine, France): a microcosm approach. Res Microbiol 160:10–18

Rodriguez-R LM, Overholt WA, Hagan C, Huettel M, Kostka JE, Konstantinidis KT (2015) Microbial community successional patterns in beach sands impacted by the Deepwater Horizon oil spill. ISME J 9:1928–1940

Romine MF, Stillwell LC, Wong KK et al (1999) Complete sequence of a 184-kilobase catabolic plasmid from *Sphingomonas aromaticivorans* F199. J Bacteriol 181:1585–1602

Sardessai Y, Bhosle S (2002) Tolerance of bacteria to organic solvents. Res Microbiol 153: 263–268

Sentchilo V, Mayer AP, Guy L et al (2013) Community-wide plasmid gene mobilization and selection. ISME J 7:1173–1186

Shahgholi H, Gholamalizadeh Ahangar A (2014) Factors controlling degradation of pesticides in the soil environment: a Review. Agriculture Science Developments 3:273–278

Simonin M, Richaume A, Guyonnet JP, Dubost A, Martins JMF, Pommier T (2016) Titanium dioxide nanoparticles strongly impact soil microbial function by affecting archaeal nitrifiers. Scientific Reports 6

Simonin M, Martins JMF, Le Roux X, Uzu G, Calas A, Richaume A (2017) Toxicity of TiO2 nanoparticles on soil nitrification at environmentally relevant concentrations: Lack of classical dose–response relationships. Nanotoxicology 1–9

Singh BK, Quince C, Macdonald CA et al (2014) Loss of microbial diversity in soils is coincident with reductions in some specialized functions. Environ Microbiol 16:2408

Smalla K, Sobecky PA (2002) The prevalence and diversity of mobile genetic elements in bacterial communities of different environmental habitats: insights gained from different methodological approaches. FEMS Microbiol Ecol 42:165–175

Sogin ML, Morrison HG, Huber JA et al (2006) Microbial diversity in the deep sea and the underexplored “rare biosphere. Proc Natl Acad Sci 103:12115–12120

Sota M, Yano H, Ono A et al (2006) Genomic and functional analysis of the IncP-9 naphthalene-catabolic plasmid NAH7 and its transposon Tn4655 suggests catabolic gene spread by a tyrosine recombinase. J Bacteriol 188:4057–4067

Sowers KR, May HD (2013) In situ treatment of PCBs by anaerobic microbial dechlorination in aquatic sediment: are we there yet? Curr Opin Biotechnol 24:482–488

Spagnuolo M, Puglisi E, Vernile P, Bari G, de Lillo E, Trevisan M, Ruggiero P (2010) Soil monitoring of pentachlorophenol by bioavailability and ecotoxicity measurements. J Environ Monit 12:1575–1581

Springael D, Top EM (2004) Horizontal gene transfer and microbial adaptation to xenobiotics: new types of mobile genetic elements and lessons from ecological studies. Trends Microbiol 12:53–58

Stalder T, Barraud O, Casellas M, Dagot C, Ploy MC (2012) Integron involvement in environmental spread of antibiotic resistance. Frontiers Microbiol 3

Stauffert M, Cravo-Laureau C, Duran R (2015a) Dynamic of sulphate-reducing microorganisms in petroleum-contaminated marine sediments inhabited by the polychaete *Hediste diversicolor*. Environ Sci Pollut Res 22:15273–15284

Stauffert M, Cravo-Laureau C, Duran R (2015b) Structure of hydrocarbonoclastic nitrate-reducing bacterial communities in bioturbated coastal marine sediments. FEMS Microbiol Ecol 89:580–593

Stauffert M, Duran R, Gassie C, Cravo-Laureau C (2014) Response of archaeal communities to oil spill in bioturbated mudflat sediments. Microb Ecol 67:108–119

Stauffert M, Cravo-Laureau C, Jézéquel R et al (2013) Impact of oil on bacterial community structure in bioturbated sediments. PLoS ONE 8

Stokes HW, Hall RM (1991) Sequence analysis of the inducible chloramphenicol resistance determinant in the TN1696 integron suggests regulation by translational attenuation. Plasmid 26:10–19

Stokes HW, Gillings MR (2011) Gene flow, mobile genetic elements and the recruitment of antibiotic resistance genes into Gram-negative pathogens. FEMS Microbiol Rev 35:790–819

Thomas CM, Smalla K (2000) Trawling the horizontal gene pool. Microbiol Today 27:24–27

Tobor-Kapłon MA, Bloem J, Römkens PFAM, Ruiter PCd (2005) Functional stability of microbial communities in contaminated soils. Oikos 111:119–129

Top EM, Springael D (2003) The role of mobile genetic elements in bacterial adaptation to xenobiotic organic compounds. Curr Opin Biotechnol 14:262–269

Tuffin IM, de Groot P, Deane SM, Rawlings DE (2005) An unusual Tn21-like transposon containing an ars operon is present in highly arsenic-resistant strains of the biomining bacterium *Acidithiobacillus caldus*. Microbiol Sgm 151:3027–3039

Ubeda C, Maiques E, Knecht E, Lasa I, Novick RP, Penades JR (2005) Antibiotic-induced SOS response promotes horizontal dissemination of pathogenicity island-encoded virulence factors in staphylococci. Mol Microbiol 56:836–844

Ufarte L, Laville E, Duquesne S, Potocki-Veronese G (2015) Metagenomics for the discovery of pollutant degrading enzymes. Biotechnol Adv 33:1845–1854

Van Der Heijden MGA, Bardgett RD, Van Straalen NM (2008) The unseen majority: soil microbes as drivers of plant diversity and productivity in terrestrial ecosystems. Ecol Lett 11:296–310

van der Meer JR, Zehnder AJB, Devos WM (1991) Identification of a novel composite transposable element, Tn5280, carrying chlorobenzene dioxygenase genes of Pseudomonas sp strain P51. J Bacteriol 173:7077–7083

Vercraene-Eairmal M, Lauga B, Saint Laurent S et al (2010) Diuron biotransformation and its effects on biofilm bacterial community structure. Chemosphere 81:837–843

Viret H, Pringault O, Duran R (2006) Impact of zinc and nickel on oxygen consumption of benthic microbial communities assessed with microsensors. Sci Total Environ 367:302–311

Volant A, Bruneel O, Desoeuvre A et al (2014) Diversity and spatiotemporal dynamics of bacterial communities: physicochemical and other drivers along an acid mine drainage. FEMS Microbiol Ecol 90:247–263

Walker A (1987) Further observations on the enhanced degradation of iprodione and vinclozolin in soil. Pestic Sci 21:219–231

Wang ZG, Liu S, Xu WH, Hu YL, Hu Y, Zhang Y (2016) The microbiome and functions of black soils are altered by dibutyl phthalate contamination. Appl Soil Ecol 99:59–69

Watanabe K (2001) Microorganisms relevant to bioremediation. Curr Opin Biotechnol 12: 237–241

Widenfalk A, Bertilsson S, Sundh I, Goedkoop W (2008) Effects of pesticides on community composition and activity of sediment microbes—responses at various levels of microbial community organization. Environ Pollut 152:576–584

Woignier T, Fernandes P, Soler A, Clostre F, Carles C, Rangon L, Lesueur-Jannoyer M (2013) Soil microstructure and organic matter: Keys for chlordecone sequestration. J Hazard Mater 262:357–364

Wu YC, Zhu QH, Zeng J, Ding QM, Gong Y, Xing P, Lin XG (2016) Effects of pH and polycyclic aromatic hydrocarbon pollution on thaumarchaeotal community in agricultural soils. J Soils Sediments 16:1960–1969

Xiao F, Gulliver JS, Simcik MF (2013) Predicting aqueous solubility of environmentally relevant compounds from molecular features: a simple but highly effective four-dimensional model based on Project to Latent Structures. Water Res 47:5362–5370

Yachi S, Loreau M (1999) Biodiversity and ecosystem productivity in a fluctuating environment: the insurance hypothesis. Proc Natl Acad Sci USA 96:1463–1468

Yano H, Garruto CE, Sota M et al (2007) Complete sequence determination combined with analysis of transposition/site-specific recombination events to explain genetic organization of IncP-7 TOL plasmid pWW53 and related mobile genetic elements. J Mol Biol 369:11–26

Zaneveld JR, Nemergut DR, Knight R (2008) Are all horizontal gene transfers created equal? Prospects for mechanism-based studies of HGT patterns. Microbiol-Sgm 154:1–15

Zaneveld JRR, Parfrey LW, Van Treuren W et al (2011) Combined phylogenetic and genomic approaches for the high-throughput study of microbial habitat adaptation. Trends Microbiol 19:472–482

Zhang C, Bennett GN (2005) Biodegradation of xenobiotics by anaerobic bacteria. Appl Microbiol Biotechnol 67:600–618

Chapter 5
Engineered Nanoparticles in the Environments: Interactions with Microbial Systems and Microbial Activity

Samuel Eduok and Frédéric Coulon

Abstract There has been a remarkable development of nanotechnology and growing interest in the application of engineered nanoparticles in several products over the last decade. Their use in several consumer products have been associated with increased concern for human and environmental health due to the potential toxicological implications of engineered nanoparticles (ENPs) released into the environment which could have adverse effect on bacteria-dependent processes. Despite the great research attention commanded by ENPs effect on biological systems in recent years, there is still a considerable challenge in the analytical procedures and evaluation. ENPs exert their antimicrobial effect through a wide range of mechanisms including the formation of reactive oxygen species, disruption of microbial physiology and metabolic processes although there is increasing evidence to suggest that ENPs could also augment microbial-mediated processes in the ecosystem. Although little is known about the environmental fate and transport of ENPs, wastewater would serve as a sink for most of the nano-enabled waste and by-products. To date, nano-ecotoxicological studies report contrasting findings on bacterial inhibition and/or stimulation, survival and death which are dose- and species-specific and analytical protocol-dependent. Further to this, studies are largely influenced by the exposure duration, the type and the composition of the environmental matrix tested, and the reactive properties of the ENPs. Without caution, the interpretation of ENPs ecotoxicological effect on microbial activity, community structure, composition and diversity by different analytical protocols can be true but misleading because ENPs are differentially toxic to diverse microorganisms in pure and mixed cultures. Therefore, the development of a general and holistic guideline for microbial nano-ecotoxicity evaluation can at best be described as work in progress.

S. Eduok · F. Coulon (✉)
School of Water, Energy and Environment, Cranfield University, Cranfield MK43 0AL, UK
e-mail: f.coulon@cranfield.ac.uk

S. Eduok
Department of Microbiology, University of Uyo, P.M.B 1017, Uyo, Nigeria

C. Cravo-Laureau et al. (eds.), *Microbial Ecotoxicology*,
DOI 10.1007/978-3-319-61795-4_5

Keywords Engineered nanoparticles · Nano-ecotoxicity · Nanotoxicity · Nano-tolerance

5.1 Introduction

Engineered nanoparticles (ENPs) are materials with a size ranging between 1 and 100 nm in at least one dimension, synthesized and incorporated into a variety of consumer products because of their novel physical and chemical properties. ENPs are intentionally produced and designed with very specific properties related to shape, size, surface properties and chemistry. These properties are reflected in different forms such as aerosols, colloids, or powders. As colloids, most ENPs are insoluble in aqueous medium, and not retained on saturated porous media against the prediction of the Derjaguin-Landau-Verwey-Overbeek (DLVO) theory in which sorbed particles are expected to remain attached to media surfaces (Eduok 2013). In relation to their optical property, conductivity and reactivity, ENPs obey the laws of quantum physics instead of colloidal chemistry which enhances their functional characteristics. Most of the ENPs are metal oxide-derived through physical, chemical and biological synthesis. Comparatively, the biologically synthesized nanoparticles are the most eco-friendly and less disruptive to the environment (Durán et al. 2010).

An inventory of ENPs-enabled product applications indicates more than 1814 products are being manufactured using ENPs and it is projected that the number of products will triple by 2020 (Fig. 5.1; Vance et al. 2015; Woodrow Wilson Database 2016). The main components of global ENPs production consist of metal oxide nanomaterials with titanium dioxide (TiO_2) and silver oxide (Ag°) nanoparticles having the highest production volumes (Fig. 5.2). Their increased use in industrial, agricultural applications, consumer products and a variety of medical applications (Fig. 5.3) contribute to their unintended release in the environment via various pathways and acting in unknown manner on soils, waters and biota (Eduok 2013; Eduok et al. 2013; Schaumann et al. 2015).

Apart from accidental release during production, it is unlikely that ENPs will remain bound to the products at the end of the product life cycle (Daughton and Ternes 1999; Moore 2006). For instance, there are empirical evidence that ENPs are present in sewage sludge (Wigger et al. 2015; Batley et al. 2012; Kim et al. 2010; Kiser et al. 2010), wastewater effluents (Kim et al. 2010; Kiser et al. 2009), and landfill leachates (Hennebert et al. 2013; Kim et al. 2010; Blaser et al. 2008; Mueller and Nowack 2008). Indeed, sewage sludge is used for various purposes such as agriculture and soil amendment (55%), thermal energy generation (25%) and solid waste landfills (20%) (EEA 2013). Thus, wastewater is a primary point source of aged-ENPs input into the environment most likely either through wastewater-sludge-digestate-soil pathway or wastewater-effluent-surface water

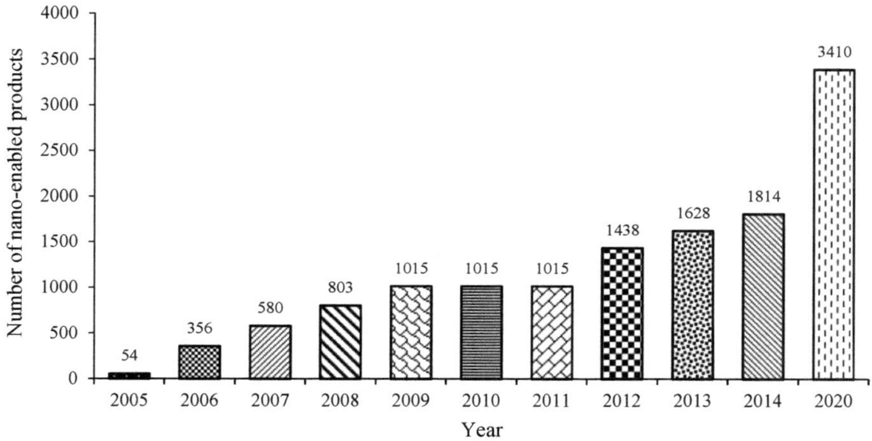

Fig. 5.1 Trends and estimated projection in the number of ENPs-enabled product applications (adapted from Woodrow Wilson Database 2016)

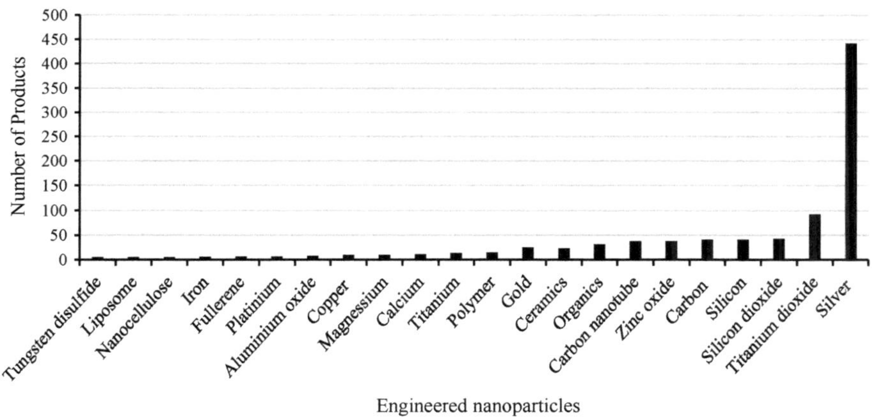

Fig. 5.2 Distribution of engineered nano-materials used in industrial and consumer products. ENPs used in less than 5 products are omitted (adapted from Woodrow Wilson Database 2016)

pathway (Fig. 5.4). Moreover, there is a marginal and restricted use of ENPs such as single- and multiwalled carbon nanotubes (S/MWCNTs), zinc, iron, phosphorus, calcium, magnesium, copper, molybdenum, silver, silica, aluminium silicate and manganese nanoparticles in biocontrol and pesticide compositions (Parisi et al. 2015; Patel et al. 2014). The ENPs are used as carriers, additives or as active ingredients in these formulations (Naderi and Danesh-Shahraki 2013) to enhance the bioavailable concentration for contact and effect. The application of

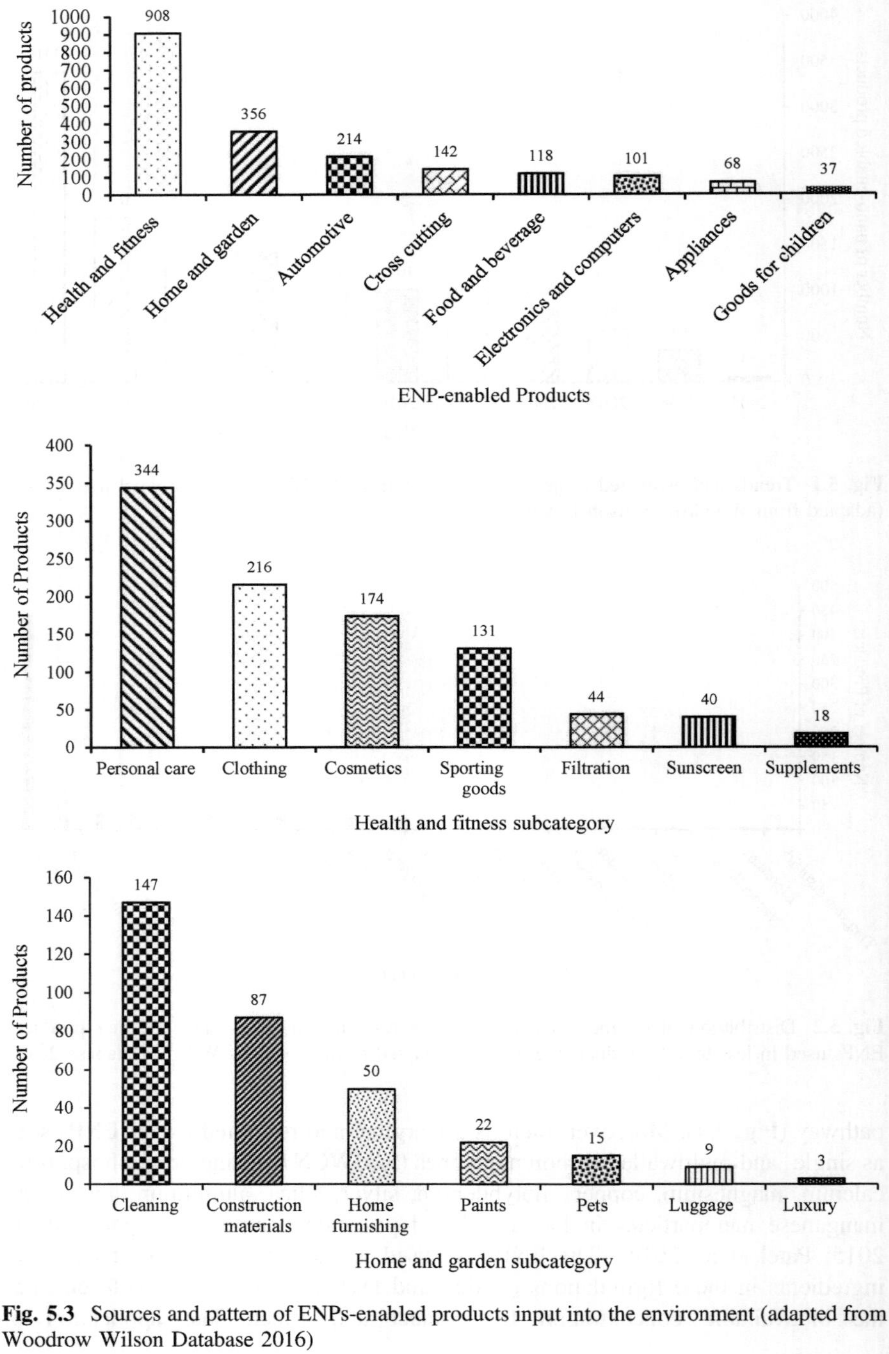

Fig. 5.3 Sources and pattern of ENPs-enabled products input into the environment (adapted from Woodrow Wilson Database 2016)

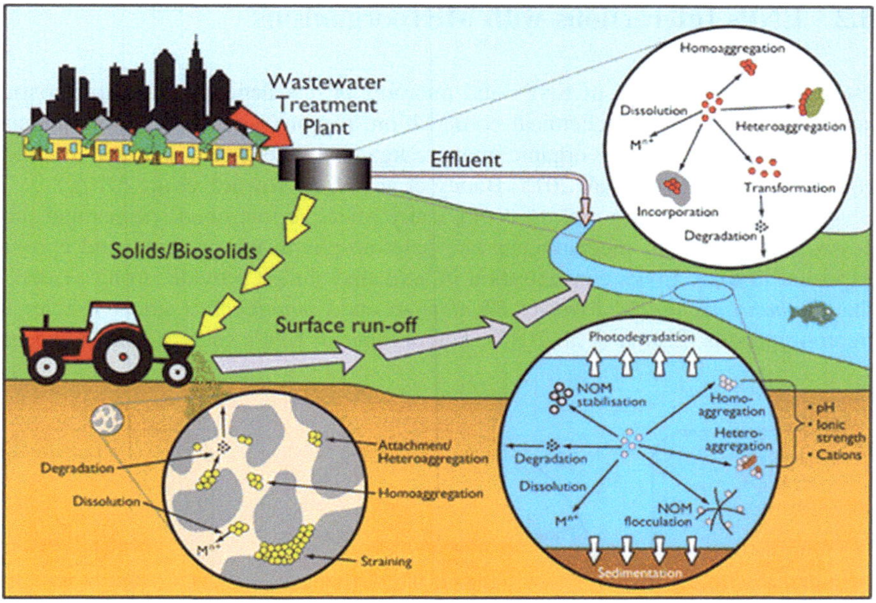

Fig. 5.4 Wastewater treatment plant as a major source distribution influencing fate and transport of engineered nanoparticles in the environment (reproduced from Batley et al. 2012)

nanotechnology in agriculture to produce and use nanopesticides for effective growth control of phytopathogenic fungi, viruses and bacteria, and as nanofertilizers to enhance plant growth and increase yield have been extensively reviewed (Mukherjee et al. 2016; Lui and Lal 2015; Servin et al. 2015; Patel et al. 2014; Jo et al. 2009). Once ENPs are released into the environment as dry powders, aerosolized sprays or associated with waste, biosolids and effluents, the soil or sediment becomes a major sink for both pristine and aged-ENPs (Cornelis et al. 2014; Keller et al. 2013; Batley et al. 2012).

Further to this, control of pathogenic microbes by antimicrobial ENPs is a promising approach to defeat pathogens exhibiting multi-resistance such as methicillin resistant *Staphylococcus aureus*. However, non-target effects on the microbial populations that play beneficial roles in the environment (i.e. nitrifiers) could have negative consequences (Simonin et al. 2017; Eduok et al. 2013). Microorganisms are the natural engines that drive biogeochemical processes in the earth's ecosystem including degradation, transformation and stabilisation of contaminants during waste treatment. Diverse microbial groups are involved in these processes either as free-living organisms or in association with other organisms including plants. Thus, it is important to understand the potential impacts of ENPs in the environment.

5.2 ENPs Interactions with Microorganisms

Overall, the interactions of ENPs and microorganisms depends on characteristics such as the size, shape, chemical composition, capping agent and environmental factors including natural organic matter, ligands, surfactants, pH and colloids (Fig. 5.5; Schaumann et al. 2015; Batley et al. 2012; Nowack et al. 2010).

Although ENP pose a risk to ecologically sensitive microbial species and processes, the growth of methanogens and heterotrophs in the presence and chronic exposure to toxic ENPs concentration in activated sludge provide strong evidence that *Methanosarcina, Acidovorax, Rhodoferax* and *Commamonas*, are nano-tolerant microorganisms (Eduok et al. 2015). Direct input of ENPs such as iron (Lacoanet

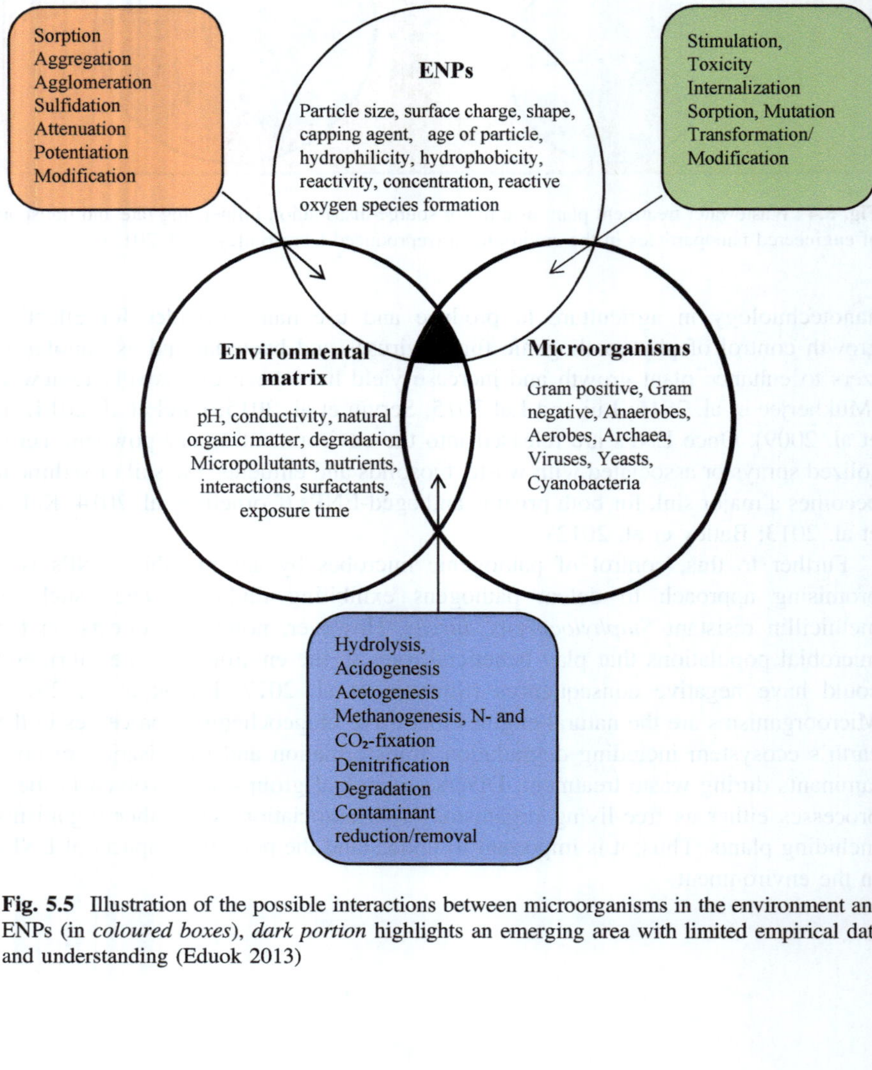

Fig. 5.5 Illustration of the possible interactions between microorganisms in the environment and ENPs (in *coloured boxes*), *dark portion* highlights an emerging area with limited empirical data and understanding (Eduok 2013)

and Wiesner 2004; Yavuz et al. 2006) and TiO_2 nanoparticles (Mach 2004) used in water treatment and environmental remediation inhibit and stimulate target organisms respectively, and at the same time exert adverse effect on non-target microorganisms and other biological systems.

ENPs are also known to associate with natural organic matter and input into the environment most likely will be through the wastewater treatment plant (WWTP)–biosolids/digestate-or-effluent pathway. However, the presence of biosolids, NOM and colloids influenced ENPs behaviour, fate and transport by either aggregation, absorption, change in oxidation state, precipitation or formation of complexes with ligands in wastewater and soil to mitigate toxic effect (Xiu et al. 2011; Liu et al. 2010). In addition, ENPs undergo transformations in the environmental matrix (Figs. 5.6 and 5.7) that attenuates or increases toxic effects, and thus is important to understand the changes for an unbiased evaluation of nano-ecotoxicity (Lowry et al. 2012). For example, silver oxide nanoparticle (Ag°) is known to react with chlorides, sulphide (Levard et al. 2011) and natural organic matter (Arvizo et al. 2010) resulting in attenuation of the toxic effect.

In contrast, zero-valent iron nanoparticles (nZVI) exerted stimulatory effect on soil microbial activities (Cullen et al. 2011), whereas other ENPs such as zinc oxide (ZnO), copper oxide (CuO), Ag° and single-walled carbon nanotubes (SWCNTs), multi-walled carbon nanotubes (MWCNTs) or fullerene soot (FS) exert varying acute and chronic toxic effects on pure cultures of microorganisms (Table 5.1) and soil microbial communities (Table 5.2). Moreover, ENPs undergo several transformations in the environment such as photochemical transformations, dissolutions,

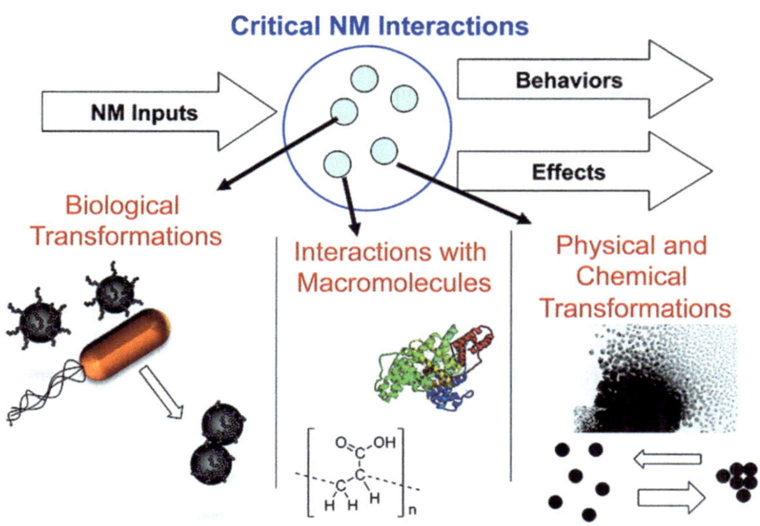

Fig. 5.6 Transformations of nanomaterials (NM) are critical processes affecting interactions such as physical and chemical transformations, biologically mediated transformations, and interactions with macromolecules and biomacromolecules (adapted from Lowry et al. 2012)

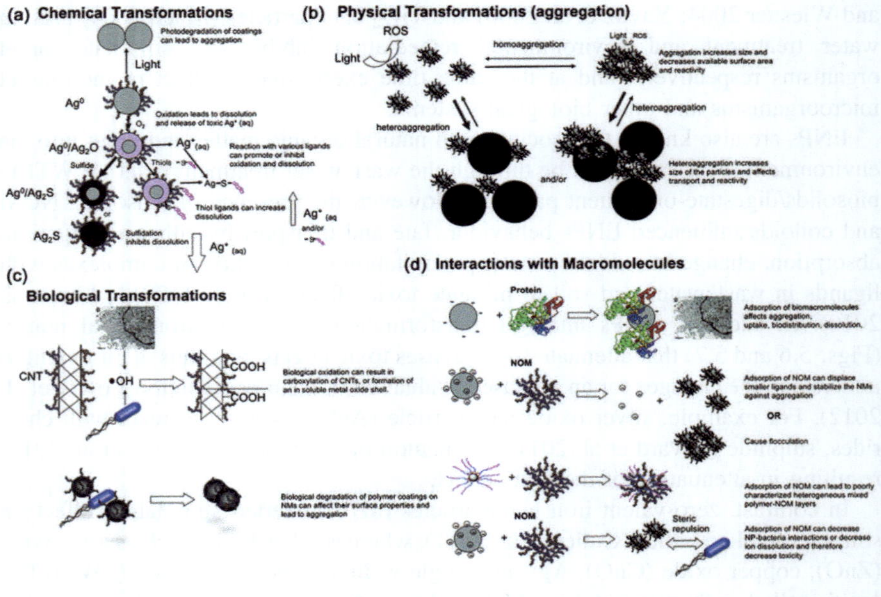

Fig. 5.7 **a** Representative chemical transformations of metal nanomaterials, the potential impacts on their behaviour and effects in the environment. AgNPs are used to exemplify the types of transformations that may occur. **b** Effects of physical transformations including aggregation and heteroaggregation on the reactivity and transport of nanomaterials. (The magnitude of *arrows* approximately correlates with potential for processes (**a** and **b**) to occur as determined from the limited data available on these processes.) **c** Biologically mediated transformations of nanomaterials and their coatings, and the subsequent impact on fate, transport, and effects. **d** Effects of nanomaterial interactions with macromolecules such as proteins and natural organic matter. Adsorbed macromolecules can affect aggregation, nanoparticle-biointeractions, bio-uptake, and fate, transport, and effects in the environment. *Arrows* do not indicate the relative potential for processes (**c** and **d**) to occur due to the limited data currently available for that assessment (adapted from Lowry et al. 2012)

precipitation, oxidation, reduction, adsorption and desorption, combustion, abrasion and biotransformations. However, information on the biotransformation of ENPs and its effect on its structure and reactivity is limited (Mitrano et al. 2015).

Other factors such as the size of ENPs (Sotiriou and Pratsinis 2010), the presence of divalent cations/anions and surface charges (Chen and Elimelech 2007; El-Badawy et al. 2010; Li et al. 2010), the bacterial cell wall composition and their charges (Sondi and Salopek-Sondi 2004; Morones et al. 2005; Jin et al. 2009), and the use of capping agents which repel ENPs by electrostatic, steric or electrosteric forces to avoid forming aggregate (Zhang et al. 2015; Hotze et al. 2010; Phenrat et al. 2008) can either enhance or attenuate ENP microbiocidal effect. It is suspected that sorption and accumulation of ENPs in sludge can adversely affect the efficiency of activated sludge and anaerobic digestion processes, which rely on diverse microbial communities. With wastewater biosolids serving as a sink and source of

Table 5.1 Effect of ENPs on pure cultures of microorganisms

ENPs	Size (nm)	Test organism	Effect on microorganisms	References
PVA- and Na_2ATP-doped Ag°	7 ± 3 40 ± 14	*Nitrosomonas europaea* ATCC19718	Size- and capping material-dependent inhibition of ammonia oxidation, disintegration of nucleoid, damage of cell wall	Yuan et al. (2013)
nZVI	20–90	*Paracoccus sp.*	Dose-dependent inactivation with low cell density, 50 mg L^{-1} promoted both cell growth and biodegradation of nitrate	Jiang et al. (2015)
ZnO	2–28	*Pseudomonas sp., Fusarium sp.*	Growth inhibition and disruption of bacterial and fungal cell wall/membrane	Sharma et al. (2010)
ZnO	70 ± 15	*Botrytic cinerea, Penicillium expansum*	3 mmol L^{-1} significantly inhibited fungal growth with *P. expanssum* as the most sensitive to the treatment. Deformation of fungal hyphae and inhibition of cellular activities in *B. cinerea*. Prevented conidiophore development in *P. expansum* resulting in the death of fungal hyphae	He et al. (2011a, b)

ENPs-enabled waste into the environment, soil microorganisms thus constitute the bulk of unintentional target of the toxic effects. Like most xenobiotic compounds, the adverse effect of ENPs on microorganisms is gradually emerging and still not well understood (Woodrow Wilson Database 2016; Pan et al. 2010; Kim et al. 2010; Liang et al. 2010; Weisner et al. 2009; Choi et al. 2008). This inference is as a result of the established antimicrobial properties of several ENPs on most pure cultures of different microorganisms (Eduok and Coulon 2017) and emerging evidence of ENPs toxic effect on soil microbial community (Table 5.2).

5.3 Influence of Capping Agents and Functional Groups

ENPs are very reactive because of their nano-size with a tendency to aggregate in aqueous medium. Generally, capping agent are used to coat the ENPs surfaces to stabilize the particles and minimize aggregation for optimum reaction. The coating agents (Tables 5.3 and 5.4) therefore becomes an integral part of the ENPs and can influence the differential effect exerted by the nanoparticles. These agents can be categorized into cationic, neutral and anionic (Zhang et al. 2015). In addition, other materials such as polyvinylpyrrolidone (PVP), citrate, ascorbic acid, sodium polyacrylate, polyoxyethylene glycerol trioleate, polyoxyethylene sorbitan

Table 5.2 Effect of ENPs on microbial community activity and composition in different environmental matrix

ENPs	Size (nm)	ENPs concentration	Environmental matrix/microbial group	Exposure time	Effect on microbial cell, community composition, diversity, activity and dynamics	References
AgO (citrate coated)	9.9	1, 10, 100, 1000 μg g^{-1}	Surface soil (0–5 cm)	7-days, microcosm	Significantly inhibited the activities of exoenzymes associated with nutrient cycle (urease, acid phosphatase, arylsulfatase, β-glucosidase) and the overall microbial activity (dehydrogenase and fluorescein diacetate hydrolase)	Shin et al. (2012)
Aged-Ag°	18.34	0.5–100 μg mL^{-1}	Rhizosphere bacteria	4 and 6-h, microcosms	Pit formation and leakage of reducing sugars. *Bacillus thuringiensis* SBURRI completely eliminated compared to *B. amyloliquefaciens* SBURR5 that was tolerant with growth stimulated by the aged-Ag°	Mirzajani et al. (2013)
Ag°	20	100 mg kg^{-1}	Rhizosphere microbial community in 30 cm sandy loam from corn field	75-days, mesocosms	Significant Ag°-induced alteration in rhizosphere bacterial community composition with slight modifications in the fungal community composition compared to the bulk soil	Sillen et al. (2015)

(continued)

Table 5.2 (continued)

ENPs	Size (nm)	ENPs concentration	Environmental matrix/microbial group	Exposure time	Effect on microbial cell, community composition, diversity, activity and dynamics	References
PVP-Ag°, Ag_2S	20–30	1, 10, 100 mg kg^{-1}	Sludge-amended sandy loam soil (10 cm) Rhizosphere microbial community	8 weeks, mesocosms	Significantly reduced arbuscular mycorrhizal colonization, microbial biomass, Gram-negative bacteria, Actinomycetes, and fungi with proportional increase in Gram-positive bacterial community structure and composition but not concentration-dependent	Judy et al. (2015)
Ag°	10 and 100	6 mg L^{-1}	Genetically modified *Escherichia coli* (JM109) and *Sinorhizobium meliloti* for overproduction of EPS	24-h, microcosm	Exopolymeric substance (EPS) enhanced tolerance and resistance resulting in higher survival and viability rates compared to non EPS producing strain. EPS trapped Ag°, prevented contact with cell wall, induced cell aggregation to reduce cell surface exposed to Ag°	Joshi et al. (2012)
Ag°	65 ± 30	0, 20, 200, 2000 µg L^{-1}	Biofilm microbial community in 4 L freshly collected seawater	4-days, mesocosms	Concentration-dependent and significant decrease in biofilm volume and biomass. Inhibition of biofilm succession and relative abundance of major bacterial	Febrega et al. (2011)

(continued)

Table 5.2 (continued)

ENPs	Size (nm)	ENPs concentration	Environmental matrix/microbial group	Exposure time	Effect on microbial cell, community composition, diversity, activity and dynamics	References
					groups with potential long-term effect on biofilm development and function	
Ag°	10	22.4 mg L^{-1}	*Pseudomonas chlororaphis* 06	72-h, microcosm	Dose-dependent inhibition of cell growth and viability with 3 mg L^{-1} as bactericidal, no evidence of cell wall disruption or leakage. Impaired growth mitigated by production of EPS	Dimkpa et al. (2011a, b)
Phosphate-stabilized Ag°	20 and 80	0.25–5 ppm	Ammonia-oxidizing bacteria (*Nitrosomonas europaea*) in 155 mL batch reactor	3-h, microcosm	Size-dependent toxic effect, decreased ammonia-oxidizing activity and destabilization of outer-membrane	Radniecki et al.(2011)
Ag°, CuO, SiO_2	20, 20, 15 respectively	50 µg mL^{-1}	Arctic soil microbial community, *Bradyrhizobium canariense* (ATCC BAA-1002™)	176-days, microcosm	Ag° was highly toxic to arctic soil microbial community. Community-identified plant-associating bacteria, *Bradyrhizobium canariense* was significantly sensitive to Ag° compared to CuO and silicon oxide (SiO_2) nanoparticle	Kumar et al. (2011)

(continued)

Table 5.2 (continued)

ENPs	Size (nm)	ENPs concentration	Environmental matrix/microbial group	Exposure time	Effect on microbial cell, community composition, diversity, activity and dynamics	References
Ag°	10–40	10, 25, 100, 250, 500 mg L^{-1}	*Bacillus pumilus, E. coli* (ATCC 13534, *E. coli* (ATCC 25922), *S. aureus* (ATCC 25923), *Micrococcus luteus* (clinical isolate)	24-h, Microcosm	Reduced amount of exopolysaccharide by *B. pumilus* after exposure to Ag°. Exopolysaccharide capped-Ag° exerted low toxic effect on test organisms	Khan et al. (2011)
Myco-synthesized Ag°	5–12	0.4 µg mL^{-1}	Beneficial gram negative rhizosphere soil bacterium: *Pseudomonas putida* KT2440	48-h, microcosm	Loss of bioluminescence correlated directly with loss of microbial viability	Gupta et al. (2015)
Ag°	10	1 and 3 mg L^{-1}	Agricultural soil (15 cm), *Pseudomonas chlororaphis* 06	4-days, microcosm	Toxic effect mitigated by humic substances in loamy soil with evidence of no cell death compared to loss of viability and cultivability in sandy soil	Calder et al. (2012)
Ag° (PVP-coated), ZnO	52 and 30 respectively	140 and 1400 mg kg^{-1} respectively	Sandy soil (0–25 cm)	Soil/sludge mixture in lysimeters	Inhibition of nitrification; reduced fungal, bacterial and methanogenic archaeal composition with no major changes in the microbial community structure	Durenkamp et al. (2016)
Ag° nanospheres Ag° nanospheres Ag° nanorods Ag° nanoplates	15 20–40 50 50–80	0.072–0.708 0.108–0.814 0.141–1.529 0.070–0.678 (all in mg L^{-1})	Deciduous soil microbial community	120-h, microcosm	Lower 96-h effective concentration (EC_{50}) (mg L^{-1}) of 0.201 (50–80 nm nanoplates), 0.212 (15 nm nanospheres), 0.222	Zhai et al. (2016)

(continued)

Table 5.2 (continued)

ENPs	Size (nm)	ENPs concentration	Environmental matrix/microbial group	Exposure time	Effect on microbial cell, community composition, diversity, activity and dynamics	References
					(20–40 nm nanosperes) than 0.342 (50 nm nanorods). Dose-response relationship exhibited but no significant difference between 15 and 20–40 nm nanospheres exposure, whereas 50 nm nanorods and 50–80 nanoplates exposures significantly reduced microbial community diversity	
$NaBH_4$-Ag°, TA-Ag°, S.cit-Ag°, AA-Ag°	30–50	5, 25, 75 μg mL^{-1}	*E. coli* and *Salmonella typhii*	24-h, microcosm	Dose-dependent increase in cellular disruption and ROS production with the influence of surface potential and cell type on cytotoxicity	Kaur and Tikoo (2013)
PVP-Ag°	<15	1, 50 and 200 mg L^{-1}	Wastewater biofilm (1.5 mm thickness) microorganisms	24-h, 4-days, microcosm	No significant effect on the survival rate of biofilm microorganisms. Slight decrease in the viability of microbial genera indicating high tolerance of wastewater biofilm microorganisms to Ag° based on EPS produced prior to exposure	Sheng and Liu (2011)

(continued)

Table 5.2 (continued)

ENPs	Size (nm)	ENPs concentration	Environmental matrix/microbial group	Exposure time	Effect on microbial cell, community composition, diversity, activity and dynamics	References
PVA- and Na_2ATP-doped Ag°	7 ± 3, 40 ± 14	0, 1, 5, and 10 mg L^{-1}	Pure nitrifying bacteria: *Nitrosomonas europaea* ATCC 19718	3-h, microcosm	Size- and capping material-dependent inhibition of ammonia oxidation, disintegration of nucleoid, damage of cell wall	Yuan et al. (2013)
CuO, Fe_3O_4	<50	0.1 and 1%	Sandy loam and Sandy clay loam (0–10 cm)	48-h, microcosm	Inhibition of bacterial hydrolytic activity, oxidative potential, changes in community composition and abundance by 1% CuO in sandy loam soil compared to sandy clay soil. Fe_3O_4 inhibited hydrolytic activity and bacterial composition in sandy loam but not in sandy clay loam. However, bacterial abundance decreased at 0.1% CuO, increased at 1% CuO in sandy loam, whereas in sandy clay loam, abundance was reduced at 1% CuO. Members of the Bacilli, Rhizobiales, Sphingobacterials were adversely affected	Frenk et al. (2013)

(continued)

Table 5.2 (continued)

ENPs	Size (nm)	ENPs concentration	Environmental matrix/microbial group	Exposure time	Effect on microbial cell, community composition, diversity, activity and dynamics	References
Mycosynthesized CeO_2, Fe_3O_4, and SnO_2	50–105, 20–30, and 61 respectively	0, 10, and 100 mg kg^{-1}	Organ-mineral horizons (A1 and A2) of Epileptic Cambisol	7 and 60-days, mesocosm	Increased microbial C/N ratio from ectomycorrhizae predominance. Increased metabolic quotient indicating microbial stress and changes in bacteria/fungal biomass ratio	Antisari et al. (2013)
CuO	40–80	0–200 mmol g^{-1}	Mineral and organic pasture soils	5–7 h, microcosm	Inhibitory effect on bacterial growth in mineral soil, toxic to bacterial community (Log $EC_{50} = 1.55 \pm 0.10$). Dose-response relationship in organic soil. Significant bacterial growth inhibition exerted by the highest dose	Rousk et al. (2012)
Pristine, carboxyl-, hydroxyl-, and amino functionalized MWCNTs and pristine SWCNTs	8–15, 1–2	0, 100, 200 mg L^{-1}	Aquatic environment created by mixing deionized water with soil	10–40-days, microcosm	Contact time-dependent and differentially toxic effect with reduced microbial population. Stronger toxicity exerted by functionalized CNTs. *Bacillus* and *Acidothiobacillus* were nano-tolerant species	Wang et al. (2015)
HMT-CeO_2, TMAOH-TiO_2, S.cit-Au°, S.cit-Ag°	12, 7.5, 20 and 30 respectively	0.64, 0.84, 0.075 and 0.13 mg mL^{-1} respectively	Ordinary heterotrophic organisms (OHO)	4-h, microcosm	100, 83 and 33% inhibitions exerted by CeO_2, TiO_2 and Ag° respectively, with no inhibitory effect exhibited by Au°. 60%	Garcia et al. (2012)

(continued)

Table 5.2 (continued)

ENPs	Size (nm)	ENPs concentration	Environmental matrix/microbial group	Exposure time	Effect on microbial cell, community composition, diversity, activity and dynamics	References
					decrease in the growth of *Pseudomonas fluorescens* with 0.002 mg mL^{-1} of Ag°	
HMT-CeO_2, TMAOH-TiO_2, S.cit-Au°, S.cit-Ag°	12, 7.5, 20 and 30 respectively	0.64, 0.84, 0.075 and 0.13 mg mL^{-1} respectively	Ammonia-oxidizing bacteria (AOB)	1 and 4-h, microcosm	No significant to low inhibitory effect (<4%) by Ag° and TiO_2. 14% inhibitory effect by Au° that remained unchanged with increased exposure time. CeO_2 exerted the most significant inhibitory effect	Garcia et al. (2012)
HMT-CeO_2, TMAOH-TiO_2, S.cit-Au°, S.cit-Ag°	12, 7.5, 20 and 30 respectively	0.64, 0.84, 0.075 and 0.13 mg mL^{-1} respectively	Mesophilic and thermophilic anaerobic microbial consortia, biogas production	50-days, mesocosm	CeO_2 inhibited 90% of the microbial group with significantly reduced biogas production, whereas TiO_2 exerted 10% increased production of biogas	Garcia et al. (2012)
C60, SWCNTs, MWCNTs, FS	1, 1, 10–15 respectively	1, 10, 100, 1000 mg kg^{-1}	Agricultural soil (Dystic cambisol), 5–20 cm	21-days, mesocosm	Dose dependent decrease in ^{14}C-glucose uptake with varying pattern and no significant effect on soil microbial activity in the short term	Oyelami and Semple (2015)

(continued)

Table 5.2 (continued)

ENPs	Size (nm)	ENPs concentration	Environmental matrix/microbial group	Exposure time	Effect on microbial cell, community composition, diversity, activity and dynamics	References
CuO, ZnO	<50, <100	20, 100, 200, 500 mg L^{-1} and 10, 20, 50, 100, 500 mg L^{-1} respectively	*Pseudomonas chlororaphis* 06	48-h, microcosms	Dose-dependent reduction in cultivability of *P. chlororaphis* 06, although highly resistant to ZnO. Presence of EPS mitigated toxicity of CuO	Dimkpa et al. (2011b)
MWNTs	30–50	Low: 10, 100, 1000 and 10,000 mg kg^{-1} (worst case scenario)	Sandy loam (0–10 cm)	90-days, microcosms	No effect on soil respiration, enzyme activity and microbial composition at 10, 100, and 1000 mg kg^{-1}. Increased fungal fatty acid methyl esters (FAMEs) at highest treatment. Decreased abundance of bacterial genera (*Derxia*, *Holophago*, *Opitutus*, *Waddlia*). Increased abundance of organisms associated with degrading recalcitrant contaminants (*Rhodococcus*, *Cellulomonas*, *Norcardiodes* and *Pseudomonas*)	Shrestha et al. (2013)
HMT-CeO_2, TMAOH-TiO_2, S.cit-Au°, S.cit-Ag°	12, 7.5, 20 and 30 respectively	0.64, 0.84, 0.075 and 0.13 mg mL^{-1} respectively	Mesophilic and thermophilic anaerobic microbial consortia, biogas production	50-days, mesocosms	CeO_2 inhibited 90% of the microbial group with significantly reduced biogas production, whereas TiO_2 exerted 10% increased production of biogas	Garcia et al. (2012)

(continued)

Table 5.2 (continued)

ENPs	Size (nm)	ENPs concentration	Environmental matrix/microbial group	Exposure time	Effect on microbial cell, community composition, diversity, activity and dynamics	References
TiO_2, CuO	20, 40	0, 100, 500 and 1000 mg kg^{-1} soil	Flooded paddy soil (0–20 cm composite sample)	90-days flooded, 20-days drought. Microcosms	Significantly reduced soil microbial biomass, total phospholipid fatty acids (PLFAs), and enzyme activity (urease, phosphatase and dehydrogenase) by CuO than TiO_2	Xu et al. (2015)
TiO_2, CMC, HM-CMC, HM-PEG, SDS/DDAB, Mo/NaO, $TiSiO_4$, CdSe/ZnS quantum dots, Au nanorods, Fe/Co Magnetic fluid	<100, ns, ns, 1, 30, 60, <50, ns, 10, 7, respectively	5,10,10, 10, 1.7, 5, 5, 0.5, 3.34 and 0.5 mg kg^{-1} respectively	Standard artificial OECD soil	30-days, microcosm	TiO_2, CMC, HM-CMC, HM-PEG, SDS/DDAB exerted negative impacts on the soil bacterial community structure and diversity	Nogueira et al. (2012)
TiO_2, ZnO	15–20, 20–30	0, 0.5, 1.0 and 2.0 mg TiO_2 g^{-1}; 0.5, 0.1, 0.5 mg ZnO g^{-1}	Loamy soil (0–10 cm)	60-days, microcosm	Stimulatory and inhibitory effects exerted on different microbial taxa. Reduced microbial diversity of organisms associated with nitrogen-fixation (Rhizobiales, Bradyrhizobiaceae and *Bradyrhizobium*) and methane oxidation (Methylobacteriaceae). Positive effects on organisms associated with degrading recalcitrant organic pollutants	Ge et al. (2012)

(continued)

Table 5.2 (continued)

ENPs	Size (nm)	ENPs concentration	Environmental matrix/microbial group	Exposure time	Effect on microbial cell, community composition, diversity, activity and dynamics	References
					(Sphingomonidaceae) and biopolymers (Streptomycetaceae and *Streptomyces*)	
TiO_2, ZnO	15–20, 20-30	0.5 mg g^{-1} soil	Loamy soil (0–10 cm)	60-days, microcosm	Reduced respiratory activity, microbial biomass, diversity and community composition. Stronger shifts in bacterial community composition exerted by ZnO than TiO_2	Ge et al. (2011)
TiO_2, CeO_2, MWCNTs	29 ± 9, <50, 20–30 respectively	2007 ± 79, 830 and 3000 mg kg^{-1} respectively	Arbuscular mycorrhizal fungi	14 weeks, mesocosm	Arbuscular mycorrhizal fungi root colonization were not adversely affected by the ENPs exposure. No negative effect on nitrogen fixation although slightly increased at 3000 mg L^{-1} of MWCNTs	Moll et al. (2015)
TiO_2	21	11 and 27 mg L^{-1}	*Rhizobium trifolii* 30141 and red clover symbiotic interaction	28-days, Hydroponic microcosm	Differential growth rate and reduced number of nodules formed	Moll et al. (2016)
TiO_2	20	1 and 500 mg kg^{-1} dry soil	Sandy loam, Loam and silty clay soils (0–20 cm)	90-days Microcosm	Low concentration of TiO_2 exerted significant adverse effect on C-mineralization in the silty clay soil with high organic matter. Adverse effect	Simonin et al. (2015)

(continued)

Table 5.2 (continued)

ENPs	Size (nm)	ENPs concentration	Environmental matrix/microbial group	Exposure time	Effect on microbial cell, community composition, diversity, activity and dynamics	References
					depended on soil pH and organic matter content in contrast to soil texture	
TiO_2	21 ± 17	0.14 mg kg^{-1} soil	Floodplain soil of Cartecay and Chewacala series (0–10 cm)	50-days replicated long-term mesocosm field experiment	52 and 27% reduced activity of microbial extracellular enzyme leucine amino and phosphatase respectively. Significant adverse effect on soil bacterial community composition	Colman et al. (2013)
SDS/DDAD, Mo/NaO, Au nanoparticle, CdSe/ZnS quantum dot and Fe/Co	30, 60, 10, 3.4 and 7 respectively	500 µL	White-rot fungi: *Trametes versicolor*, *Lentinus sajor caju*, *Pleurotus ostreatus*, and *Phanerochaete chrysosporium*	3 and 8-days, Microcosm	All ENPs significantly inhibited fungal growth with pronounced effect on the chemical composition of the mycelium, although Gold and CdSe/ZnS exerted higher adverse effect compared to other ENPs	Galindo et al. (2013)
TiO_2, Fe_3O_4	22–25	0, 100, 200 mg per 1 kg of soil	Silt loam rhizosphere soil (10 cm), rhizobia and AMF	8 weeks, mesocosms	Microbial community structure of rhizobia and arbuscular mycorrhizae fungi root colonization of soybean plants (*Glycine max*) not adversely affected by nanoparticle type, concentration or charge	Burke et al. (2015)

(continued)

Table 5.2 (continued)

ENPs	Size (nm)	ENPs concentration	Environmental matrix/microbial group	Exposure time	Effect on microbial cell, community composition, diversity, activity and dynamics	References
Organic (SS/DDBA, Mo/NaO) and inorganic (TiO_2, $TiSiO_4$, QDs, CdSe/ZnS 530, Fe/Co magnetic fluid, Au nanorods)	30, 60, <100, 50, 7, and 10 respectively	1.7, 5, 5, 5, 0.5, 0.5, and 3.35 respectively	Aged-ENPs spiked soil elutriates, *Salmonella typhimurium* (TA98 and TA100 strains) and *Vibrio fischeri*	2-h and 30-days Elutriate	Solid Phase Microtox® toxic effect on bioluminescence of *Vibrio fischeri,* differential effect of ageing with increased EC_{50} after 30-days. Elutriates from soil spiked with organic and inorganic nanomaterials were genotoxic and mutagenic to *S. typhimurium* in 30-days	Pereira et al. (2011)
$NaBH_4$-nZVI	<50	0.1 g L^{-1}	Chlorinated hydrocarbon groundwater	20-days, batch microcosm	Time-dependent inhibition and stimulation of heterotrophic microbial biomass. Decreased *Dehalococcoides* population and inhibition of biological dechlorination. Stimulates methanogenesis and anaerobic sulfate reduction	Ronavari et al. (2016)
PAA-coated nZVI	12.5 ± 3	10 mg kg^{-1}	Rowland series soil, 5-20 cm	28-days, microcosm	Disrupts soil bacterial community composition with reduced activity of chloroaromatic mineralizing microorganisms	Tilston et al. (2013)
$NaBH_4$-nZVI	ns	50–1000 mg L^{-1}	Basal medium in 50 mL serum bottles	20-days, microcosm	Dose-dependent stimulation and inhibition of bacterial growth and biodenitrification. Promotes both cell growth and nitrate biodegradation, adsorbs to cell wall	Jiang et al. (2015)

(continued)

Table 5.2 (continued)

ENPs	Size (nm)	ENPs concentration	Environmental matrix/microbial group	Exposure time	Effect on microbial cell, community composition, diversity, activity and dynamics	References
nZVI	50	0.3, 1.0, 5 g L^{-1}	Soil yeast: *Trichosporon cutaneum* (strain CCY 30.5.10)	48-h, microcosm	Humic acid acted as electrosteric barrier, hindered interaction between fungi and nZVI and mitigated the toxic effects of nZVI sorbed to the cell surface	Padrova et al. (2016)
nZVI	<50	10 µg L^{-1} to 1 g L^{-1}	*Pseudomonas putida* G7	72-h, microcosm	Loss of viability causing 60% microbial death at 0.1 mg L^{-1}. Dose-dependent and increased turning event characterized by 4-fold higher turning event	Ortega-Calvo et al. (2016)
nZVI	<50	34 mg g^{-1} soil	Bulk agricultural field soil	72-h, microcosm	Little effect on the microbial cellular and biological activity. No bactericidal effect on bulk soil microbial community based on biomarker genes *narG*, *nirS* and *gyrA*. Significant changes in soil microbial population structure and composition based on fluorescent in situ hybridisation (FISH)	Fajardo et al. (2011)
nZVI	12.5	10 mg g^{-1} soil	Rowland series silt loam soil	14-days, microcosm	Inhibited ammonia oxidation potential, stimulated dehydrogenase activity with minimal effect on hydrolase activity	Cullen et al. (2011)

(continued)

Table 5.2 (continued)

ENPs	Size (nm)	ENPs concentration	Environmental matrix/microbial group	Exposure time	Effect on microbial cell, community composition, diversity, activity and dynamics	References
ZnO	70 ± 15	0, 3, 6 and 12 mmol L^{-1}	Plant pathogenic fungi: *Botrytis cinerea* and *Penicillium expansum*	12-days, microcosm	Significantly inhibited fungal growth with *P. expansum* as the most sensitive to the treatments. Deformation of fungal hyphae and inhibiting cellular activities in *B. cinerea*. Prevented conidiophore development of *P. expansum* resulting in death of hyphae	He et al. (2011a, b)
ZnO	20	0–200 mmol g^{-1}	Mineral and organic pasture soils	5–7 h, microcosm	Significant dose-response effect in mineral soil (R^2 = 0.83) and organic soil (R^2 = 0.68). Pronounced inhibitory effect in mineral soil (Log EC_{50} = 1.81 ± 0.10) than organic soil (LogEC_{50} = 2.27 ± 0.14)	Rousk et al. (2012)
ZnO	<50	3, 10, and 300 mM,	Biofilm microbial community: *Pseudomonas aeruginosa* and planktonic cells	12-h, microcosm	Pronounced inhibition of biofilm formation and pyocyanin production, quinolone signal (PQS), pyochelin and hemolytic activity of *Pseudomonas*. Increased cellular hydrophilicity of *Pseudomonas*. No observable effect on growth of planktonic cells	Lee et al. (2014)

(continued)

Table 5.2 (continued)

ENPs	Size (nm)	ENPs concentration	Environmental matrix/microbial group	Exposure time	Effect on microbial cell, community composition, diversity, activity and dynamics	References
ZnO	30, 300	238, 381, 610, 976, 1520 and 2500 mg kg^{-1} dw soil	Forest soil	7-days, microcosm	Dose mediated shifts correlated positively with soil pH, pronounced changes in bacterial community composition. Stimulatory effect on *Proteobacteria*, reduced population growth of *Acidobacteria*	Read et al. (2016)
ZnO	90 ± 15	0, 400, 800, 1600, 3200 mg kg^{-1}	Arbuscular mycorrhizal fungi (AMF): *Glomus versiforme* and *G. caledonium*) and maize plant symbiotic interaction	8 weeks, microcosm	AMF mitigates ZnO induced phototoxicity and enhance tolerance of maize plant to high ZnO doses by increased antioxidant activities and reduced oxidative damage of biomolecules	Wang et al. (2016a, b)

AA ascorbic acid, *Ag°* silver oxide nanoparticle, *Au* Gold nanoparticle, *$BaTiO_3$* barium titanate, *Brij-76* polyoxyethylene stearyl ether, *CNTs* carbon nanotubes, CdSe/ZnS quantum (Lumidot™), *CuO* copper oxide nanoparticle, *Cu* copper, *CeO_2* cerium (iv) oxide nanoparticle, *CMC* carboxylmethyl-cellulose, *EPS* extracellular polymeric substances, *EC_{50}* effective concentration of substance that generate 50% reduction in bioluminescence, *FS* fullerene soot, *HM-PEG* hydrophobically modified polyethylgycol, *HM-CMC* hydrophobically modified CMC, *HMT* hexamethylenetetramine, *LOEC* lowest observed effect concentration, *MWNTs* multi-walled nanotubes, *MWCNTs* multi-walled carbon nanotubes, *Mo/NaO* monoolein/sodium oleate, *MIC* minimum inhibitory concentration, *MBC* minimum bactericidal concentration, *$NaBH_4$* sodium borohydride, *Na_2ATP* adenosine triphosphate disodium, *NOEC* no observed effect concentration, *PVA* polyvinyl alcohol, *PVP* polyvinylpyrrolidone, *PAA* poly-acrylic acid; *QDs* quantum dots, *SWCNTs* single-walled carbon nanotubes, *SDS/DDAB* sodium dodecylsulfate/didodecyl dimethylammonium bromide, *SDS* sodium dodecyl sulfate, *$TiSiO_4$* titanium silicon oxide, *S.cit.* sodium citrate, *TMAOH* tetramethylammonium hydroxide, *TA* tannic acid, *TiO_2* titanium dioxide nanoparticles, *nZVI* zero-valent iron nanoparticles, *ns* not specific, *ZnO* zinc oxide nanoparticles

Table 5.3 Main capping agents/coatings used with engineered nanomaterials

Nanomaterial	Typical capping agents/coatings	
	Inorganic and small organic molecules	Synthetic and organic macromolecules
Zinc oxide	2-mercaptoethanol, triethoxycarprylsilane, triethanolamine, acetate	Polyvinylpyrrolidone (PVP), polysaccharides
Silver	Citrate, decanethiol, tannic acid, ethylenediaminetetraacetic acid (EDTA)	Polyethylene glycol (PEG), PVP, gum arabic
Gold	Citrate, octanethiol, cetyltrimethyl ammonium bromide (CTAB), cysteine, tannic acid	Biotin, bovine serum albumin (BSA), polypeptides
Cerium oxide	Oleic acid	PVP, poly(acrylic acid)-octyl amine
Titanium dioxide	Oleic acid	Poly(acrylic acid)
Quantum dots (CdSe, CdS)	Silica (inorganic), zinc sulfide (inorganic), citrate, mercaptopropionic acid	PEG, aminodextran
Iron oxide	Dodecylamine, oleic acid	BSA, poly(acrylic acid), poly (methacrylic acid), PEG
Zerovalent iron (ZVI)	Au, Pd, Pt, Ni	Carboxymethyl cellulose, xanthan gum, polypropylene glycol

Adapted from Lowry et al. (2012)

monolaurate, acrylic/acrylate copolymer and polycarboxylate ether are usually employed as surface coatings (Zhai et al. 2016; Durenkamp et al. 2016; Whitley et al. 2013; Benoit et al. 2013; Jaiswal et al. 2010). The presence of long aliphatic chains in most surface coating agents especially in the cationic and anionic species enhance ENPs toxic effects (Zhang et al. 2015).

The surface capping agent are associated with positive or negative influence on the toxicity of the ENPs (Fig. 5.8) on pure and mixed microbial cultures (Table 5.1), biosolid amended and unamended soils (Table 5.2). For instance, in the absence of sewage sludge (soil unamended with sewage biosolid), PVP and citrate had pronounced effects on silver nanoparticle aggregation state and partitioning into pore water, whereas these effects were mitigated during incubation of the ENPs with sewage sludge amended soil (Whitley et al. 2013). Also, the addition of a functional group enhances the toxic potential of ENPs to cell cultures and whole organisms. For instance, $C_{60}HxC_{70}Hx$—a fullerene derivative used in aquatic bioassay test was more toxic with higher impact on the hopping behaviour of *Daphnia* species compared with the conventional nano-fullerene (nano-C_{60}) (Lovern et al. 2007). Similarly, hydroxyl-, carboxyl- and amino-functionalized carbon nanotubes (CNTs) were more toxic than pristine CNTs on aquatic microbial community composition (Wang et al. 2015), whereas sodium citrate coated-silver oxide nanoparticle exerted low (<4%) inhibitory effect on heterotrophic, mesophilic

Table 5.4 Chemical nature of selected surface coating agents used on nanoparticles

Cationic	Neutral	Anionic
Oleylamine	p-Aminothiophenol	Sodium dodecyl sulfate
Tetraoctylammonium bromide	Triphenylphosphate	Myristic acid
Hexadecylamine	Dodecanethiol	Tetradecylphosphonic acid
Cetyltrimethylammonium bromide	Tributylphosphine	

Adapted from Zhang et al. (2015)

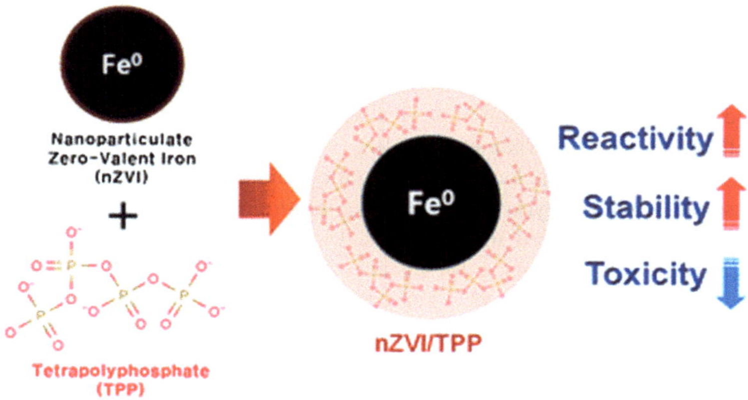

Fig. 5.8 Zero-valent iron nanoparticle (nZVI) coated with polyphosphate organic compounds and its stability and bacterial toxicity. The TPP coating made nZVI stable, prevent the nanoparticles forming aggregate or sediments. Besides, nZVI and its oxidation products were unable to inactivate *Escherichia coli* and indicate that the toxic effect and damage to the cell membrane was mitigated (adapted from Kim et al. 2017)

and thermophilic anaerobic bacteria and ammonia-oxidizing bacteria (AOB) compared to the >90% inhibition by hexamethylenetetramine capped cerium oxide (HMT-CeO_2) and 2–14% inhibitory effect on AOB biomass with 10 mM tetramethylammonium hydroxide (TMAOH) (Garcia et al. 2012). These results are clearly consistent with the perception that nanoparticle stabilizer solutions influence ENPs toxicity and should be considered during the toxicity evaluations.

5.4 Nanotoxicology and Nano-ecotoxicology of ENPs

The study of the toxicity of nanomaterials known as nanotoxicology and the assessment of potential effect of engineered nanomaterials in the environment termed as nanoecotoxicology seem to be relatively new terms used to describe a phenomenon and scenario that has been known for over a century (Nowack et al. 2010).

From a critical review of the relevant scientific literature compiled in 2010, only 12 studies were identified that can actually be classified as ecological studies (i.e. studies considering the complexity of natural ecosystem). The use of silver oxide as a typical bulk and nanosized particles aptly exemplifies the toxicity of most substances in nanoecotoxicology. Historically, silver-derived nanoparticles in its colloidal form at a regulated and appropriate dose was administered safely as medication for almost a century without much concern on the adverse effect. Moreover, about 53% of biocidal products/applications registered by the US Environmental Protection Agency (USEPA) such as water filters, algicides and antimicrobials were nano-silver impregnated and were used since 1954 in the United States (Nowack et al. 2010). Concerns raised by the detection of nanomaterials such as nano-silver in environmental matrices are based on the notion that regarded nano-silver as a new substance with unique and unknown properties. These nano-scale substances are known to react and exert adverse effect on biological systems and the environment in ways yet to be fully elucidated (Nowack et al. 2010).

Until now, however, the harmful effects of colloidal silver at high concentrations were associated with argyria—a cosmetic condition that caused human skin discoloration—which occurred in the 1930s. Usually, the toxic effect of silver are exerted at high concentrations, for instance, the lethal dose (LD_{50}) for rats was higher than 1600 mg kg^{-1} $days^{-1}$ for orally administered doses (Nowack et al. 2010; Wijnhoven et al. 2009; Drake and Hazelwood 2005). At the moment, the scientific community is faced with a dilemma in part to decide whether nanoparticles, for instance, nano-silver is fundamentally a new substance. In part, if the addition of prefix 'nano' to known compounds with small size range can change its nomenclature and properties to level in a borderline of confusion or arbitrariness. The challenge is exacerbated in environmental regulatory standards in which ionic silver is used, whereas most silver in the environment exist in the nano form or small clusters (Nowack et al. 2010). Thus, while we largely understand the properties of bulk materials and/or chemicals at the molecular level, there are new properties of materials being discovered in the zone between "molecule" and "bulk"—that is the nanoscale. When bulk materials are made into smaller and smaller pieces of matter their surface chemistry changes and chemical reactivity increases. It is the reaction between the highly increased reactive surfaces of nanomaterials, due to the increased surface-to-volume ratio, and "wet" biochemistry that is the focus of attention in nanotoxicology and nano-ecotoxicology. Since then, investigations into the toxicological potential of nanomaterials have been constantly trying to catch up with the rapid growth of nanotechnology.

5.4.1 *Toxicity Varies with ENPs Concentrations*

Microbial bioassay for nano-toxicity measurement has gained prominence because the tests are simple to reproduce, rapid and cost effective (Parvez et al. 2006).

Furthermore, the diversity of microorganisms used in nano-toxicity evaluation has increased exponentially. This increase is due in part because microbes are differentially sensitive, resistant or tolerant to ENPs due to either their cell wall composition, presence of divalent ions or natural organic matter in the medium (Eduok and Coulon 2017). However caution is needed as the assertive assumption that ENPs with small particle size are more toxic than large particle size ENPs could be misleading (Table 5.5).

ENPs also exert in vitro toxic effect on pure culture of diverse bacterial cells with concentrations that vary from those inhibiting mixed culture of microbial communities in the soil or other complex matrices (Table 5.2).

The bioavailable concentration of ENPs exerting antimicrobial effects constitutes a certain level of uncertainty in nano-ecotoxicity assessment because toxic concentrations are inconsistent and unpredictable. For example, ZnO nanoparticle inhibits different microorganisms with varying and inconsistent concentrations such as: Minimum inhibitory concentration (MIC) and minimum bactericidal concentration (MBC) of 0.1 and 0.8 $\mu g\ mL^{-1}$ on *E. coli* K88 (Wang et al. 2012), 3 and ≥ 12 mmol L^{-1} inhibited the growth of *E. coli* O157:H7 (Liu et al. 2009), 179 and 1790 $\mu g\ mL^{-1}$ antibacterial effect on *S. aureus* (Soderberg et al. 1990), MIC of 500 ± 306.18 $\mu g\ mL^{-1}$ and MBC of 500 $\mu g\ mL^{-1}$ on *Streptococcus mutans* (Hernandez-Sierra et al. 2008).

In complex environmental matrix such as the soil, ZnO concentrations that ranged from 0 to 200 mmol g^{-1} (Rousk et al. 2012), 140 to 1400 mg kg^{-1} (Durenkamp et al. 2016) and 238 to 2500 mg kg^{-1} (Read et al. 2016) at different exposures exerted inhibitory effect on microbial community activities. Already, extensive review on the toxic effect of ENPs (Eduok and Coulon 2017) seek in varying degree to validate and convey the present and perceived risk to specific and non-target microbial communities (Table 5.2). However, interpretation of toxic effect based on the concentration can potentially mislead because a particular ENPs dose in the soil matrix can stimulate microbial community activity whereas individual organisms or groups are inhibited (Ge et al. 2012). In addition, different outcomes are exhibited by diverse microbes that interact with varying types of ENPs and their concentrations. Typically, low concentrations of ENPs can exhibit divergent outcomes including hormesis in complex environment. For instance, 0–2.0 mg g^{-1} TiO_2 stimulated and also inhibited microbes in loamy soil (Ge et al. 2012), whereas 0.072–0.708 mg L^{-1} of Ag° was toxic to microbes in deciduous soil (Zhai et al. 2016). Thus, the interpretation of the bioavailable dose of different ENPs in the environment and the associated biotic responses in simple and complex media vary due to factors such as the presence of natural organic matter (NOM), colloids, physicochemical and biological transformations, complexation reaction with ligands, physicochemical properties of the ENPs, and contact time (Strigul et al. 2009; Barrena et al. 2009; Crane et al. 2008; Heinlaan et al. 2008; Nowack and Bucheli 2007).

Table 5.5 Percentage growth inhibition of microbial cells by ENPs suspension

Organism	ENPs (nm)	Growth inhibition (%) at different concentrations (ppm)						
		10	50	100	500	1000	2000	5000
B. subtilis	TiO_2 (330)	nd	0	0	0	75 ± 6.6	99 ± 0.9	nd
	SiO_2 (205)	nd	0	0	0	7 ± 4.7	84 ± 9.9	99 ± 1.8
	ZnO (480)	90 ± 4.4	98 ± 0.8	98 ± 1.4	98 ± 0.8	nd	nd	nd
E. coli	TiO_2 (330)	nd	0	0	15 ± 4.2	44 ± 7.0	46 ± 11.3	72 ± 9.4
	SiO_2 (205)	nd	0	0	15 ± 6.4	19 ± 8.3	32 ± 10.1	48 ± 8.5
	ZnO (480)	14 ± 3.5	22 ± 6.5	28 ± 4.9	38 ± 8.9	48 ± 7.7	nd	nd

Adapted from Adams et al. (2006)
Nanoparticles at the advertised size in suspension were either bactericidal or non-toxic to the test organisms (±1 standard deviation, n = 6), *nd* not determined

5.4.2 Acute Versus Chronic Exposure and Effects on Microbial Community

The adverse effect of ENPs on representative pure culture microorganisms (Table 5.1) in most studies are based on in vitro exposure to shock doses, whereas ENPs are expected to be repeatedly released into the soil or aquatic environments. Therefore ENPs ecotoxic assessment must be evaluated based on repeated exposures that are likely to occur in the natural environment. However, there are some inherent challenges posed by repeated exposure compared to single exposures. For example in soil microcosms, TiO_2 dose applied significantly inhibited nitrifying activity, ammonia-oxidizing *Archaea* and bacterial population during repeated and chronic exposures which was at the end more damaging to nitrifying microbial communities in the soil than acute exposure (Simonin et al. 2016). Similarly, Choi and Hu (2008) showed that repeated exposure to Ag° was more toxic to ammonia-oxidizing bacterial (AOB) biomass than a single acute exposure.

Also interpreting the effect of ENPs on aquatic organisms and food web is challenging, because the dissolution of nanoparticles in the aqueous matrix will affect the reactive nature of the nanoparticles by changing the size, surface charges and release of ions, thus influencing the toxic effects (Misra et al. 2012). It is also important to note that single species under laboratory conditions *vs* in situ conditions will respond differently as microbial cells serve as food for grazers and filter-feeders and therefore ENPs are likely to have effect on the prey-predator relationship in the aquatic environment (Baptista et al. 2015). Notably, bacterioplankton and phytoplankton populations had their photosynthetic efficiency significantly reduced when exposed to $\geq 500\ \mu g\ L^{-1}$ of Ag°. This is consistent with the assertion that ENPs in complex medium such as soil (Mirzajani et al. 2013; Judy et al. 2015) and activated sludge (Eduok et al. 2015) exerts selective toxic effect on the different microbial groups and species (Fig. 5.9).

5.4.3 Mechanism of ENPs Toxicity

The release of ion—such as superoxide anions (O_2^-), hydrogen peroxide (H_2O_2), and hydroxyl free radicals (OH)—and indirect change in soil water availability are prominent mechanisms to exert toxic effect by ENPs on pure and mixed microbial communities (Ge et al. 2013; Navarro et al. 2008). These ions can be influenced by confounding factors that mitigate or enhance their toxic properties. Notably, the hydroxyl free radical is the most damaging because changes in the redox potential of ions in the environment causes lipid peroxidation creating holes in the cell wall and membrane and generates other toxic radicals (Devlin 2006). Also, ENPs generate reactive oxygen species (ROS) such as free radicals (OH^-), singlet oxygen ($_1O^2$) and superoxides (O_2^-) which exerts several adverse effects on microorganisms including disruption of cell wall leading to leakage of cytoplasmic content,

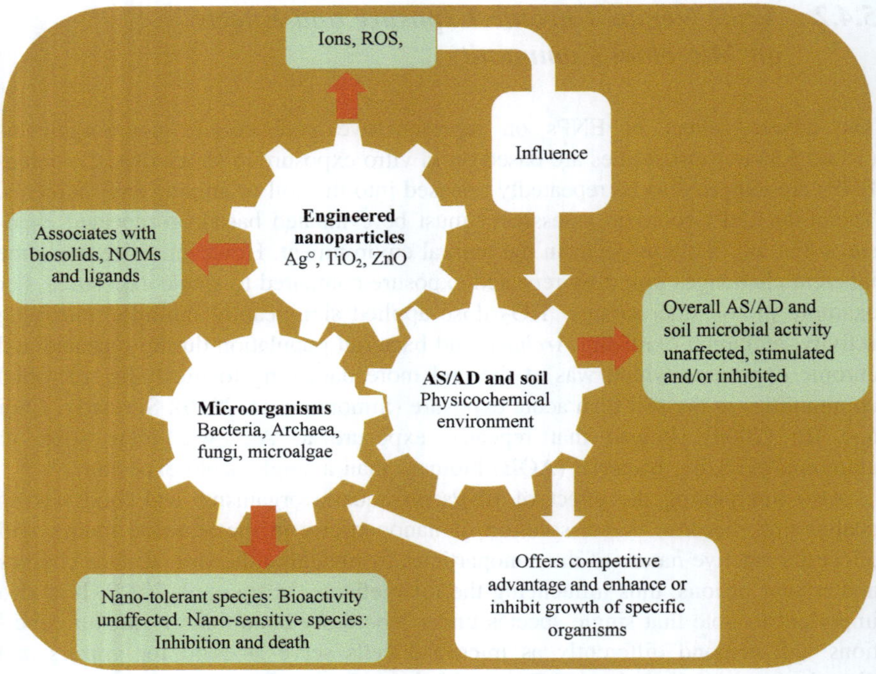

Fig. 5.9 Effect of ENPs on the microbial community in activated sludge (AS), anaerobic digester (AD) and soil microbial activity (adapted from Eduok et al. 2015)

damage of DNA/RNA, lipid peroxidation, oxidative stress, photo-oxidizing and photo-catalytic effect with interaction that enhanced membrane permeability, loss of proton motive force, inhibition of exopolysaccharide and biofilm formation (Sirelkhatim et al. 2015; Eduok et al. 2015; Ge et al. 2013; Wang et al. 2012; Pelletier et al. 2010). In addition, the abrasive surface texture of some ENPs such as ZnO enhanced antibacterial activity (Sirelkhatim et al. 2015).

5.4.3.1 Photo-Induced Toxic Effect of ENPs

Most ENPs exert toxic effects under exposure to visible or ultraviolet lights, and inhibitory effect are more pronounced in the presence of light than in its absence. Indeed, a 2.5-fold higher inhibition of *B. subtilis* compared to 1.8-fold for *E. coli* exerted by TiO_2 in the presence of light have been reported (Adams et al. 2006), and indicates that TiO_2 toxic effects are associated with photocatalytic conditions that generate ROS (Cabiscol et al. 2000; Maness et al. 1999). Also, photo-reactivity increased on exposure of copper sulfate (CuS) nanomaterial (Li et al. 2010) and TiO_2 to UV light which enhances the production of damaging reactive oxygen species (ROS) capable of causing oxidative stress and DNA damage (Ivask et al.

2010; Dunford et al. 1997). However, contrasting evidence exist because Cu-doped TiO_2 enhanced survival rate of *Shewanella oneidensis* MR-1 exposed to ultraviolet light after 24-h incubation at 30 °C (Wu et al. 2011) which suggests that the coating agent mitigated the adverse effect of a supposedly toxic ENPs under the same exposure scenario.

5.4.3.2 Production of Ions

The toxic mechanism of action from dissolved or suspended solutions of ENPs on various organisms such as bacteria, *Daphnia*, algae and fish (Griffitt et al. 2008), wastewater microorganisms (Eduok et al. 2015; Blaser et al. 2008) have been reported and attributed to release of ions (Beer et al. 2012). Notably, silver (Ag°), TiO_2, and ZnO nanoparticles exert toxic effect in varying ways but the release of ions is consistently associated with the antimicrobial properties demonstrated against diverse organisms such as *Bacillus subtilis* (Hsueh et al. 2015) with *E. coli* (Wen-Ru et al. 2010) and wastewater microorganisms (Eduok et al. 2015).

However, the ion-induced toxicity of ENPs are dependent on various factors prevailing in the medium or environment (Fig. 5.10) such as cooperative entry into the cell (Wang et al. 2014), presence of divalent cation, surface charges and microbial cell wall composition (Lyon et al. 2005), pH of the medium (El-Badawy et al. 2010), concentration and contact time (Beer et al. 2012). The adverse effects include the leaking of reducing sugars and proteins, enzyme inhibition; cell disruption, and scattered vesicles which slowly dissolve thus inhibiting cellular respiration and cell growth (Fig. 5.11) (Wen-Ru et al. 2010).

However, the response of microorganisms exposed to ENPs in different environmental matrices will vary based on the physicochemical reactions taking place (Figs. 5.6 and 5.7). More empirical evidence exist to reinforce the suggestion that toxicity is influenced when ENPs are transformed (Fig. 5.12) in the soil and aquatic environments (Dale et al. 2015). For example, the dissolution and aggregation of ZnO in synthetic freshwater and natural water were influenced due to difference in water chemistry including pH, ion, dissolved organic matter, and hardness. This resulted in reduced toxic effect of ZnO on *Escherichia coli* in natural water which was directly correlated with the presence of divalent ions such as calcium (Ca^{2+}) and magnesium (Mg^{2+}) in the natural water (Li et al. 2013).

5.5 Challenges in Assessing and Interpreting Microbial Nano-ecotoxicity

Nano-ecotoxicology is still in its infancy and determining the endpoints of ENPs toxic effect on microorganisms are yet to be further investigated as to date there are inconsistent findings in relation to the varying factors that can influence the

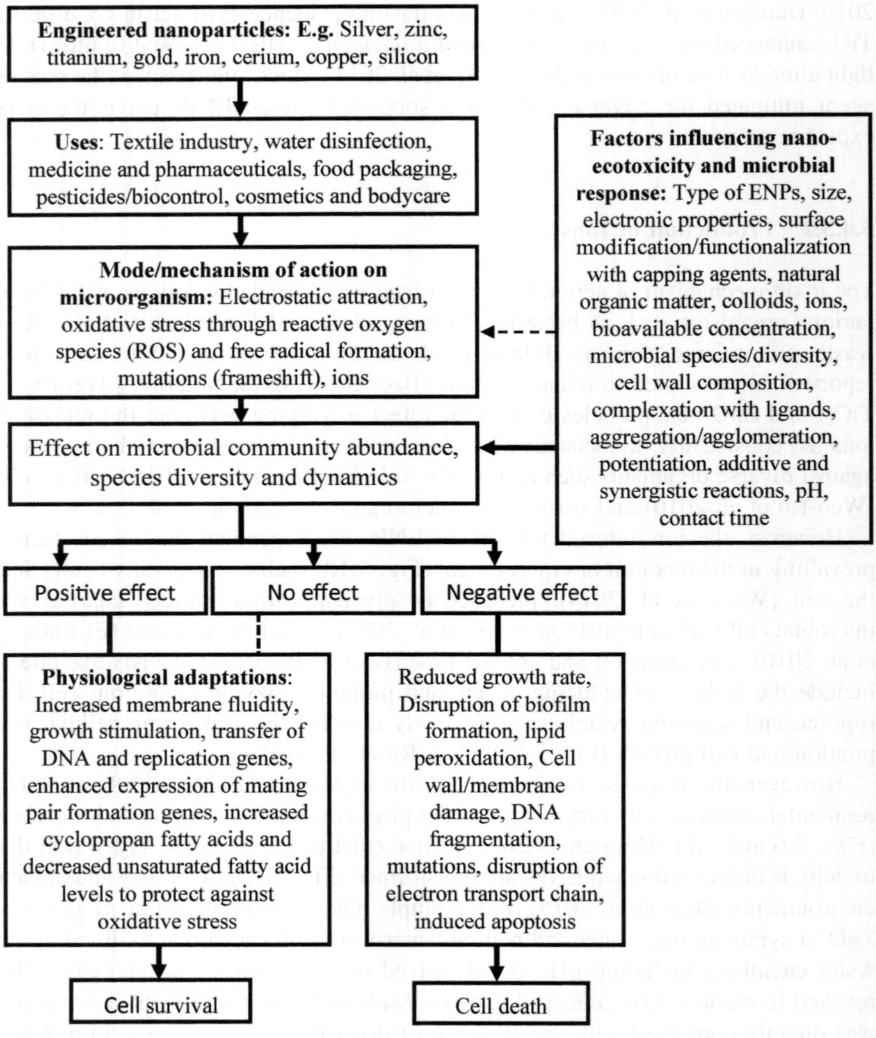

Fig. 5.10 Schematic diagram of the outcomes of engineered nanoparticles toxic effect and microbial species response in environmental media. *Broken lines* indicates a likely indirect influence of most of the factors on ENPs and microbial response

toxicology assessment On one hand, there is currently no evidence that ENPs pose a significant threat to the environment; on the other hand, many gaps in our knowledge remain with regard to ENPs ecotoxicity. This lack of evidence however should by no means be interpreted to imply that environmental damage cannot occur. Further to this, most of the ecotoxicological studies are conducted under controlled laboratory conditions involving cell cultures or model organisms

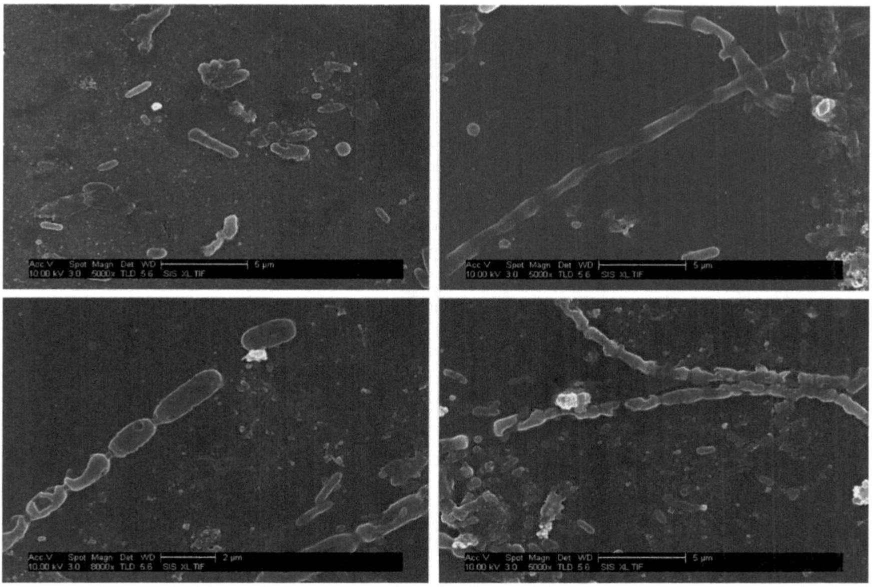

Fig. 5.11 SEM micrographs showing cell wall and membrane peroxidation and damage in activated sludge bacteria exposed to mixed ENPs (Ag°, TiO_2 and ZnO)

(see Table 5.1) and one of the main critiques is the use of unrealistically high dose of nanoparticles. Such high doses, also termed "overdose" are often necessary to trigger some effect or microbial response. However, they can lead to analytical artefacts as it has been shown that ENPs can form large aggregates that can alter the bioavailability and thus the toxicity of a nanomaterial. Moreover the physico-chemical composition and complexity of natural ecosystems is not taken into account and the concentrations used often are well above realistic exposure scenarios. For example, risk assessment for nanosilver show that maximum concentrations in waters are currently probably about 0.1 μg/L. Despite this, most ecotoxicological studies use nanosilver in concentrations in the mg/L range.

In the absence of appropriate mechanism for endpoint determination to date, the conventional concepts used in ecotoxicology have been applied in nano-ecotoxicology and although with significant caveats, there are still problematic and generally inappropriate. For instance, the descriptive concepts and hypothesis-based concentration-effect models used in ecotoxicological assessments such as no observed effect concentration (NOEC), no observed adverse effect concentration (NOAEC), lowest observed effect concentration (LOEC), and the maximum acceptable toxicant concentration (MATC—the geometric mean of NOEC and LOEC) do not indicate whether the ENPs concentration considered is biologically safe. Also, the use of point estimate toxicity such as lethal concentration (LC), effective concentration (EC), or inhibition concentration (IC) are

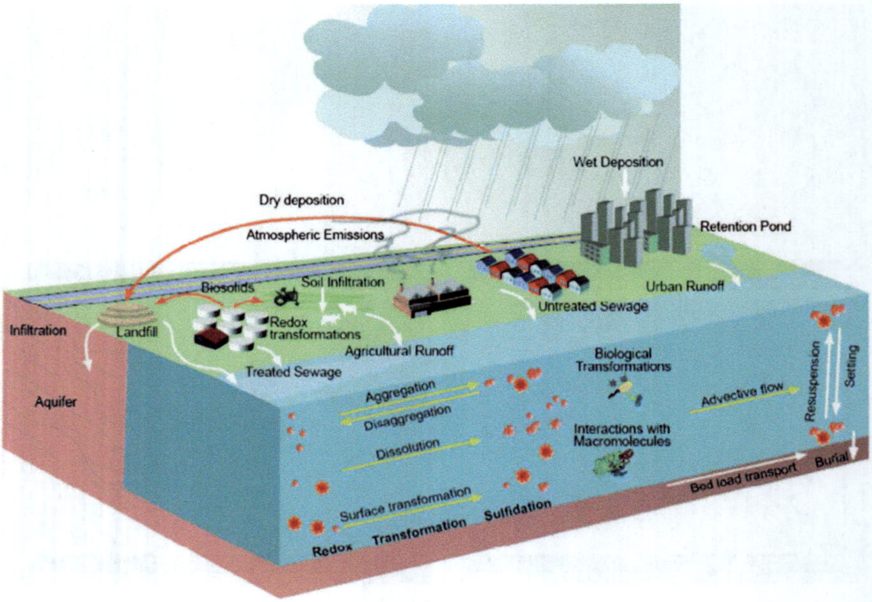

Fig. 5.12 Schematic representation of sources and flow of nanomaterials in the environment, and the key processes determining the fate and behaviour of nanomaterials in aquatic environments (adapted from Dale et al. 2015)

influenced by several factors already enumerated which invalidate result interpretation. Thus, the biological concept of safe concentration in relation to the statistically determined NOEC, NOAEC and MATC are considered to be biased, inconsistent and misleading (Newman 2008; Moore and Caux 1997; Noppert et al. 1994) and without any biologically meaningful effect level (Crane and Newman 2000; Weber et al. 1989) because of the reactivity and bioavailability of ENPs. Besides, ENPs with catalytic potentials such as TiO_2 exposed to UV light are usually photoinduced with enhanced production of damaging reactive oxygen species (ROS) to exert toxic effect (Ivask et al. 2010; Dunford et al. 1997). Thus, the endpoint of toxic effect based on the quantitative ENPs dose (mg kg^{-1}) or point estimate of toxicity can be fundamentally flawed and misleading. For instance, a study on the effect of ENPs in wastewater treatment plant on microbial bioluminescence, the widely known Microtox® acute and solid phase test (SPT) assay was less sensitive to detect aged-ENPs toxic effect from wastewater effluent and digestate (Eduok et al. 2013). Moreover, the NOAEC in the digestate indicated that ENPs toxic effect interpreted using this bioassay can lead towards a wrong conclusion. This is because the bioavailable dose of a substance is closely related to its adverse effect (Crane et al. 2008). Notably, low concentration of aged-ENPs ions in the filtrate reduced bioluminescence of *Vibrio fischeri* compared to the high concentrations in the digestate cake. Thus, apart from varying osmotic conditions

(Gutierrez et al. 2002), redox and sorption potential, confounding factors such as lipophilic properties and uncertainty of bioavailable dose makes Microtox® a less suitable assay for assessing ENPs toxicity in digestate. This assertion raises the need for developing a more responsive and sensitive bioanalytical methods able to detect biological activity below the present bulk chemical limit of detection—using microbial indicators relevant to the particular environmental matrix for nano-ecotoxicity assessment.

Another aspect challenging our understanding and interpretation of ENPs toxic effect is the occurrence of synergistic adverse and/or stimulatory effects on microbial community (McKee and Filser 2016). Contrasting outcomes and effects are usually observed depending on the type of analytical protocol employed indicating that only a holistic assessment of microbial response can minimise the misleading or interpretation error in ENP eco-toxicity. For instance, adverse effect was exerted by zero-valent iron nanoparticle (nZVI) on soil microbial structure and composition based on FISH analysis, whereas biomarker genes indicated no bactericidal effect on the same soil organisms (Fajardo et al. 2011).

Even though scientific uncertainties still exist, the precautionary principle should be applied in the sense of preventive risk minimization. Further ecotoxicological research is also needed to determine both the starting and available dose in the environmental matrices considered given the number of biotic and abiotic transformations that nanoparticles can undertake in the environment. Increasingly, ecotoxicological research should focus on the environmental relevance of the nanomaterials and consider the complexity of the natural systems when assessing microbial nano-ecotoxicity. As it is to date difficult to relate or compare effect on pure microbial cultures to those on indigenous microbial communities, mainly due to the following reasons:

- Few ENPs have been studied in full details to date;
- Limited number of microorganisms involved in biogeochemical processes are exposed to ENPs toxic assay in contrast to the diverse organisms in the environment;
- Confounding factors such as toxic endpoint determination vary widely for different ENPs and microorganisms;
- Interferences and pre-treatments alter exposure-response outcomes;
- Modifications such as aggregation, agglomeration, addition, potentiation, synergism influence and change or mitigate toxic effects;
- Different diluents with varying effects on ENPs are used in exposure assay and therefore inducing a bias for the nano-ecotoxicity assessment;
- Exposure response vary for different organisms;
- Liquid and solid phase toxicity assay are inconsistent;
- Bioavailable dose is difficult to determine and toxic concentrations are not uniform.

5.6 Conclusion

Several opportunities and challenges are presented in nano-ecotoxicology as a result of the increased ENPs production and subsequent release into the environment. ENPs are selectively toxic to different species and groups of microorganisms influenced by exposure matrix and time, bioavailable concentration and confounding factors. Microbial ecotoxicity data are obtained with species commonly used in laboratory toxicity and pathogenicity testing but they often do not cover the most relevant taxonomic groups involved in biogeochemical transformations, degradation and other microbiologically mediated processes. Although eco-nanotoxicology is developing, more studies are needed for a holistic understanding of the mechanism, transformation and reactivity of ENPs in the environment as well as for unbiased interpretation of their ecotoxicological effect on microorganisms. Also long-term studies would be necessary to assess retarded environmental impacts of ENPs and to help determine potential adaptive mechanisms. More studies on bioaccumulation in the food chain as well as on the interaction of ENPs with other pollutants in the environment would also be necessary.

References

Adams LK, Lyon DY, Alvarez PJJ (2006) Comparative eco-toxicity of nanoscale TiO^2, SiO_2, and ZnO water suspensions. Water Res 40:3527–3532

Antisari LV, Carbone S, Gatti A, Vianello G, Nannipeiro P (2013) Toxicity of metal oxides (CeO_2, Fe_3O_4, SnO_2) engineered nanoparticles on soil microbial biomass and their distribution in soil. Soil Biol Biochem 60:87–94

Arvizo RR, Miranda OR, Thompson MA, Pabelick CM, Bhattacharya R, Robertson JD, Rotello VM, Prakash YS, Mukherjee P (2010) Effect of nanoparticle surface charge at the plasma membrane and beyond. Nano Lett 10:2543–2548

Baptista MS, Miller RJ, Halewood ER, Hanna SK, Almeida CMR, Vasconcelos VM, Keller AA, Lenihan HS (2015) Impacts of silver nanoparticles on a natural estuarine plankton community. Environ Sci Technol 49:12968–12974

Barrena R, Casals E, Colon J, Font X, Sanchez A, Puntes V (2009) Evaluation of the ecotoxicity of model nanoparticles. Chemos 75:850–857

Batley GE, Kirby JK, McLaughlin MJ (2012) Fate and risks of nanomaterials in aquatic and terrestrial environments. Acc Chem Res 46:854–864

Beer C, Foldbjerg R, Hayashi Y, Sutherland DS, Autrup H (2012) Toxicity of silver nanoparticles —Nanoparticle or silver ions? Toxicol Lett 202:286–292

Benoit R, Wilkinson KJ, Sauve S (2013) Partitioning of silver and chemical speciation of free Ag in soils amended with nanoparticles. Chem Cent J 7:75

Blaser SA, Scheringer M, MacLeod M, Hungerbuhler K (2008) Estimation of cumulative aquatic exposure and risk due to silver: contribution of nano-functionalized plastics and textiles. Sci Total Environ 390:396–409

Burke DJ, Pietrasiak N, Situ SF, Abenojar EC, Porche M, Kraj P, Lakliang Y, Samia ACS (2015) Iron oxide and titanium dioxide nanoparticle effects on plant performance and root associated microbes. Int J Mol Sci 16:23630–23650

Cabiscol E, Tamarit J, Ros J (2000) Oxidative stress in bacteria and protein damage by reactive oxygen species. Int Microbiol 3:3–8
Calder AJ, Dimkpa CO, McLean JE, Britt DW, Johnson W, Anderson AJ (2012) Soil component mitigates the antibacterial effects of silver nanoparticles towards a beneficial soil bacteria, *Pseudomonas chlororaphis* 06. Sci Tot Environ 429:215–222
Chen KL, Elimelech M (2007) Influence of humic acid on the aggregation kinetics of fullerene (C-60) nanoparticles in monovalent and divalent electrolytic solutions. J Colloid Interface Sci 309:126–134
Choi O, Hu Z (2008) Size dependent and reactive oxygen species related nanosilver toxicity to nitrifying bacteria. Environ Sci Technol 42:4583–4588
Choi O, Deng KK, Kim N-J, Ross L Jr, Surampalli RY, Hu Z (2008) The inhibitory effect of silver nanoparticles, silver ions, and silver chloride colloids on microbial growth. Water Res 42:3066–3074
Colman BP, Arnaout CL, Anciaux S, Gunsch CK, Hochella MF, Kim B, Lowry GV, McGill BM, Reinsch BC, Richardson CJ, Unrine JM, Wright JP, Yin L, Bernhardt ES (2013) Low concentrations of silver nanoparticles in biosolids cause adverse ecosystem responses under realistic field scenario. PLoS ONE 8(2):e57189
Cornelis G, Hund-Rinke K, Kuhlbusch T, van den Brink N, Nickel C (2014) Fate and bioavailability of engineered nanoparticles in soils: a review. Crit Rev Environ Sci Technol 44:2720–2764
Crane M, Newman MC (2000) What level of effect is a no observed effect? Environ Toxicol Chem 19:516–519
Crane M, Handy RD, Garrod J, Owen R (2008) Ecotoxicity test methods and environmental hazard assessment for engineered nanoparticles. Ecotoxicology 17:421–437
Cullen LG, Tilson EL, Mitchell GR, Collins CD, Shaw LJ (2011) Assessing the impact of nano- and micro-scale zerovalent iron particles on soil microbial activities: particle reactivity interferes with assay conditions and interpretation of genuine microbial effects. Chemosphere 82:1675–1682
Dale AL, Casman EA, Lowry GV, Lead JR, Viparelli E, Baalousha M (2015) Modeling nanomaterial environmental fate in aquatic systems. Environ Sci Technol 49:2587–2593
Daughton CG, Ternes TA (1999) Pharmaceutical and personal care products in the environment: agents of subtle change? Environ Health Perspect 107(6):907–938
Devlin TM (2006) Textbook of biochemistry with clinical correlations. Wiley-Liss, Hoboken, 574 pp
Dimkpa CO, Calder A, Gajjar P, Merugu S, Huang W, Britt DW, McLean JE, Johnson WP, Anderson AJ (2011a) Interaction of silver nanoparticles with environmentally beneficial bacterium *Pseudomonas chlororaphis*. J Hazard Mater 188:428–435
Dimkpa CO, Calder A, Britt DW, McLean JE, Anderson AJ (2011b) Response of a soil bacterium, *Pseudomonas chlororaphis* 06 to commercial metal oxide nanoparticles compared to responses to metal ions. Environ Pollut 159:1749–1756
Drake PL, Hazelwood KJ (2005) Exposure-related health effects of silver and silver compounds: a review. Ann Occup Hyg 49:575–588
Dunford R, Salinaro A, Cai L, Serpone N, Horikoshi S, Hidaka H, Knowland J (1997) Chemical oxidation and DNA damage catalysed by inorganic sunscreen ingredients. FEBS Lett 418: 87–90
Durán N, Marcato PD, De Conti R, Alves OL, Costab FTM, Brocchib M (2010) Potential use of silver nanoparticles on pathogenic bacteria, their toxicity and possible mechanisms of action. J Braz Chem Soc 21:949–959
Durenkamp M, Pawlett M, Ritz K, Harris JA, Neal AL, McGrath SP (2016) Nanoparticles within WWTP sludge have minimal impact on leachate quality and soil microbial community structure and function. Environ Pollut 211:399–405
Eduok S (2013) Evaluation of the impact of engineered nanoparticles on the operation of wastewater treatment plant. PhD thesis, Cranfield University, 197 pp. http://dspace.lib.cranfield.ac.uk/handle/1826/8261

Eduok S, Coulon F (2017) Microbiological toxicity of nanoparticles, Chap 6. In: Busquets R (ed) Emerging nanotechnologies in food science, 1st edn. Elsevier, Amsterdam. ISBN 9780323429801 (in press)

Eduok S, Martin B, Nocker A, Villa R, Jefferson B, Coulon F (2013) Evaluation of engineered nanoparticle toxic effect on wastewater microorganisms: current status and challenges. Ecotox Environ Safe 95:1–9

Eduok S, Hendry C, Ferguson R, Martin B, Villa R, Jefferson B, Coulon F (2015) Insights into the effect of mixed engineered nanoparticles on activated sludge performance. FEMS Microb Ecol 91:1–9; fiv082

El-Badawy AM, Luxton TP, Silva RG, Scheckel KG, Suidan MT, Tolaymat TM (2010) Impact of environmental conditions (pH, ionic strength, and electrolyte type) on the surface charge and aggregation of silver nanoparticles suspensions. Environ Sci Technol 44:1260–1266

European Environmental Agency, EEA (2013) EEA annual report 2013 and environmental statement. Technical report No 12/2013. http://www.eea.europa.eu/publications/air-quality-in-europe-2013. Accessed Sept 2013

Fajardo C, Oritz LT, Rodriguez-Membibre ML, Nande M, Lobo MC, Martin M (2011) Assessing the impact of zero-valent iron (ZVI) nanoparticles on soil microbial structure and functionality: a molecular approach. Chemos 86:802–808

Febrega J, Zhang R, Renshaw JC, Liu W-T, Lead JR (2011) Impact of silver nanoparticles on natural marine biofilm bacteria. Chemosphere 85:961–966

Frenk S, Ben-Moshe T, Dror I, Berkowitz B, Minz D (2013) Effect of metal oxide nanoparticles on microbial community structure and function in two different soil types. PLoS ONE 8(12): e84441

Galindo TPS, Pereira R, Freitas AC, Santos-Rocha TAP, Rasteiro MG, Antunes F, Rodrigues D, Soares AMVM, Gonçalves F, Duarte AC, Lopes I (2013) Toxicity of organic and inorganic nanoparticles to four species of white-rot fungi. Sci Tot Environ 458–460:290–299

Garcia A, Delgado L, Tora JA, Casals E, Gonzalez E, Puntes V, Font X, Carrera J, Sanchez A (2012) Effects of cerium oxide, titanium dioxide, silver and gold nanoparticles on the activity of microbial communities intended for wastewater treatment. J Hazard Mater 199–200:64–72

Ge Y, Schimel JP, Holden PA (2011) Evidence for negative effects of TiO_2 and ZnO nanoparticles on soil bacterial communities. Environ Sci Tech 45:1659–1664

Ge Y, Schimel JP, Holden PA (2012) Identification of soil bacteria susceptible to TiO_2 and ZnO nanoparticles. Appl Environ Microbiol 78:6749–6758

Ge Y, Priester JH, Van De Werfhorst LC, Schimel JP, Holden PA (2013) Potential mechanisms and environmental controls of TiO_2 nanoparticle effects on soil bacterial communities. Environ Sci Technol 47:14411–14417

Griffitt RJ, Luo J, Gao J, Bonzongo JC, Barber DS (2008) Effects of particle composition and species on toxicity of metallic nanomaterials in aquatic organisms. Environ Toxicol Chem 27:1972–1978

Gupta IR, Anderson AJ, Rai M (2015) Toxicity of fungal-generated silver nanoparticles to soil-inhabiting *Pseudomonas putida* KT2440, a rhizobium bacterium responsible for plant protection and bioremediation. J Hazard Mater 286:48–54

Gutierrez M, Extebarria J, de las Fuentes L (2002) Evaluation of wastewater toxicity: comparative study between Microtox® and activated sludge oxygen uptake inhibition. Water Res 36: 919–924

He S, Feng Y, Ren H, Zhang Y, Gu N, Lin X (2011a) The impact of iron oxide magnetic nanoparticles on the soil bacterial community. J Soils Sediments 11:1408–1417

He L, Liu Y, Mustapha A, Lin M (2011b) Antifungal activity of zinc oxide nanoparticles against *Botrytis cinerea* and *Penicillium expansum*. Microbiol Res 166:207–215

Heinlaan M, Ivask A, Blinova I, Dubour-guier H-C, Kahru A (2008) Toxicity of nanosized and bulk ZnO, CuO and TiO_2 to bacteria *Vibrio fischeri* and crustaceans *Daphnia magna* and *Thamnocephalus platyurus*. Chemosphere 71:1308–1316

Hennebert P, Avellan A, Yan J, Aguerre-Cheriol O (2013) Experimental evidence of colloids and nanoparticles presence from 25 waste leachates. Waste Manage 33:1870–1881

Hernandez-Sierra J, Ruiz F, Pena D, Martinez-Gutierrez F, Martinez AE, Guillen AJ, Tapia-Perez H, Castanon GM (2008) The antimicrobial sensitivity of *Streptococcus mutans* to nanoparticles of silver, zinc oxide, and gold. Nanomed-Nanotechnol 4:237–240

Hotze EM, Phenrat T, Lowry GV (2010) Nanoparticle aggregation: challenges to understanding transport and reactivity in the environment. J Environ Qual 39:1909–1924

Hsueh Y-H, Lin K-S, Ke W-J, Hsieh C-T, Chiang C-L, Tzou D-Y, Liu S-T (2015) The antimicrobial properties of silver nanoparticles in *Bacillus subtilis* are mediated by released Ag^+ ions. PLoS ONE 10(12):e0144306

Ivask A, Bondarenko O, Jepihhina N, Kahru A (2010) Profiling of the reactive oxygen species-related ecotoxicity of CuO, ZnO, TiO_2, silver and fullerene nanoparticles using a set of recombinant luminescent *Escherichia coli* strains: differentiating the impact of particles and solubilised metals. Anal Bioanal Chem 398:701–716

Jaiswal S, Duffy B, Jaiswal AK, Stobie N, McHale P (2010) Enhancement of the antibacterial properties of silver nanoparticles using β-cyclodextrin as a capping agent. Int J Antimicrob Agent 36:280–283

Jiang C, Xu X, Meghary M, Naidu R, Chen Z (2015) Inhibition or promotion of biodegradation of nitrate by *Paracoccus* sp. in the presence of nanoscale zero-valent iron. Sci Tot Environ 530–531:241–246

Jin T, Sun D, Su JY, Zhang H, Sue HJ (2009) Antimicrobial efficacy of zinc oxide quantum dots against *Listeria monocytogenes*, *Salmonella enteritidis* and *Escherichia coli* 0157:H7. J Food Sci 74:46–52

Jo Y-K, Kim BH, Jung G (2009) Antifungal activity of silver ions and silver nanoparticles on phytopathogenic fungi. Plant Dis 93:1037–1043

Joshi N, Ngwenya BT, French CE (2012) Enhanced resistance to nanoparticle toxicity is conferred by overproduction of extracellular polymeric substances. J Hazard Mater 241–242:363–370

Judy JD, Kirby JK, Creamer C, McLaughlin MJ, Fubiger C, Wright C, Cavagnaro TR, Bertsch PM (2015) Effect of silver sulfide nanomaterials on mycorrhizal colonization of tomato plants and soil microbial communities on biosolid-amended soil. Environ Pollut 206:256–263

Kaur J, Tikoo K (2013) Evaluating cell specific cytotoxicity of differentially charged silver nanoparticles. Food Chem Toxicol 51:1–14

Keller AA, Mcferran S, Lazareva A, Suh S (2013) Global life cycle releases of engineered nanomaterials. J Nanopart Res 15:1692

Khan S, Mukherje A, Chandrasekaran N (2011) Silver nanoparticle tolerant bacterium from sewage. J Environ Sci 32:346–352

Kim KT, Klaine SJ, Lin S, Ke PC, Kim SD (2010) Acute toxicity of a mixture of copper and single-walled carbon nanotubes to *Daphnia magna*. Environ Toxicol Chem 29:122–126

Kim H-H, Kim MS, Kim H-E, Lee H-L, Jang M-H, Choi J, Hwang Y, Lee C (2017) Nanoparticulate zero-valent iron coupled with polyphosphate: the sequential redox treatment of organic compounds and its stability and bacterial toxicity. Environ Sci Nano 4:396–405. doi:10.1039/c6en00502k

Kiser MA, Westerhoff P, Benn T, Wang Y, Perez-Rivera J, Hristovski K (2009) Titanium nanomaterial removal and release from wastewater treatment plants. Environ Sci Technol 43:6757–6763

Kiser MA, Ryu H, Jang H, Hristovski K, Westerhoff P (2010) Biosorption of nanoparticles to heterotrophic wastewater biomass. Water Res 44:4105–4114

Kumar N, Shah V, Walker VK (2011) Perturbation of an arctic soil microbial community by metal nanoparticles. J Hazard Mater 190:816–822

Lacoanet HF, Wiesner MR (2004) Velocity effect on fullerene and oxide nanoparticle deposition in porous media. Environ Sci Technol 38:4377–4382

Lee J-H, Kim Y-G, Cho MH, Lee J (2014) ZnO nanoparticles inhibit *Pseudomonas aeruginosa* biofilm formation and virulence factors production. Microbiol Res 160:888–896

Levard C, Reinsch BC, Michael FM, Oumahi C, Lowry GV, Brown GE (2011) Sulfidation process of PVP-coated silver nanoparticles in aqueous solution: impact on dissolution rate. Environ Sci Technol 45:526–5266

Li F, Wu J, Qin Q, Li Z, Huang X (2010) Controllable synthesis, optical and photocatalytical properties of CuS nanomaterials with hierarchical structures. Powder Technol 198:267–274

Li M, Lin D, Zhu L (2013) Effect of water chemistry on the dissolution of ZnO nanoparticles and their toxicity to *Escherichia coli*. Environ Pollut 173:97–102

Liang Z, Das A, Hu Z (2010) Bacterial response to a shock load of nanosilver in an activated sludge treatment system. Water Res 44:5432–5438

Liu Y, He L, Mustapha A, Li H, Hu ZQ, Lin M (2009) Antibacterial activities of zinc oxide nanoparticles against *Escherichia coli* O157:H7. J Appl Microbiol 107:1193–1201

Liu JY, Sonshine DA, Shervani S, Hurt RH (2010) Controlled release of biologically active silver from nanosilver surfaces. ACS Nano 4:6903–6913

Lovern SB, Strickler JR, Klaper R (2007) Behavioral and physiological changes in *Daphnia magna* when exposed to nanoparticle suspensions (titanium dioxide, nano-C, and CHxCHx). Environ Sci Technol 41:4465–4470

Lowry GV, Gregory KB, Apte SC, Lead JR (2012) Transformations of nanomaterials in the environment. Environ Sci Technol 46:6893–6899

Lui R, Lal R (2015) Potentials of engineered nanoparticles as fertilizers for increasing agronomic productions. Sci Tot Environ 514:131–139

Lyon DY, Fortner JD, Sayes CM, Colvin VL, Hughes JB (2005) Bacterial cell association antimicrobial activity of a C-60 water suspension. Environ Toxicol Chem 24:2757–2762

Mach R (2004) Nanoscale treatment of groundwater. Federal remedial technology roundtable: naval facilities engineering command. Available at https://frtr.gov/pdf/meetings/l–mach_09jun04.pdf. Accessed Feb 2017

Maness PC, Smolinski S, Blake DM, Huang Z, Wolfrum EJ, Jacoby WA (1999) Bactericidal activity of photocatalytic TiO_2 reaction: towards an understanding of its killing mechanism. Appl Environ Microbiol 65:4094–4098

McKee MS, Filser J (2016) Impacts of metal-based engineered nanomaterials on soil communities. Environ Sci: Nano 3:506–533

Mirzajani F, Askari H, Hamzelou S, Farzaneh M, Ghassempour A (2013) Effect of nanosilver on *Oryza sativa* L. and its rhizosphere bacteria. Ecotoxicol Environ Safety 88:48–54

Misra SK, Dybowska A, Berhanu D, Luoma SN, Valsami-Jones E (2012) The complexity of nanoparticles dissolution and its importance in nanotoxicological studies. Sci Tot Environ 438:225–232

Mitrano DM, Motellier S, Clavaguera S, Nowack B (2015) Review of nanomaterials aging and transformation through the life cycle of nano-enhanced products. Environ Int 77:132–147

Moll J, Gogos A, Bucheli TD, Widmer F, van der Heijden MGA (2015) Effect of nanoparticles on red clover and its symbiotic microorganisms. J Nanobiotechnol 14:36

Moll J, Okupnik A, Gogos A, Knauer K, Bucheli TD, van der Heijden MGA, Widmer F (2016) Effects of titanium dioxide nanoparticles on red clover and its rhizobial symbiont. PLoS ONE 11(5):e0155111

Moore MN (2006) So nanoparticles present ecotoxicological risks for the health of the aquatic environment? Environ Int 32:967–976

Moore DRJ, Caux P-Y (1997) Estimating low toxic effects. Environ Toxicol Chem 16(4):784–801

Morones JR, Elechiguerra JL, Camacho A, Holt K, Kouri JB, Ramirez JT, Yacaman MJ (2005) The bactericidal effect of silver nanoparticles. Nanotechnology 16:2346–2353

Mueller NC, Nowack B (2008) Exposure modelling of engineered nanoparticles in the environment. Environ Sci Technol 42:4447–4453

Mukherjee A, Majumdar S, Servin AD, Pagano L, Dhankher OP, White JC (2016) Carbon nanomaterials in agriculture: a critical review. Front Plant Sci 7:172. doi:10.3389/fpls.2016.00172

Naderi MR, Danesh-Shahraki A (2013) Nanofertilizers and their roles in sustainable agriculture. Int J Agric Crop Sci 5(19):2229–2232

Navarro E, Baun A, Behra R, Hartmann NB, Filser J, Miao AJ, Quigg A, Santschi PH, Sigg L (2008) Environmental behavior and ecotoxicity of engineered nanoparticles to algae, plants, and fungi. Ecotoxicology 17(5):378–386

Newman MC (2008) What exactly are you inferring? A closer look at hypothesis testing. Environ Toxicol Chem 27:1013–1019

Nogueira V, Lopes I, Rocha-Santos T, Santos AL, Rasteiro GM, Antunes F, Gonçalves F, Soares AMVM, Cunha A, Almeida A, Gomes NNCM, Ruth P (2012) Impact of organic and inorganic nanomaterials in the soil microbial community structure. Sci Tot Environ 424: 344–350

Noppert F, Van der Hoeven N, Leopold A (1994) How to measure no effect? Towards a new measure of chronic toxicity in ecotoxicology. Netherlands Working Group on Statistics and Ecotoxicology, Delft, The Netherlands

Nowack B, Bucheli TD (2007) Occurrence, behaviour and effects of nanoparticles in the environment. Environ Pollut 150:5–22

Nowack B, Krug HF, Height M (2010) 120 years of nanosilver history: implications for policy makers. Environ Sci Technol 45:1177–1183

Ortega-Calvo JJ, Sanchez CJ, Pratarolo P, Pullin H, Scott TB, Thompson IP (2016) Tactic response of bacteria to zero-valent iron nanoparticles. Environ Pollut 213:438–445

Oyelami AO, Semple KT (2015) Impact of carbon nanomaterials on microbial activity in soil. Soil Biol Biochem 86:172–180

Padrova K, Mat'atkova O, Sikova M, Fuzik T, Masak J, Cejkova A, Jirku V (2016) Mitigation of Fe° nanoparticles toxicity to *Trichosporon cutaneum* by humic substances. New Biotechnol 33:144–152

Pan X, Redding JE, Wiley PA, Wen L, McConnell JS, Zhang B (2010) Mutagenicity evaluation of metal oxide nanoparticles by bacterial reverse mutation assay. Chemosphere 79:113–116

Parisi C, Viani M, Rodriguez-Cerezo E (2015) Agricultural nanotechnologies: what are the current possibilities? Nano Today 10:124–127

Parvez S, Venkataraman C, Mukheriji S (2006) A review on advantages of implementing luminescence inhibition test (*Vibrio fischeri*) for acute toxicity prediction of chemicals. Environ Int 32:265–268

Patel N, Desai P, Patel N, Jha A, Gautam HK (2014) Agronanotechnology for plant fungal disease management: a review. Int J Curr Microbiol Appl Sci 3:71–84

Pelletier DA, Suresh AK, Holton GA, McKeown CK, Wang W, Gu B, Mortensen NP, Allison DP, Joy DC, Allison MR, Brown SD, Phelps TJ, Doktycz MJ (2010) Effects of engineered cerium oxide nanoparticles on bacterial growth and viability. Appl Environ Microbiol 76:7981–7989

Pereira R, Rocha-Santos TAP, Antunes FE, Rasteiro MG, Rubeiro R, Goncalves F, Soares AMVM, Lopes I (2011) Screening evolution of the ecotoxicity and genotoxicity of soils contaminated with organic and inorganic and inorganic nanoparticles: the role of ageing. J Hazard Mater 194:345–354

Phenrat T, Saleh N, Sirk K, Kim HJ, Tilton RD, Lowry GV (2008) Stabilization of aqueous nanoscale zerovalent iron dispersions by anionic polyelectrolytes: adsorbed anionic polyelectrolyte layer properties and their effect on aggregation and sedimentation. J Nanopart Res 10:795–814

Radniecki TS, Stankus DP, Neigh A, Nason JA, Semprini L (2011) Influence of liberated silver nanoparticles on nitrification inhibition of *Nitrosomonas europaea*. Chemosphere 85:43–49

Read DS, Matzke M, Gweon HS, Newbold LK, Heggelund L, Ortiz MD, Lahive E, Spurgeon D, Svendsen C (2016) Soil pH effects on the interactions between dissolved zinc, non-nano- and nano-ZnO with soil bacterial communities. Environ Sci Pollut Res 23:4120–4128

Ronavari A, Balazs M, Tolmacsov P, Molnar C, Kiss I, Kukovecz A, Konya Z (2016) Impact of the morphology and reactivity of nanoscale zero-valent iron (NZVI) on dechlorinating bacterium. Water Res 95:165–173

Rousk J, Ackermann K, Curling SF, Jones DL (2012) Comparative toxicity of nanoparticulate CuO and ZnO to soil bacterial communities. PLoS ONE 7(3):e34197

Schaumann GE, Philippe A, Bundschuh M, Metrevelia G, Klitzke S, Rakcheev D, Grün A, Kumahor SK, Kühn M, Baumann T, Lang F, Manz W, Schulz R, Vogel H-J (2015) Understanding the fate and biological effects of Ag- and TiO_2-nanoparticles in the environment: the quest for advanced analytics and interdisciplinary concepts. Sci Tot Environ 535:3–19

Servin A, Elmer W, Mukherjee A, De la Torre-Roche R, Hamdi H, White JC, Bindraban P, Dimkpa C (2015) A review of the use of engineered nanomaterials to suppress plant disease and enhance crop yield. J Nanopart Res 17:92

Sharma D, Rajput J, Kaith BS, Kaur M, Sharma S (2010) Synthesis of ZnO nanoparticles and study of their antimicrobial and antifungal properties. Thin Solid Films 519:1224–1229

Sheng Z, Liu Y (2011) Effect of silver nanoparticles on wastewater biofilms. Water Res 45:6039–6050

Shin YJ, Kwak JI, An Y-J (2012) Evidence for the inhibitory effect of silver nanoparticles on the activities of soil exoenzymes. Chemosphere 88:524–529

Shrestha B, Acosta-Martinez V, Cox SB, Green MJ, Li S, Canas-Carrell JE (2013) An evaluation of the impact of multiwalled carbon nanotubes on soil microbial community structure and functioning. J Hazard Mater 261:188–197

Sillen WMA, Thijs S, Abbamondi GR, Janssen J, Weyens N, White JC, Vangronsveld J (2015) Effect of silver nanoparticles on soil microorganisms and maize biomass are linked in the rhizosphere. Soil Biol Biochem 91:14–22

Simonin M, Guyonnet JP, Martins JMF, Ginot M, Richaume A (2015) Influence of soil properties on the toxicity of TiO_2 nanoparticles on carbon mineralization and bacterial abundance. J Hazard Mater 285:529–535

Simonin M, Martins JMF, Uzu G, Vince E, Richaume A (2016) Combined study of titanium dioxide nanoparticle transport and toxicity on microbial nitrifying communities under single and repeated exposures in soil columns. Environ Sci Technol 50:10693–10699

Simonin M, Martins JM, Le Roux X, Uzu G, Calas A, Richaume A (2017) Toxicity of TiO_2 nanoparticles on soil nitrification at environmentally relevant concentrations: lack of classical dose-response relationships. Nanotoxicology. doi:10.1080/17435390.2017.1290845

Sirelkhatim A, Mahmud S, Seeni A, Kaus NHM, Ann LC, Siti Khadijah Mohd Bakhori SKM, Hasan H, Mohamad D (2015) Review on zinc oxide nanoparticles: antibacterial activity and toxicity mechanism. Nano-Micro Lett 7:219–242

Soderberg TA, Sunzel B, Holm S, Elmros T, Hallmans G, Sjoberg S (1990) Antibacterial effect of zinc oxide *in vitro*. Scand J Plast Reconstr Surg Hand Surg 24:193–197

Sondi I, Salopek-Sondi B (2004) Silver nanoparticles as antimicrobial agent: a case study on *E. coli* as a model for Gram-negative bacteria. J Colloid Interface Sci 275:177–182

Sotiriou GA, Pratsinis SE (2010) Antibacterial activity of nanosilver ions and particles. Environ Sci Technol 44:5649–5654

Strigul N, Vaccaria L, Galduna C, Waznea M, Liua X, Christodoulatosa C, Jasinkiewiczb K (2009) Acute toxicity of boron, titanium dioxide, and aluminum nanoparticles to *Daphnia magna* and *Vibrio fischeri*. Desalination 248:771–782

Tilston EL, Collins CD, Mitchell GR, Princivalle J, Shaw LJ (2013) Nanoscale zerovalent iron alters soil bacteria community structure and inhibits chloroaromatic biodegradation potential in Arochlor 1242-contaminated soil. Environ Pollut 173:36–46

Vance ME, Kuiken T, Vejerano EP, McGinnis SP, Hochella MF Jr, Rejeski D, Hull MS (2015) Nanotechnology in the real world: redeveloping the nanomaterial consumer products inventory. Beilstein J Nanotech 6:1769–1780

Wang C, Liu L-L, Zhang A-T, Xie P, Lu J-J, Zou X-T (2012) Antibacterial effects of zinc oxide nanoparticles on *Escherichia coli* K88. African J Biotechnol 11:10248–10254

Wang J, Yao H, Hi X (2014) Cooperative entry of nanoparticles into the cell. J Mech Phys Solids 73:151–165

Wang F, Yao J, Liu H, Liu R, Chen H, Yi Z, Yu Q, Ma L, Xing B (2015) Cu and Cr enhanced the effect of various carbon nanotubes on microbial communities in aquatic environment. J Hazard Mater 292:137–145

Wang P, Menzies NW, Dennis PG, Guo J, Forstner C, Sekine R, Lombi E, Kappen P, Bertsch PM, Kopittke PM (2016a) Silver nanoparticles entering soils via the wastewater–sludge–soil pathway pose low risk to plants but elevated Cl concentrations increase Ag bioavailability. Environ Sci Technol 50:8274–8281

Wang F, Liu X, Shi Z, Tong R, Adam CA, Shi X (2016b) Arbuscular mycorrhizae alleviates negative effects of zinc oxide nanoparticles and zinc accumulation in maize plant—a soil microcosm experiment. Chemosphere 147:88–97

Weber CI, Peltier WH, Norberg-King TJ, Horning WB, Kessler FA, Menkedick JR, Neiheisel TW, Lewis PA, Klemm DJ, Pickering QH, Robinson EL, Lazorchak JM, Wymer LJ, Freyberg RW (1989) Short-term methods for estimating the chronic toxicity of effluents and receiving waters to freshwater organisms, EPA/600/4-89/001. Environmental Monitoring Systems Laboratory, Environmental Protection Agency, Cincinnati, OH, 315 pp

Weisner MR, Lowry GV, Jones KL, Hochella MF, Di Guilio RT, Casman E, Bernhardt ES (2009) Decreasing uncertainties in assessing exposure, risk and ecological implication of nanomaterials. Environ Sci Technol 43:6458–6462

Wen-Ru L, Xiao-Bao X, Qing-Shan S, Hai-Yan Z, You-Sheng O-Y, Yi-Ben C (2010) Antibacterial activity and mechanism of silver nanoparticles on *Escherichia coli*. Appl Microbiol Biotechnol 85:1115–1122

Whitley AR, Levard C, Oostveen E, Bertsch PM, Matocha CJ, von der Kammerd F, Unrine JM (2013) Behavior of Ag nanoparticles in soil: effects of particle surface coating, aging and sewage sludge amendment. Environ Pollut 182:141–149

Wigger H, Hackmann S, Zimmermann T, Koser J, Thoming J, von Gleich A (2015) Influences of use activities and waste management on environmental releases of engineered nanomaterials. Sci Tot Environ 535:160–171

Wijnhoven SWP, Peijnenburg WJGM, Herberts CA, Hagens WI, Oomen AG, Heugens EHW, Roszek B, Bisschops J, Gosens I, Van De Meent D, Dekkers S, De Jong WH, van Zijverden M, Sips AJAM, Geertsma RE (2009) Nano-silver—a review of available data and knowledge gaps in human and environmental risk assessment. Nanotoxicology 3:109–138

Woodrow Wilson Database (2016) The project on emerging nanotechnologies: consumer products inventory. Woodrow Wilson International Centre for Scholars, Washington DC. (http://www.nanoproduct.org/inventories/consumer/. Accessed 11 Nov 2016

Wu B, Zhuang W-R, Sahu M, Bismas P, Tang YJ (2011) Cu-doped TiO_2 nanoparticles enhance survival of *Shewanella oneidensis* MR-1 under ultraviolet (UV) light exposure. Sci Tot Environ 409:4635–4639

Xiu Z-M, Ma J, Alvarez PJJ (2011) Differential effect of common ligands and molecular oxygen on antimicrobial activity of silver nanoparticles versus silver ions. Environ Sci Technol 45:9003–9008

Xu C, Peng C, Sun L, Zhang S, Huang H, Chen Y, Shi J (2015) Distinctive effects of TiO_2 and CuO nanoparticles on soil microbes and their community structures in flooded paddy soil. Soil Biol Biochem 86:24–33

Yavuz CT, Mayo JT, Yu WW, Prakash A, Falkner JC, Yean S, Cong L, Shipley HJ, Kan A, Tomson M, Natelson D, Colvin VL (2006) Low-field magnetic separation of monodisperse Fe_3O_4 nanocrystals. Science 314:964–967

Yuan Z, Li J, Cui L, Xu B, Zhang H, Yu C-P (2013) Interaction of silver nanoparticles with pure nitrifying bacteria. Chemosphere 90:1404–1411

Zhai Y, Hunting ER, Wouters M, Peijnenburg WJGM, Vijver MG (2016) Silver nanoparticles, ions, and shape governing soil microbial functional diversity: nano shapes micro. Front Microbiol 7:1123–1132

Zhang Y, Newton B, Lewis E, Fu PP, Kafoury R, Ray PC, Yu H (2015) Cytotoxicity of organic surface coating agents used for nanoparticles synthesis and stability. Toxicol In Vitro 29: 762–768

Chapter 6
Marine Microbial Community Adaptation and Resiliency to Anthropogenic Stresses Through Horizontal Gene Transfer

Suja Rajan and Patricia A. Sobecky

Abstract Horizontal gene transfer (HGT) in prokaryotic lineages is an evolutionary, dynamic mechanism known to promote adaptation to novel habitats by acquisition of genes and alteration of the genetic composition of an organism. The inherent complexity of the microbial species concept exists due to the difficulty involved in quantifying HGT events that results in extensive genome differences even among closely related organisms. In marine ecosystems, HGT leads to an enrichment of microbial populations comprising degradation pathways that enable adaptation to multiple anthropogenic stressors caused by coastal development, oil and gas industry, and aquaculture. Genetic elements such as plasmids, phages and integrons are frequently associated with the dissemination of xenobiotic degradation genes that facilitates microbial adaptation to stress within marine environments. The factors involved in microbial adaptation to stress and their connection to resilience of ecosystems after an anthropogenic disturbance are being increasingly reported and can have implications on coastal recovery and restoration. The increased incidence of metagenomic studies in recent years has also facilitated a better understanding of genetic elements and their contribution to microbial diversity in marine environments. This chapter discusses HGT processes, resistance and resilience patterns found in marine microbial communities along with recent findings about genetic elements that respond to the selective pressures associated with coastal and ocean ecosystems.

Keywords Mobile genetic elements · Horizontal gene transfer · Lateral gene transfer · Marine ecosystems · Bioremediation · Marine bacteria · Microbial adaptation · Anthropogenic stress

---

S. Rajan · P.A. Sobecky (✉)
Department of Biological Sciences, University of Alabama, Tuscaloosa, AL, USA
e-mail: psobecky@ua.edu

C. Cravo-Laureau et al. (eds.), *Microbial Ecotoxicology*,
DOI 10.1007/978-3-319-61795-4_6

6.1 Introduction

6.1.1 Anthropogenic Stresses in Marine Environments

Marine microbial communities are continuously challenged by dynamic and rapidly changing environmental conditions posed by numerous parameters such as changes in temperature, oxygen, salinity, nutrients, predators, and anthropogenic stresses. Microbial growth is influenced by complex combinations of biotic and abiotic factors that act as adaptation drivers through selection. Thus, genetic adaptation by horizontal gene transfer (HGT) is one mechanism by which microorganisms rapidly respond to changing environmental conditions and is a well-documented process resulting in microbial evolution and diversification. This chapter will focus on the genetic adaptability of microbial communities in marine offshore, nearshore and deep sea environments.

Marine microorganisms account for nearly half of the global primary production and play pivotal roles in global nutrient cycling (Arrigo 2005). They are also known to possess complex adaptation mechanisms, which provide selective advantages for microbial growth, persistence and survival in the continuously changing marine ecosystem. Marine microbes exhibit a wealth of metabolic activities including bioremediation capabilities (Dash et al. 2013), biofim formation (Poli et al. 2010), bioactive compound synthesis (Debbab et al. 2010), and biosurfactant production (Banat et al. 2010). As human population growth increases, marine environments are subject to increasing anthropogenic stresses. For example, human-induced activities such as nearshore, land-based industrial, and residential operations add to increasing levels of organic and inorganic pollution into coastal waters while deeper ocean-based drilling operations pose significant risks to marine ecosystem health due to shipping and resource extraction activities. Such risks to deep-sea marine ecosystems have recently been highlighted by the detection of additional oil-impacted coral communities up to 30 km from the 2010 *Deepwater Horizon* spill Macondo wellhead site (Fisher et al. 2014). Thus, as has been previously reported for nearshore estuarine and coastal habitats, pollutants such as hydrocarbons (and newly introduced chemical dispersants to marine habitats) persist for extended periods of time and compromise the health of the shallow and deep-water marine food web on multiple trophic levels (Islam and Tanaka 2004).

The fate and persistence of pollutants in marine environments depends upon the nature of the pollutant, the bioavailability of the pollutant, and the conditions for promoting microbial activity at the site of contamination (Nogales et al. 2011). The type of pollutant compounds in marine environments are varied and they originate from point and non-point sources such as, atmospheric deposition, discharges from pipelines and ships, riparian input, and runoff from nonpoint sources on land (Islam and Tanaka 2004). The type of pollutants that are known to cause degradation of global marine environments are: (1) heavy metals associated with sewage effluents and other nonpoint sources; (2) petroleum hydrocarbons from oil spills; (3) organochlorine-based pesticides used for domestic and agricultural purposes;

(4) radionuclides associated with discharges from nuclear power plants and sea-dumping of nuclear waste; (5) synthetic polymers (plastics) associated with marine litter originating from municipal waste streams and other nonpoint sources. Additionally, agriculture and livestock farming introduces pollutants such as inorganic fertilizers and antibiotics into the marine environment causing eutrophication, hypoxia and the emergence of antibiotic resistant microbial pathogens (Kennish 1996).

The toxicity posed by pollutants in marine environments result in the selection and blooms of microorganisms that can utilize the pollutants for carbon and/or energy sources. Thus, as a function of microbial metabolism, biodegradation is an important mechanism by which pollutants are transformed to less or non-toxic products in marine habitats. Microorganisms undergo changes in their genetic composition that enables them to metabolize the pollutant using various degradation enzymes (Shade et al. 2012). The processes involved in genetic adaptations have been relatively well documented and are briefly described below.

6.1.2 HGT Processes Involved in Genetic Adaptations of Marine Bacteria

HGT can be defined as the physical transfer of DNA into a recipient genome to enable stable inheritance (Stokes and Gillings 2011). The adaptation of microbial communities by HGT was first recognized when multi-drug antibiotic resistance in emerging pathogens was detected (Freeman 1951). With the advent of molecular biology, unequivocal evidence showed that mobile genetic elements such as plasmids, transposons and phages were contributing significantly to bacterial adaptability, diversity as well as genome instability by facilitating the horizontal transfer of genes (Darmon and Leach 2014). Bacterial species typically share a core genome and differ in their assembly of the pan genome due to the presence or absence of the collection of variable genes exchanged by HGT (Lerat et al. 2005).

The marine environment consists of various selection pressures along with large quantities of organic matter that can be utilized by a diverse array of microorganisms thereby creating reservoirs of gene transfer. Although gene exchange typically occurs more frequently between closely related organisms, inter-phylum genetic exchange has been reported in marine environments (Zhaxybayeva et al. 2009; Nelson-Sathi et al. 2012; Caro-Quintero and Konstantinidis 2015). Figure 6.1 briefly illustrates the major components and processes of HGT prevalent in marine organisms (Lang et al. 2012; Brochier-Armanet and Moreira 2015). Although other mechanisms of HGT such as nanotubes (Dubey and Ben-Yehuda 2011) and membrane vesicles (Mashburn-Warren and Whiteley 2006) are known to occur among bacteria, studies focused on these mechanisms in marine organisms are limited. A recent study reported that marine cyanobacteria, *Prochlorococcus*, releases DNA within membrane vesicles suggesting that vesicles may serve as vectors for HGT in marine ecosystems (Biller et al. 2014).

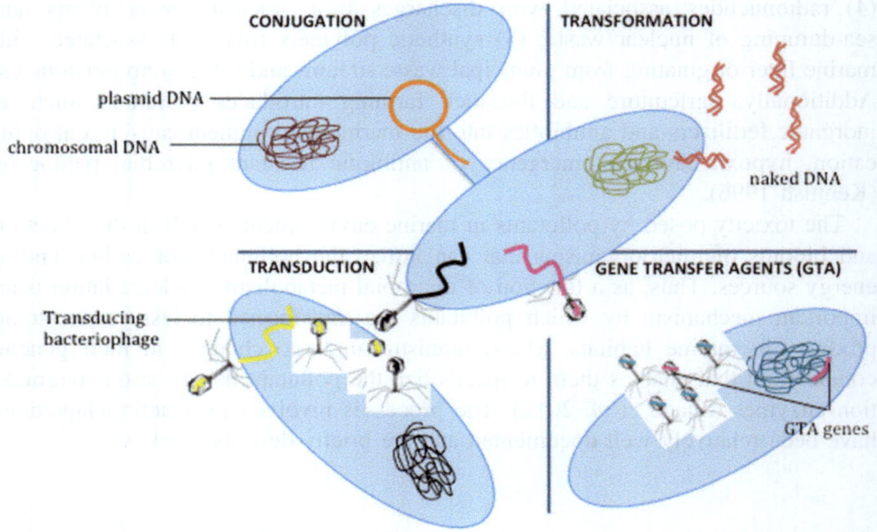

Fig. 6.1 Principal processes of horizontal gene transfer in marine organisms: Conjugation involves transfer of DNA between donor and recipient cells by cell-to-cell contact. **Transformation** involves the uptake of DNA by recipient cells in the presence or absence of donor cells. **Transduction** involves the transfer of DNA between donor and recipient cells via viral vectors. **Gene transfer agents (GTA)** are bacteriophage-like particles that carry random pieces of bacterial DNA from the donor cell and transfer it to the recipient cell upon lysis of the donor cell. Contrary to transducing bacteriophages, GTA-particles contain random segments of host DNA that is insufficient to transfer GTA-producing ability to recipient cells. Modified from reference 160 (von Wintersdorff et al. 2016)

In the past decade, studies have increasingly found GTA clusters in marine bacterioplankton genomes, suggesting the importance of GTA-mediated gene transfer in the ocean (Lang and Beatty 2007; Biers et al. 2008; Lang et al. 2012). Moreover, frequencies of GTA-mediated HGT was found to be between thousand to a million times higher than earlier HGT estimates which implies that GTA-mediated gene transfer could be an adaptive mechanism to maintain metabolic stability in constantly changing marine environments (McDaniel et al. 2010).

6.1.3 Gene Transfer Agent (GTA)-Mediated Gene Transfer in Marine Environments

A phage like gene transfer agent (now known as RcGTA) was first discovered in the purple photosynthetic bacterium *Rhodobacter capsulatus* (Marrs 1974) and is the most well studied GTA till date. The cluster of genes encoding the RcGTA was identified decades later (Lang and Beatty 2000) and it was then found to be widely present in Rhodobacterales (class Alphaproteobacteria) which are well documented marine bacterioplankton (Lang and Beatty 2007; Biers et al. 2008). GTA production has also been identified in members of marine Rhodobacterales species such as *Ruegeria pomeroyi* and *Roseovarius nubinhibens* and DNA transfer was found to be directed to heterologous recipients. Additionally, antibiotic resistance gene transfer rates were found to be a million-fold higher in *Roseovarius nubinhibens* than earlier gene transfer rates recorded in marine environments. Other known GTAs identified in bacterial cells are VSH-1 in the spirochaete *Brachyspira hyodysenteriae*, and Dd1 in the deltaproteobacterium *Desulfovibrio desulfuricans* (Lang et al. 2012). GTAs are not limited to bacteria and were found to be associated with gene transfer in archaea, e.g., *Methanococcus voltae* (Bertani 1999). Most GTAs are tailed phage-like particles that need not necessarily harbor any GTA-encoding gene and instead, contains random pieces of the genome of the GTA-producing cell. Genes encoding the GTA are present on the host chromosome and result in the occasional formation of a particle containing GTA-encoding genes when expressed. The GTAs are then released into the environment by the lysis of the GTA-producing cell and transferred to the recipient cell. Contrastingly, transducing phage particles are produced due to the expression of phage genes and replication of phage genome within the host cell which subsequently results in either, packaging of the complete phage genome or, occasionally, bits of both phage and host DNA (Lang et al. 2012). Recent studies have suggested that GTAs use a combination of transduction and transformation mechanisms to initiate genetic exchange between the GTA particle and recipient cells (Brimacombe et al. 2015).

GTA-mediated transfer events have been proposed to have a number of advantages over the traditional gene transfer mechanisms in that, the GTA particles protect DNA from the damaging effects of the environment as compared to naked DNA that undergoes transformation (Stanton 2007). Moreover, the frequency of GTA-mediated events was found to be significantly higher in ocean environments when compared to earlier reports of other HGT mechanisms (McDaniel et al. 2010). In this study, GTA particles were incubated with recipient cells in culture as well as seawater samples from natural environments such as open ocean, coastal, coral reef and estuaries, to test GTA-mediated gene transfer. All environmental experiments were conducted on a cruise except for the coral reef that was conducted *in situ*. Antibiotic resistance, GTA production and GTA-transfer frequencies were found to be significantly higher than the frequency of transduction previously measured in the marine environment (Jiang and Paul 1998).

Marine viral communities could be using GTA to transfer a considerable number of host-adaptation genes to bacterial cells. However, our current understanding about the prevalence of GTA-mediated transfer within marine viral communities and the role of marine viruses in conferring adaptive mechanisms to bacterial cells is inadequate. GTA-mediated gene transfer studies need to be expanded to address critical information regarding the various components involved in transfer and their contribution to genetic adaptation mechanisms in marine environments.

6.2 Genetic Elements Involved in Adaptation

6.2.1 *Plasmids*

Plasmids are among the best-studied categories of mobile genetic elements. Catabolic plasmids are powerful tools and vectors in the dissemination of antibiotic resistance and xenobiotic degrading genes among marine bacterial communities (Sobecky and Hazen 2009). Plasmids have been categorized into different types based on incompatibility groups determined by their replication and partitioning systems (Shintani et al. 2015). However, plasmid classification based on replicon typing can be difficult as plasmids may contain multiple replicons. Classification of plasmids based on mobility type has been found to overcome this limitation but was deemed to be inappropriate for non-transmissable plasmids (Garcillán-Barcia et al. 2011). It has been reported that globally, approximately one fourth of the plasmids are conjugative, as many are mobilizable, and half of all plasmids examined to date are non-mobilizable (Smillie et al. 2010). Selection pressures that occur in natural environments facilitate plasmid maintainance within microbial communities. Plasmid "addiction" due to toxin-antitoxin systems, a set of closely linked genes encoding a toxin and anti-toxin, has been shown to stabilize plasmids in the host by inhibiting the growth of daughter cells lacking the plasmid, thereby making them persistent even in the absence of selective pressures (Gerdes et al. 1986; Pal et al. 2015).

The incidence of xenobiotic degradation plasmids occurring in marine coastal ecosystems has been reported since the 1980s. More recently, advances in next-generation sequencing (NGS) technology have been instrumental in the detection of novel plasmids and plasmid families in bacterial communities present in marine environments (Palenik et al. 2009; Ma et al. 2012). Marine sediment bacteria were found to contain antibiotic resistance genes similar to those found in human pathogens making them potentially important players in the global spread of antibiotic resistance (Yang et al. 2013). For instance, the occurrence of *Enterococcus feacium,* found in coastal marine sediment that harbored multi-drug antibiotic resistance plasmids (Hegstad et al. 2010; Palmer et al. 2010), was enhanced by the release of antibiotics in the aquatic ecosystem as a result of aquaculture (Di Cesare et al. 2013).

The alphaproteobacterial *Roseobacter* clade constitutes 25% of the marine microbial community promoting global carbon and sulfur cycling (Wagner-Döbler and Biebl 2006). *Roseobacter* are capable of thriving on algal metabolites and hence, can degrade dimethylsulfopropionate (DMSP), an osmoprotectant that is released in large amounts during algal blooms (Moran et al. 2007). Genes encoded on plasmids in the *Roseobacter* clade include key genes for biochemical cycles in addition to genes for the degradation of aromatic compounds (Petersen et al. 2013). Plasmids in *Roseobacters* can constitute up to 20% of the genome content and is found to be transferred via conjugation or GTAs (Swingley et al. 2007; Biers et al. 2008; Buchan and González 2010). RepA-I-type plasmids have also been detected in *Roseobacter* species which have been shown to promote biofim formation thus contributing to *Roseobacter* spp. colonizing submerged surfaces (Michael et al. 2016). RepA-I-type plasmids are found to be predominant in Rhodobacteraceae with the rhamnose operon playing a crucial role in biofilm formation. Elimination of the 65-kb RepA-I-type plasmid from *P. inhibens* DSM 17395 to understand their functional role within the host bacterium, resulted in the loss of genes for polysaccharide metabolism and subsequently, the capacity for biofilm synthesis. Additionally, the comparative analyses of >30 RepA-I-type plasmids facilitated the identification of an exopolysaccharide transport system that was found on all tested biofilm plasmids within the *Roseobacter* group.

6.2.2 Genomic Islands (GEIs)

Research findings on genetic elements continue to add new insights and key concepts considering novel information from recent studies. Attempts have been made in the past to unify various classes of mobile genetic elements involved in transfer of resistance genes through HGT into integrative conjugative elements, also known as conjugative transposons (Ochman et al. 2000; Darmon and Leach 2014; Johnson and Grossman 2015), and through the scope of genomic islands (GEIs) (van der Meer and Sentchilo 2003; Juhas et al. 2009). GEIs commonly denote a wide range of mobile DNA segments that are integrated into the chromosome or other replicons via integrase or transposase genes (Peters et al. 2014). Figure 6.2 illustrates the categorization of the various classes of genetic elements within genomic islands. GEIs tend to carry genes distinct from the rest of the chromosome that offer selective advantageous traits. Although GEIs were first discovered in pathogenic bacteria (Hacker et al. 1990), the detection of adaptive genes in non-pathogenic bacteria (Dobrindt et al. 2004; Coleman et al. 2006) suggested that some of these genes were involved in increasing general bacterial fitness and adaptability. These adaptive genes account for functions such as virulence, compound degradation, and heavy metal resistance to the host organism (Juhas et al. 2009). For example, GEIs were recently found to confer resistance to copper and oxidative stress in a coastal marine *Synechococcus* sp. strain (Stuart et al. 2013). GEIs constituting genes associated with secondary metabolite biosynthesis were also found in marine Actinobacteria, *Salinospora*, suggesting possible links between secondary metabolism and functional adaptation (Penn et al. 2009).

Since GEIs are often linked to integrative elements such as conjugative plasmids and phages, their transfer often occurs through conjugation or transduction (Juhas et al. 2009) and occasionally through transformation. Bacteria that undergo transformation naturally acquire competence by certain DNA-uptake systems found in marine ecosystems such as Type-IV pili systems. Type-IV pili systems activated by chitin, have been found to play a key role in establishing natural competence, colonization, and biofilm formation in *Vibrio* species from marine and brackish water ecosystems (Antonova and Hammer 2015). Bacterial secretion systems such

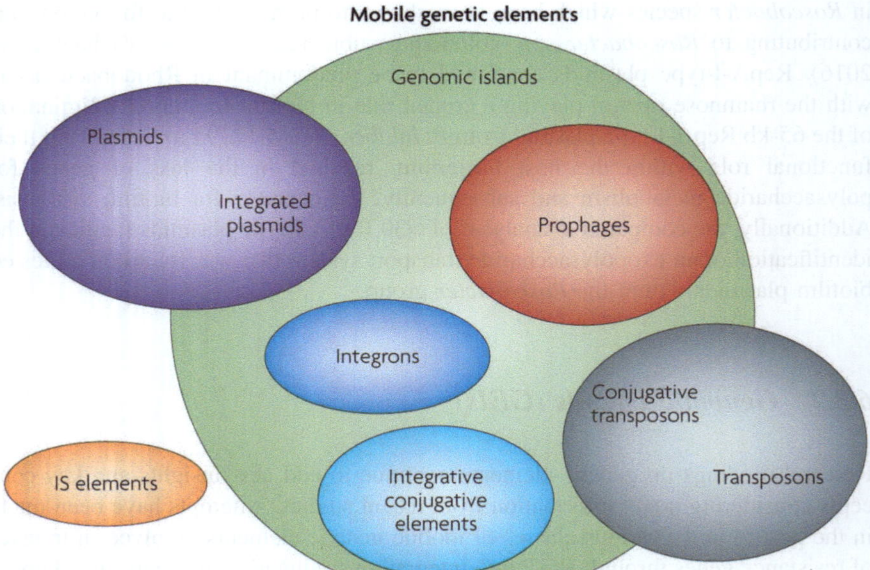

Fig. 6.2 Mobile genetic elements. A general schematic of the class structure of mobile genetic element (MGE) definitions. The broad definition of genomic islands (GIs), which are large genomic regions with probable horizontal origins, allows several other MGEs to be grouped within GIs and illustrates the fact that many GI prediction methods can be applied to other MGEs. IS, insertion sequence. *Image* Langille et al. (2010: 374). Retrieved May 4, 2017, from Academic Search Premier Database

as Type-IV secretion systems (T4SS) were found to play an integral part in genomic island-mediated HGT of resistance and virulence genes in *Haemophilus* and *Pseudomonas* (Juhas et al. 2007a, b, 2008; Morales-Espinosa et al. 2012). Additionally, T4SS genes related to conjugation transfer systems were detected in surface waters of the Sargasso Sea (Venter et al. 2004). Bacteriophages also allow transfer of bacterial DNA along with GEIs through their genomes via transduction. A comprehensive analyses of GEI transduction in marine environments is relatively limited due to the under representation of majority of viral metagenome sequences in genomic databases (Brum and Sullivan 2015). Although this problem is rapidly resolving with the increase in NGS, further studies are required to better understand phage-host interactions and mechanisms of HGT in marine environments through transduction.

GEIs in environmental bacteria have been associated with the presence of xenobiotic degradation pathways (van der Meer and Sentchilo 2003; Dobrindt et al. 2004; Tang et al. 2012). Genomic island ICE *clc* encoding two catabolic pathways associated with 3- and 4-chlorocatechol (*clc*) and 2-aminophenol (*amn*) degradation were found in *Pseudomonas* sp. strain B13 isolated from a sewage plant (Dorn et al. 1974; Gaillard et al. 2006; Miyazaki et al. 2015). Gaillard et al. reported the *clc* element to be 102,784 bp with *attR* and *attL* at the right and left ends respectively. The first gene located next to the *attR* sequence was the *IntB13* integrase gene that was found to facilitate chromosomal integration of the *clc* element. The right half of the *clc* element consisted of previously known genes encoding catabolic properties and the left half of the element primarily consisted of genes encoding hypothetical proteins. GEIs have also been found in *Magnetospirillum gryphiswaldense* that encoded genes for magnetite biomineralization (Ullrich et al. 2005). The putative genomic magnetosome island was approximately 130 kb, rich in insertion sequences and underwent frequent transposition and subsequent deletion upon exposure to stressful conditions such as increased oxygenation and high iron concentrations.

6.2.3 Phages

Viruses are known to be the most abundant group of organisms in the marine environment with most them being bacteriophages (Breitbart 2012), although accurate information regarding current phage abundance estimation and distribution in the marine environment are limited (Aziz et al. 2015). Phages influence bacterial diversity, community composition and function through HGT and are known to carry and encode a wide range of host genes. Environmental factors such as temperature, salinity, nutrients and UV radiation impacts phage and phage-host dynamics (Mojica and Brussaard 2014) which in turn influences microbial abundance, community structure and biogeochemical cycling in marine habitats. Also, anthropogenic pollutants were found to cause prophage induction in marine microbial communities thereby suggesting a role for lysogeny in influencing bacterial host diversity (Jiang and Paul 1996; Cochran and Paul 1998). For example, exposure to relatively low concentrations (10 μg/mL) of heavy oil was found to increase phage production in marine bacterial populations causing changes in community composition (Yoshida and Suzuki 2014). Lysogeny has also been suggested as a phage-host survival mechanism under adverse environmental conditions subsequently contributing to phage diversity (Paul 2008; Wang et al. 2010). A recent study used metagenomics to demonstrate that prophages induced from natural microbe assemblages were unique and lower in diversity when compared to ambient viral populations (McDaniel et al. 2014).

The advent of NGS technologies have made characterization of marine phages independent of the tedious steps involved in phage isolation and cultivation. This methodology development allows researchers to build strategies that will advance our understanding of phage-host interactions and resistance mechanisms.

6.2.4 Integrons

Integrons are distinct genetic elements that facilitate adaptation in bacteria and are frequently associated with the dissemination of antibiotic resistance genes (Gillings 2014). It is now known that integrons are commonly embedded in chromosomes, plasmids or transposons and can be found in both pathogenic and environmental bacteria contributing to bacterial diversity and evolution (Mazel 2006; Boucher et al. 2007). Integrons are typically composed of three key features; (1) an integrase gene *intI*, (2) recombination sites *attI*, and (3) a promoter P_c located upstream of the integration site. The integrase gene can recombine circular DNA molecules called gene cassettes generally consisting of an *attC* recombination site and a single promoterless open reading frame. The *IntI* integrase gene, encoding a site-specific tyrosine recombinase, recognizes the *attC* site on the gene cassette and recombines with the *attI* site behind the promoter P_c in the integron, expressing the newly integrated gene. Consecutive insertion of integron cassettes results in the formation of a cassette array that promotes adaptive functional roles and depicts functions that were valuable to the cell in the past (Mazel 2006; Darmon and Leach 2014). Integrons are not considered mobile on their own as they depend on physical links to recombinase and transposase genes, commonly present with their gene cassettes or adjacent to the integrase gene, for movement between chromosomes (Gillings 2014). Integrons detected in clinical environments were divided into Classes 1, 2, 3, based on the sequence similarity of integrase genes *intI1*, *intI2* and *intI3*, respectively (Abella et al. 2015). The cassette promoter P_c was confirmed in Class 1 and 3 integrons and is found within the integrase gene or *attI* recombination site (Collis et al. 2002; Boucher et al. 2007). Other integrases such as *IntI4*, *IntI5*, *IntI6*, *IntI7* and *IntI8* along with hundreds of other integrase types were subsequently detected in other environments giving rise to new integron classes (Mazel et al. 1998; Nield et al. 2001; Rapa and Labbate 2013; Gillings 2014). Integrons are widespread in bacterial communities and have been estimated to constitute 17% of all sequenced bacterial genomes, with the vast majority of integrons present in Gamma Proteobacteria (Cambray et al. 2010; Abella et al. 2015). Class 1 integrons were detected in 40–70% of gram negative pathogenic bacteria (Martinez-Freijo et al. 1998; van Essen-Zandbergen et al. 2007) and from pathogens isolated from livestock (Goldstein et al. 2001; Ebner et al. 2004). Although Class I integrons are typically found in clinical environments, a group of Class I integrons were found in environmental bacteria associated with estuarine environments (Wright et al. 2008).

A major focus of recent studies characterizing integrons has been associated with integron elements carrying antibiotic resistance determinants in freshwater and soil impacted by anthropogenic stresses (Byrne-Bailey et al. 2011; Gaze et al. 2011; Stalder et al. 2014). Studies addressing non-clinical/environmental integrons in estuarine or marine ecosystems impacted by anthropogenic stresses have increased in the past decade as they have been progressively found to have roles in microbial adaptation and genome evolution. (Wright et al. 2008; Rodríguez-Minguela et al. 2009; Elsaied et al. 2011; Uyaguari et al. 2013). In fact, Class 1 integrons have been recently suggested to serve as a marker for anthropogenic pollution due to its widespread occurrence in environmental bacteria, physical link to genes associated with xenobiotic resistance, rapid generation times, and response to stressors (Gillings et al. 2015). Most genes within the integron gene cassettes isolated from marine environments have, as yet, unidentified functions. The genes that have been identified can be broadly classified into two categories: (1) genes encoding xenobiotic-degrading enzymes (Gillings et al. 2009; Koenig et al. 2009; Elsaied and Maruyama 2011) such as catechol 1,4-dioxygenase and carboxymuconolactone decarboxylase implicated in the degradation of 1,4-dichlorobenzene degradation as well as benzoate degradation, and (2) metabolism related genes encoding iron transport proteins and proteins involved in sulfur metabolism such as thioredoxin reductase (Elsaied et al. 2007; Elsaied and Maruyama 2011). Novel integrase genes, attachment sites and associated gene cassettes have been detected in metagenomic DNA extracted from marine sediments polluted with high loads of petroleum wastes, and sediments contaminated with a mixture of industrial domestic wastes such as chlorinated polycyclic aromatic hydrocarbons and perfluorinated compounds in Suez and Tokyo bays (Elsaied et al. 2011). Gene cassettes were found to encode enzymes such as haloacid hydrolases that facilitate the degradation of organochlorinated compounds. Gene cassettes encoding proteins homologous to enzymes associated with hydrocarbon and toxic metal pollution such as disulphide isomerases and esterases were also detected at the sites. In deep-sea environments, novel integrons and associated gene cassettes were found in hydrothermal vents with genes that encoded metabolism-related enzymes such as transferases and thioredoxin reductases (Elsaied et al. 2007). Integrons with gene cassettes encoding oxidoreductases and alkyl transferases were also identified in submarine gas hydrate cores in these deep-sea environments (Elsaied et al. 2014).

6.3 Environmental Dissemination of Antibiotic Resistance Genes (ARGs)

While the global dissemination of ARGs due to the widespread use of antibiotics in agriculture, human and veterinary medicine are well documented (Rysz and Alvarez 2004; Pruden et al. 2006), studies focused on the spread of ARGs in marine environments due to antibiotic use in aquaculture and waste effluents are relatively limited. The fact that 95% of consumed antibiotics can be excreted in an unaltered state, greatly contributes to the rise of antibiotics as emerging environmental contaminants (Pruden et al. 2006). Also, wastewater treatment plants are not equipped to remove micropollutants such as antibiotics further exacerbating the problem (Janssens et al. 1997). Antibiotic concentrations between 10 and 1000 ng L^{-1} has been detected in secondary effluents depending on antibiotic type (Le-Minh et al. 2010). Although freshwater and marine aquatic ecosystems are increasingly recognized as reservoirs of ARGs (Taylor et al. 2011; Marti et al. 2014, Li et al. 2015a, b), studies characterizing and quantifying ARGs in non-clinical settings are relatively limited (Berendonk et al. 2015). ARGs cannot be readily eliminated from natural environments, as they tend to perpetuate even in the absence of selective pressure. Contrary to popular notion that harboring ARGs reduces bacterial host fitness due to metabolic costs associated with maintaining them, certain studies have demonstrated that ARGs may instead increase the bacterial fitness even in the absence of selective pressures (Salyers and Amabile-Cuevas 1997; Enne et al. 2004; Luo et al. 2005). This has been reported mainly in clinical studies where the fitness cost incurred because of antibiotic resistance is compensated by a secondary mutation that increases the fitness without affecting the resistance capability. For example, whole genome sequencing of rifampicin-resistant *M. tuberculosis* revealed compensatory mutations in RNA polymerase increasing their fitness in vitro as well as in vivo as determined by their frequency within clinical populations (Comas et al. 2012).

Plasmids are known to play an important role in the dissemination of antibiotic resistant determinants in the environment. Wastewater treatment plants were found to be important reservoirs of diverse types of plasmids encoding antibiotic resistance genes (Rizzo et al. 2013; Marti et al. 2014). Plasmids recovered from wastewater treatment plants confer resistance to multiple drugs including tetracycline, vancomycin, sulphonamides, and erythromycin among others. Plasmid metagenomes extracted from influent, activated and digested sludge in two different wastewater treatment plants were shown to contain multi-drug resistant ARGs and heavy metal resistance genes (Li et al. 2015a, b). This study, in addition to several others (Zhang et al. 2011; Rahube et al. 2014), further demonstrates the significant role of HGT and plasmids in the dissemination of ARGs in the environment.

Integrons present on plasmids and phages also play important roles in the horizontal transfer of ARGs. Several studies have reported an increase in integrons at anthropogenically impacted sites (Elsaied et al. 2011; Gaze et al. 2011; Gillings et al. 2015). Industrial effluents and wastewater treatment plants have also been identified as hotspots for the dissemination of mobile integrons (Stalder et al. 2007). The *IntI1* gene of Class 1 integrons has been suggested as a possible indicator of antibiotic resistance in the environment (Berendonk et al. 2015). Clinical class 1 integrons are found in over 80% of Enterobacteriaceae in humans and animals (Liu et al. 2013; Marchant et al. 2013). Class 1 integrons are also increasingly detected in ground water and fresh water sediments indicating their predisposition to mobilization and contribution to antibiotic selective pressure within aquatic environments (Stokes et al. 2006; Gillings et al. 2008). *IntI* carrying bacteria are highly abundant in manure and the *IntI* gene has been found to have a slow decay rate (Burch et al. 2014). Strong correlations between the abundance of *IntI* and sulphonamide resistance genes in reclaimed water, wastewater treatment plants, and animal feeding operations have been reported (Pruden et al. 2012; Wang et al. 2014). The activation of the bacterial SOS response, an inducible DNA repair system, caused by stress was shown to result in the overexpression of integrase leading to an increase in recombination events (Guerin et al. 2009; Baharoglu et al. 2010; Cambray et al. 2011). Guerin et al. analyzed integrase expression of integrons in *V. cholerae* and *E. coli*. The results showed that induction of SOS response increased the expression of a beta-galactosidase reporter of integrase transcription by 4.5 fold in *E. coli* and 37 fold in *V. cholera*. The frequency of cassette excision was also measured which revealed that integrase expression results in functional cassette recombination and the excision rates increased by 141 fold in *E. coli* and 340 fold in *V. cholerae*.

Biofilm formation has been known to be an effective resistance mechanism to antibiotics and heavy metals (Baker-Austin et al. 2006). In recent years, the finding that biofilms are reservoirs of ARGs and that the abundance of mobile GEs is directly proportional to ARG-transfer within biofilms (Balcázar et al. 2015) augments earlier reports of resistance within biofilms. Exposure to metals and antibiotics can stimulate the production of extracellular polymeric substances resulting in cell adhesion and subsequent biofilm formation. 'Persister' cells within the biofilm have been postulated to be responsible for promoting tolerance to antimicrobial agents and heavy metals (Teitzel and Parsek 2003; Harrison et al. 2005; Baker-Austin et al. 2006). Biofilms consist of a high density of microbial cells close to each other within the matrix and the incidence of mobile GEs are found to be elevated in these environments (Gillings et al. 2008; Farkas et al. 2013).

Integrons conferring a specific adaptation can be carried by transposons with a different adaptive capability such that the recipient of the combined larger element gains both adaptive traits. This results in co-resistance of more than one adaptive traits due to the physical linkage of the genetic elements. The selection of one of these adaptive genes maintains both adaptive traits in a phenomenon called co-selection. Genetic elements such as integrons and plasmids found in polluted environments can facilitate co-selection as they usually carry gene cassettes that confer resistance to antibiotics. The marine environment, often contaminated with a mixture of compounds from anthropogenic sources, facilitates co-selection of resistance genes and phenotypes. Although aquatic systems have been hypothesized to play important ecological roles in the emergence and spread of antimicrobial resistance among bacteria (Taylor et al. 2011), fewer studies, to date, focused on the incidence of co-resistance within marine ecosystems (Allen et al. 1977; Timoney et al. 1978; Rasmussen and Sørensen 1998).

6.4 The Occurance of Co-resistance in Marine Environments

Co-resistance emerges when two or more genes responsible for resistance phenotypes are located next to each other on a single mobile genetic element (Chapman 2003). The acquisition of multiple resistance genes offers a selective advantage to bacteria and the selection typically occurs via HGT or mutational events. For instance, the increase of metal concentration in the environment has been positively correlated to the abundance of antibiotic resistance genes (Belliveau et al. 1991;

Di Cesare et al. 2016; Henriques et al. 2016). This poses as an environmental hazard since heavy metal remediation is a slow process and the consequential persistence of heavy metals in the marine environment provides constant selective pressure to resistant bacteria.

The main mechanisms by which resistance is conferred are: (1) efflux of the antimicrobial agent through the cell membrane; (2) modification and/or breakdown of the toxin within the cell; (3) sequestration within the cell; and (4) reduced sensitivity of the target within the cell (Cloete 2003). These mechanisms can be acquired by HGT from other bacteria resulting in clonal expansion of the resistant organism. The use and overuse of antibiotics are major promoters of multidrug resistant bacteria harboring resistance genes that can accumulate in certain bacteria taxa resulting in complex phenotypes. In addition to antibiotic use, heavy metal contamination is widespread due to agricultural and aquaculture practices as metals occur in fertilizers, pesticides, fish feed and anti-fouling products (Seiler and Berendonk 2007). The toxicity of heavy metals depends on their bioavailability which is influenced by abiotic factors such as pH, redox potential, and concentration of organic matter in sediment or water. Environmental pollution with heavy metals can promote the co-selection of antibiotic resistance genes along with heavy metal resistance due to physical linkage of the resistance genes, subsequently decreasing antibiotic susceptibility (Baker-Austin et al. 2006; Seiler and Berendonk 2007). Examples of shared structural and functional characteristics of bacterial antibiotic and metal resistance mechanisms have been well documented (Baker-Austin et al. 2006; Stepanauskas et al. 2006; Seiler and Berendonk 2007; Ji et al. 2012). Heavy metal pollution, due to their widespread and recalcitrant nature was found to be an important component in dissemination of antibiotic resistant bacteria in freshwater habitats (Stepanauskas et al. 2006). Co-resistance to heavy metals such as chromate and cobalt along with antibiotics was reported in the marine bacterium *Halomonas* due to the close arrangement of resistance genes on plasmid pMA21 (Osman et al. 2010). Pal et al. (2015) has demonstrated that plasmids from environments with a higher probability of acquiring multiple resistance genes were significantly more likely to be conjugative than plasmids without them. Co-resistances to multiple anthropogenic pollutants coupled with conjugative transfer capabilities further reaffirms that HGT contributes to bacterial adaptation as well as the persistence of resistant bacterial populations in the environment.

Predicting the structural and functional response of a microbial community to anthropogenic disturbances in marine environments is a strategy that is critical to assessing the impacts of similar perturbations in the future. Microbial communities can recover from a disturbance through physiological and genetic adaptation, and microbial resilience to disturbances need to be further examined to gain better insight into the consequences of anthropogenic changes on coastal, near-shore and off-shore marine ecosystems.

6.5 Resilience of Microbial Communities Against Perturbations

Ecological studies have consistently demonstrated that microbial community composition and function are often significantly altered upon perturbations. The resilience of microbial communities refers to the rate of recovery of the microbial community to its original composition (or state) after the disturbance (Allison and Martiny 2008). Although microbial communities within salt marsh sediments degraded oil released from the *Deepwater Horizon* spill (Mahmoudi et al. 2013), shoreline oiling and its effects on vegetation reduced the resilience of Louisiana salt marshes suggesting the existence of thresholds that determine the resilience of an ecosystem (Silliman et al. 2012, 2016). Phenomena such as cellular dormancy and biofilm formation are important in limiting the impacts of anthropogenic stresses of some microorganisms (Hall-Stoodley et al. 2004; Lennon and Jones 2011) thereby conferring resilience to the disturbance. The dispersal of microbial cells has also been reported to contribute to increased resistance and resilience of naturally occurring microbial communities experiencing disturbance events (Altermatt et al. 2011; Steiner et al. 2011). Metabolic flexibility as a result of HGT events, and functional redundancy in the midst of significant perturbations, such as oil spills, is another means by which microbial populations recover from anthropogenic impacts (Fernandez et al. 2000; Allison and Martiny 2008). For example, purple nonsulfur bacteria act as phototrophs in anoxic conditions and heterotrophs under aerobic conditions, thereby maximizing their competitive advantages, in the event of a disturbance (Shade et al. 2012). Additionally, the strength, frequency, spatial and temporal scales of the disturbance will also determine the resilience of the microbial communities. 'Pulse' disturbances are considered short-term events and 'presses' are more continuous, long-term events (Bender et al. 1984). Strategies such as metabolic dormancy are used for maintaining microbial community stability during pulse disturbances. Increased growth rates is yet another type of strategy for pulse disturbances to limit or ameliorate alterations to community composition and ultimately ecosystem function (Shade et al. 2012).

6.6 Conclusion

The complex relationships between genetic adaptation of microbial communities and microbial community diversity and function continue to be a subject of investigations when events such as the largest oil spill in US history were occurring in 2010 (Hazen et al. 2010; Beazley et al. 2012; Horel et al. 2012). Methods such as high throughput next-generation sequencing technologies have greatly facilitated insights into short-term and long-term temporal changes to microbial community structure, metabolic capabilities, and gene functions within coastal and deep sea environments. (Hazen et al. 2010; Bælum et al. 2012; Mason et al. 2012; Kimes et al. 2014; King et al. 2015). Analyses of microbial community and their adaptation to anthropogenic stresses will provide key insights into identifying taxonomic and functional information of relevant organisms thereby unraveling mechanisms underlying microbial community resilience.

References

Abella J, Bielen A, Huang L, Delmont TO, Vujaklija D, Duran R, Cagnon C (2015) Integron diversity in marine environments. Environ Sci Pollut Res 22:15360–15369

Allen D, Austin B, Colwell R (1977) Antibiotic resistance patterns of metal-tolerant bacteria isolated from an estuary. Antimicrob Agents Chemother 12:545–547

Allison SD, Martiny JB (2008) Resistance, resilience, and redundancy in microbial communities. Proc Natl Acad Sci 105:11512–11519

Altermatt F, Bieger A, Carrara F, Rinaldo A, Holyoak M (2011) Effects of connectivity and recurrent local disturbances on community structure and population density in experimental metacommunities. PLoS ONE 6:e19525

Antonova ES, Hammer BK (2015) Genetics of natural competence in *Vibrio cholerae* and other vibrios. Microbiol Spectr 3(3):0010–2014

Arrigo KR (2005) Marine microorganisms and global nutrient cycles. Nature 437:349–355

Aziz RK, Dwivedi B, Akhter S, Breitbart M, Edwards RA (2015) Multidimensional metrics for estimating phage abundance, distribution, gene density, and sequence coverage in metagenomes. Front Microbiol 6:381

Bælum J, Borglin S, Chakraborty R, Fortney JL, Lamendella R, Mason OU, Auer M, Zemla M, Bill M, Conrad ME (2012) Deep-sea bacteria enriched by oil and dispersant from the Deepwater Horizon spill. Environ Microbiol 4(9):2405–2416

Baharoglu Z, Bikard D, Mazel D (2010) Conjugative DNA transfer induces the bacterial SOS response and promotes antibiotic resistance development through integron activation. PLoS Genet 6:e1001165

Baker-Austin C, Wright MS, Stepanauskas R, McArthur J (2006) Co-selection of antibiotic and metal resistance. Trends Microbiol 14:176–182

Balcázar JL, Subirats J, Borrego CM (2015) The role of biofilms as environmental reservoirs of antibiotic resistance. Front Microbiol 6:1216

Banat IM, Franzetti A, Gandolfi I, Bestetti G, Martinotti MG, Fracchia L, Smyth TJ, Marchant R (2010) Microbial biosurfactants production, applications and future potential. Appl Microbiol Biotechnol 87:427–444

Beazley MJ, Martinez RJ, Rajan S, Powell J, Piceno YM, Tom LM, Andersen GL, Hazen TC, Van Nostrand JD, Zhou J (2012) Microbial community analysis of a coastal salt marsh affected by the Deepwater Horizon oil spill. PLoS ONE 7:e41305

Belliveau B, Starodub M, Trevors J (1991) Occurrence of antibiotic and metal resistance and plasmids in *Bacillus* strains isolated from marine sediment. Can J Microbiol 37:513–520

Bender EA, Case TJ, Gilpin ME (1984) Perturbation experiments in community ecology: theory and practice. Ecology 65:1–13

Berendonk TU, Manaia CM, Merlin C, Fatta-Kassinos D, Cytryn E, Walsh F, Bürgmann H, Sørum H, Norström M, Pons M-N (2015) Tackling antibiotic resistance: the environmental framework. Nat Rev Microbiol 13:310–317

Bertani G (1999) Transduction-like gene transfer in the methanogen *Methanococcus voltae*. J Bacteriol 181:2992–3002

Biers EJ, Wang K, Pennington C, Belas R, Chen F, Moran MA (2008) Occurrence and expression of gene transfer agent genes in marine bacterioplankton. Appl Environ Microbiol 74: 2933–2939

Biller SJ, Schubotz F, Roggensack SE, Thompson AW, Summons RE, Chisholm SW (2014) Bacterial vesicles in marine ecosystems. Science 343:183–186

Boucher Y, Labbate M, Koenig JE, Stokes H (2007) Integrons: mobilizable platforms that promote genetic diversity in bacteria. Trends Microbiol 15:301–309

Breitbart M (2012) Marine viruses: truth or dare. Annu Rev Mar Sci 4:425–448

Brimacombe CA, Ding H, Johnson JA, Beatty JT (2015) Homologues of genetic transformation DNA import genes are required for *Rhodobacter capsulatus* gene transfer agent recipient capability regulated by the response regulator CtrA. J Bacteriol 197:2653–2663

Brochier-Armanet C, Moreira D (2015) Horizontal gene transfer in microbial ecosystems. In: Environmental microbiology: fundamentals and applications. Springer, pp 445–481

Brum JR, Sullivan MB (2015) Rising to the challenge: accelerated pace of discovery transforms marine virology. Nat Rev Microbiol 13:147–159

Buchan A, González J (2010) Roseobacter. In: Handbook of hydrocarbon and lipid microbiology. Springer, pp 1335–1343

Burch TR, Sadowsky MJ, LaPara TM (2014) Fate of antibiotic resistance genes and class 1 integrons in soil microcosms following the application of treated residual municipal wastewater solids. Environ Sci Technol 48:5620–5627

Byrne-Bailey K, Gaze WH, Zhang L, Kay P, Boxall A, Hawkey PM, Wellington EM (2011) Integron prevalence and diversity in manured soil. Appl Environ Microbiol 77:684–687

Cambray G, Guerout A-M, Mazel D (2010) Integrons. Annu Rev Genet 44:141–166

Cambray G, Sanchez-Alberola N, Campoy S, Guerin É, Da Re S, González-Zorn B, Ploy M-C, Barbé J, Mazel D, Erill I (2011) Prevalence of SOS-mediated control of integron integrase expression as an adaptive trait of chromosomal and mobile integrons. Mob DNA 2:2–15

Caro-Quintero A, Konstantinidis KT (2015) Inter-phylum HGT has shaped the metabolism of many mesophilic and anaerobic bacteria. ISME J 9:958–967

Chapman JS (2003) Disinfectant resistance mechanisms, cross-resistance, and co-resistance. Int Biodeterior Biodegradation 51:271–276

Cloete TE (2003) Resistance mechanisms of bacteria to antimicrobial compounds. Int Biodeterior Biodegradation 51:277–282

Cochran PK, Paul JH (1998) Seasonal abundance of lysogenic bacteria in a subtropical estuary. Appl Environ Microbiol 64:2308–2312

Coleman ML, Sullivan MB, Martiny AC, Steglich C, Barry K, DeLong EF, Chisholm SW (2006) Genomic islands and the ecology and evolution of *Prochlorococcus*. Science 311:1768–1770

Collis CM, Kim M-J, Partridge SR, Stokes H, Hall RM (2002) Characterization of the class 3 integron and the site-specific recombination system it determines. J Bacteriol 184:3017–3026

Comas I, Borrell S, Roetzer A, Rose G, Malla B, Kato-Maeda M, Galagan J, Niemann S, Gagneux S (2012) Whole-genome sequencing of rifampicin-resistant *Mycobacterium tuberculosis* strains identifies compensatory mutations in RNA polymerase genes. Nat Genet 44:106–110

Darmon E, Leach DR (2014) Bacterial genome instability. Microbiol Mol Biol Rev 78:1–39

Dash HR, Mangwani N, Chakraborty J, Kumari S, Das S (2013) Marine bacteria: potential candidates for enhanced bioremediation. Appl Microbiol Biotechnol 97:561–571

Debbab A, Aly AH, Lin WH, Proksch P (2010) Bioactive compounds from marine bacteria and fungi. Microb Biotechnol 3:544–563

Di Cesare A, Luna GM, Vignaroli C, Pasquaroli S, Tota S, Paroncini P, Biavasco F (2013) Aquaculture can promote the presence and spread of antibiotic-resistant *Enterococci* in marine sediments. PLoS ONE 8:e62838

Di Cesare A, Eckert EM, D'Urso S, Bertoni R, Gillan DC, Wattiez R, Corno G (2016) Co-occurrence of integrase 1, antibiotic and heavy metal resistance genes in municipal wastewater treatment plants. Water Res 94:208–214

Dobrindt U, Hochhut B, Hentschel U, Hacker J (2004) Genomic islands in pathogenic and environmental microorganisms. Nat Rev Microbiol 2:414–424
Dorn E, Hellwig M, Reineke W, Knackmuss H-J (1974) Isolation and characterization of a 3-chlorobenzoate degrading pseudomonad. Arch Microbiol 99:61–70
Dubey GP, Ben-Yehuda S (2011) Intercellular nanotubes mediate bacterial communication. Cell 144:590–600
Ebner P, Garner K, Mathew A (2004) Class 1 integrons in various *Salmonella enterica* serovars isolated from animals and identification of genomic island SGI1 in *Salmonella enterica* var. Meleagridis. J Antimicrob Chemother 53:1004–1009
Elsaied H, Maruyama A (2011) Diversity and role of bacterial integron/gene cassette metagenome in extreme marine environments. Handbook of molecular microbial ecology II: metagenomics in different habitats. Wiley, Hoboken, NJ, USA, pp 309–318
Elsaied H, Stokes H, Nakamura T, Kitamura K, Fuse H, Maruyama A (2007) Novel and diverse integron integrase genes and integron-like gene cassettes are prevalent in deep-sea hydrothermal vents. Environ Microbiol 9:2298–2312
Elsaied H, Stokes HW, Kitamura K, Kurusu Y, Kamagata Y, Maruyama A (2011) Marine integrons containing novel integrase genes, attachment sites, attI, and associated gene cassettes in polluted sediments from Suez and Tokyo Bays. ISME J 5:1162–1177
Elsaied H, Stokes HW, Yoshioka H, Mitani Y, Maruyama A (2014) Novel integrons and gene cassettes from a Cascadian submarine gas-hydrate-bearing core. FEMS Microbiol Ecol 87:343–356
Enne VI, Bennett PM, Livermore DM, Hall LM (2004) Enhancement of host fitness by the sul2-coding plasmid p9123 in the absence of selective pressure. J Antimicrob Chemother 53:958–963
Farkas A, Butiuc-Keul A, Ciatarâş D, Neamţu C, Crăciunaş C, Podar D, Drăgan-Bularda M (2013) Microbiological contamination and resistance genes in biofilms occurring during the drinking water treatment process. Sci Total Environ 443:932–938
Fernandez AS, Hashsham SA, Dollhopf SL, Raskin L, Glagoleva O, Dazzo FB, Hickey RF, Criddle CS, Tiedje JM (2000) Flexible community structure correlates with stable community function in methanogenic bioreactor communities perturbed by glucose. Appl Environ Microbiol 66:4058–4067
Fisher CR, Hsing P-Y, Kaiser CL, Yoerger DR, Roberts HH, Shedd WW, Cordes EE, Shank TM, Berlet SP, Saunders MG, Larcom EA, Brooks JM (2014) Footprint of deepwater horizon blowout impact to deep-water coral communities. Proceedings of the National Academy of Sciences 111(32):11744–11749
Freeman VJ (1951) Studies on the virulence of bacteriophage-infected strains of *Corynebacterium diphtheriae*. J Bacteriol 61:675
Gaillard M, Vallaeys T, Vorhölter FJ, Minoia M, Werlen C, Sentchilo V, Pühler A, van der Meer JR (2006) The clc element of *Pseudomonas* sp. strain B13, a genomic island with various catabolic properties. J Bacteriol 188:1999–2013
Garcillán-Barcia MP, Alvarado A, de la Cruz F (2011) Identification of bacterial plasmids based on mobility and plasmid population biology. FEMS Microbiol Rev 35:936–956
Gaze WH, Zhang L, Abdouslam NA, Hawkey PM, Calvo-Bado L, Royle J, Brown H, Davis S, Kay P, Boxall AB (2011) Impacts of anthropogenic activity on the ecology of class 1 integrons and integron-associated genes in the environment. ISME J 5:1253–1261
Gerdes K, Rasmussen PB, Molin S (1986) Unique type of plasmid maintenance function: postsegregational killing of plasmid-free cells. Proc Natl Acad Sci 83:3116–3120
Gillings MR (2014) Integrons: past, present, and future. Microbiol Mol Biol Rev 78:257–277
Gillings M, Boucher Y, Labbate M, Holmes A, Krishnan S, Holley M, Stokes H (2008) The evolution of class 1 integrons and the rise of antibiotic resistance. J Bacteriol 190:5095–5100
Gillings MR, Labbate M, Sajjad A, Giguere NJ, Holley MP, Stokes H (2009) Mobilization of a Tn402-like class 1 integron with a novel cassette array via flanking miniature inverted-repeat transposable element-like structures. Appl Environ Microbiol 75:6002–6004
Gillings MR, Gaze WH, Pruden A, Smalla K, Tiedje JM, Zhu Y-G (2015) Using the class 1 integron-integrase gene as a proxy for anthropogenic pollution. ISME J 9:1269–1279

Goldstein C, Lee MD, Sanchez S, Hudson C, Phillips B, Register B, Grady M, Liebert C, Summers AO, White DG (2001) Incidence of class 1 and 2 integrases in clinical and commensal bacteria from livestock, companion animals, and exotics. Antimicrob Agents Chemother 45:723–726

Guerin É, Cambray G, Sanchez-Alberola N, Campoy S, Erill I, Da Re S, Gonzalez-Zorn B, Barbé J, Ploy M-C, Mazel D (2009) The SOS response controls integron recombination. Science 324:1034

Hacker J, Bender L, Ott M, Wingender J, Lund B, Marre R, Goebel W (1990) Deletions of chromosomal regions coding for fimbriae and hemolysins occur in vitro and in vivo in various extra intestinal *Escherichia coli* isolates. Microb Pathog 8:213–225

Hall-Stoodley L, Costerton JW, Stoodley P (2004) Bacterial biofilms: from the natural environment to infectious diseases. Nat Rev Microbiol 2:95–108

Harrison JJ, Turner RJ, Ceri H (2005) Persister cells, the biofilm matrix and tolerance to metal cations in biofilm and planktonic *Pseudomonas aeruginosa*. Environ Microbiol 7:981–994

Hazen TC, Dubinsky EA, DeSantis TZ et al (2010) Deep-sea oil plume enriches indigenous oil-degrading bacteria. Science 330:204–208

Hegstad K, Mikalsen T, Coque T, Werner G, Sundsfjord A (2010) Mobile genetic elements and their contribution to the emergence of antimicrobial resistant *Enterococcus faecalis* and *Enterococcus faecium*. Clin Microbiol Infect 16:541–554

Henriques I, Tacão M, Leite L, Fidalgo C, Araújo S, Oliveira C, Alves A (2016) Co-selection of antibiotic and metal (loid) resistance in gram-negative epiphytic bacteria from contaminated salt marshes. Mar Pollut Bull 109(1):427–434

Horel A, Mortazavi B, Sobecky PA (2012) Responses of microbial community from northern Gulf of Mexico sandy sediments following exposure to deepwater horizon crude oil. Environ Toxicol Chem 31:1004–1011

Islam MS, Tanaka M (2004) Impacts of pollution on coastal and marine ecosystems including coastal and marine fisheries and approach for management: a review and synthesis. Mar Pollut Bull 48:624–649

Janssens I, Tanghe T, Verstraete W (1997) Micropollutants: a bottleneck in sustainable wastewater treatment. Water Sci Technol 35:13–26

Ji X, Shen Q, Liu F, Ma J, Xu G, Wang Y, Wu M (2012) Antibiotic resistance gene abundances associated with antibiotics and heavy metals in animal manures and agricultural soils adjacent to feedlots in Shanghai; China. J Hazard Mater 235:178–185

Jiang SC, Paul JH (1996) Occurrence of lysogenic bacteria in marine microbial communities as determined by prophage induction. Mar Ecol Prog Ser 142:27

Jiang SC, Paul JH (1998) Gene transfer by transduction in the marine environment. Appl Environ Microbiol 64:2780–2787

Johnson CM, Grossman AD (2015) Integrative and conjugative elements (ICEs): what they do and how they work. Annu Rev Genet 49:577–601

Juhas M, Crook DW, Dimopoulou ID, Lunter G, Harding RM, Ferguson DJ, Hood DW (2007a) Novel type IV secretion system involved in propagation of genomic islands. J Bacteriol 189:761–771

Juhas M, Power PM, Harding RM, Ferguson DJ, Dimopoulou ID, Elamin AR, Mohd-Zain Z, Hood DW, Adegbola R, Erwin A (2007b) Sequence and functional analyses of *Haemophilus* spp. genomic islands. Genome Biol 8:R237

Juhas M, Crook DW, Hood DW (2008) Type IV secretion systems: tools of bacterial horizontal gene transfer and virulence. Cell Microbiol 10:2377–2386

Juhas M, van der Meer JR, Gaillard M, Harding RM, Hood DW, Crook DW (2009) Genomic islands: tools of bacterial horizontal gene transfer and evolution. FEMS Microbiol Rev 33: 376–393

Kennish MJ (1996) In: Practical handbook of estuarine and marine pollution. CRC Press, pp 1–2

Kimes NE, Callaghan AV, Suflita JM, Morris PJ (2014) Microbial transformation of the Deepwater Horizon oil spill—past, present, and future perspectives. Front Microbiol 5:603

King G, Kostka J, Hazen T, Sobecky P (2015) Microbial responses to the Deepwater Horizon oil spill: from coastal wetlands to the deep sea. Annu Rev Mar Sci 7:377–401
Koenig JE, Sharp C, Dlutek M, Curtis B, Joss M, Boucher Y, Doolittle WF (2009) Integron gene cassettes and degradation of compounds associated with industrial waste: the case of the Sydney tar ponds. PLoS ONE 4:e5276
Lang AS, Beatty J (2000) Genetic analysis of a bacterial genetic exchange element: the gene transfer agent of *Rhodobacter capsulatus*. Proc Natl Acad Sci 97:859–864
Lang AS, Beatty JT (2007) Importance of widespread gene transfer agent genes in α-proteobacteria. Trends Microbiol 15:54–62
Lang AS, Zhaxybayeva O, Beatty JT (2012) Gene transfer agents: phage-like elements of genetic exchange. Nat Rev Microbiol 10:472–482
Langille MG, Hsiao WW, Brinkman FS (2010) Detecting genomic islands using bioinformatics approaches. Nat Rev Microbiol 8:373–382
Le-Minh N, Khan S, Drewes J, Stuetz R (2010) Fate of antibiotics during municipal water recycling treatment processes. Water Res 44:4295–4323
Lennon JT, Jones SE (2011) Microbial seed banks: the ecological and evolutionary implications of dormancy. Nat Rev Microbiol 9:119–130
Lerat E, Daubin V, Ochman H, Moran NA (2005) Evolutionary origins of genomic repertoires in bacteria. PLoS Biol 3:e130
Li A-D, Li L-G, Zhang T (2015a) Exploring antibiotic resistance genes and metal resistance genes in plasmid metagenomes from wastewater treatment plants. Front Microbiol 6:1025
Li B, Yang Y, Ma L, Ju F, Guo F, Tiedje JM, Zhang T (2015b) Metagenomic and network analysis reveal wide distribution and co-occurrence of environmental antibiotic resistance genes. ISME J 9:2490–2502
Liu H, Wang H, Huang M, Mei Y, Gu B, Wu R, Huang Y, Chen Y, Xu Y, Wang T (2013) Analysis of antimicrobial resistance and class 1 integrons among strains from upper respiratory tract of healthy adults. J Thoracic Dis 5:149–155
Luo N, Pereira S, Sahin O, Lin J, Huang S, Michel L, Zhang Q (2005) Enhanced in vivo fitness of fluoroquinolone-resistant *Campylobacter jejuni* in the absence of antibiotic selection pressure. Proc Natl Acad Sci U S A 102:541–546
Ma Y, Paulsen IT, Palenik B (2012) Analysis of two marine metagenomes reveals the diversity of plasmids in oceanic environments. Environ Microbiol 14:453–466
Mahmoudi N, Porter TM, Zimmerman AR, Fulthorpe RR, Kasozi GN, Silliman BR, Slater GF (2013) Rapid degradation of deepwater horizon spilled oil by indigenous microbial communities in Louisiana saltmarsh sediments. Environ Sci Technol 47:13303–13312
Marchant M, Vinué L, Torres C, Moreno MA (2013) Change of integrons over time in *Escherichia coli* isolates recovered from healthy pigs and chickens. Vet Microbiol 163:124–132
Marrs B (1974) Genetic recombination in *Rhodopseudomonas capsulata*. Proc Natl Acad Sci 71:971–973
Marti E, Variatza E, Balcazar JL (2014) The role of aquatic ecosystems as reservoirs of antibiotic resistance. Trends Microbiol 22:36–41
Martinez-Freijo P, Fluit A, Schmitz F, Grek V, Verhoef J, Jones M (1998) Class I integrons in Gram-negative isolates from different European hospitals and association with decreased susceptibility to multiple antibiotic compounds. J Antimicrob Chemother 42:689–696
Mashburn-Warren LM, Whiteley M (2006) Special delivery: vesicle trafficking in prokaryotes. Mol Microbiol 61:839–846
Mason OU, Hazen TC, Borglin S et al (2012) Metagenome, metatranscriptome and single-cell sequencing reveal microbial response to Deepwater Horizon oil spill. ISME J 6:1715–1727
Mazel D (2006) Integrons: agents of bacterial evolution. Nat Rev Microbiol 4:608–620
Mazel D, Dychinco B, Webb VA, Davies J (1998) A distinctive class of integron in the *Vibrio cholerae* genome. Science 280:605–608
McDaniel LD, Young E, Delaney J, Ruhnau F, Ritchie KB, Paul JH (2010) High frequency of horizontal gene transfer in the oceans. Science 330:50

McDaniel LD, Rosario K, Breitbart M, Paul JH (2014) Comparative metagenomics: natural populations of induced prophages demonstrate highly unique, lower diversity viral sequences. Environ Microbiol 16:570–585

Michael V, Frank O, Bartling P, Scheuner C, Göker M, Brinkmann H, Petersen J (2016) Biofilm plasmids with a rhamnose operon are widely distributed determinants of the 'swim-or-stick'lifestyle in Roseobacters. ISME J 10:2498–2513

Miyazaki R, Bertelli C, Benaglio P, Canton J, De Coi N, Gharib WH, Gjoksi B, Goesmann A, Greub G, Harshman K (2015) Comparative genome analysis of *Pseudomonas knackmussii* B13, the first bacterium known to degrade chloroaromatic compounds. Environ Microbiol 17:91–104

Mojica KD, Brussaard CP (2014) Factors affecting virus dynamics and microbial host–virus interactions in marine environments. FEMS Microbiol Ecol 89:495–515

Morales-Espinosa R, Soberón-Chávez G, Delgado-Sapién G, Sandner-Miranda L, Méndez JL, González-Valencia G, Cravioto A (2012) Genetic and phenotypic characterization of a *Pseudomonas aeruginosa* population with high frequency of genomic islands. PLoS ONE 7: e37459

Moran M, Belas R, Schell M, González J, Sun F, Sun S, Binder B, Edmonds J, Ye W, Orcutt B (2007) Ecological genomics of marine Roseobacters. Appl Environ Microbiol 73:4559–4569

Nelson-Sathi S, Dagan T, Landan G, Janssen A, Steel M, McInerney JO, Deppenmeier U, Martin WF (2012) Acquisition of 1,000 eubacterial genes physiologically transformed a methanogen at the origin of Haloarchaea. Proc Natl Acad Sci 109:20537–20542

Nield BS, Holmes AJ, Gillings MR, Recchia GD, Mabbutt BC, Nevalainen KH, Stokes HW (2001) Recovery of new integron classes from environmental DNA. FEMS Microbiol Lett 195:59–65

Nogales B, Lanfranconi MP, Piña-Villalonga JM, Bosch R (2011) Anthropogenic perturbations in marine microbial communities. FEMS Microbiol Rev 35:275–298

Ochman H, Lawrence JG, Groisman EA (2000) Lateral gene transfer and the nature of bacterial innovation. Nature 405:299–304

Osman O, Tanguichi H, Ikeda K, Park P, Tanabe-Hosoi S, Nagata S (2010) Copper-resistant halophilic bacterium isolated from the polluted Maruit Lake, Egypt. J Appl Microbiol 108:1459–1470

Pal C, Bengtsson-Palme J, Kristiansson E, Larsson DJ (2015) Co-occurrence of resistance genes to antibiotics, biocides and metals reveals novel insights into their co-selection potential. BMC Genom 16:1

Palenik B, Ren Q, Tai V, Paulsen I (2009) Coastal *Synechococcus* metagenome reveals major roles for horizontal gene transfer and plasmids in population diversity. Environ Microbiol 11:349–359

Palmer KL, Kos VN, Gilmore MS (2010) Horizontal gene transfer and the genomics of enterococcal antibiotic resistance. Curr Opin Microbiol 13:632–639

Paul JH (2008) Prophages in marine bacteria: dangerous molecular time bombs or the key to survival in the seas? ISME J 2:579–589

Penn K, Jenkins C, Nett M, Udwary DW, Gontang EA, McGlinchey RP, Foster B, Lapidus A, Podell S, Allen EE (2009) Genomic islands link secondary metabolism to functional adaptation in marine Actinobacteria. ISME J 3:1193–1203

Peters JE, Fricker AD, Kapili BJ, Petassi MT (2014) Heteromeric transposase elements: generators of genomic islands across diverse bacteria. Mol Microbiol 93:1084–1092

Petersen J, Frank O, Göker M, Pradella S (2013) Extrachromosomal, extraordinary and essential—the plasmids of the *Roseobacter* clade. Appl Microbiol Biotechnol 97:2805–2815

Poli A, Anzelmo G, Nicolaus B (2010) Bacterial exopolysaccharides from extreme marine habitats: production, characterization and biological activities. Mar Drugs 8:1779–1802

Pruden A, Pei R, Storteboom H, Carlson KH (2006) Antibiotic resistance genes as emerging contaminants: studies in northern Colorado. Environ Sci Technol 40:7445–7450

Pruden A, Arabi M, Storteboom HN (2012) Correlation between upstream human activities and riverine antibiotic resistance genes. Environ Sci Technol 46:11541–11549

Rahube TO, Viana LS, Koraimann G, Yost CK (2014) Characterization and comparative analysis of antibiotic resistance plasmids isolated from a wastewater treatment plant. Front Microbiol 5:558

Rapa RA, Labbate M (2013) The function of integron-associated gene cassettes in Vibrio species: the tip of the iceberg. Front Microbiol 4:385

Rasmussen LD, Sørensen SJ (1998) The effect of longterm exposure to mercury on the bacterial community in marine sediment. Curr Microbiol 36:291–297

Rizzo L, Manaia C, Merlin C, Schwartz T, Dagot C, Ploy M, Michael I, Fatta-Kassinos D (2013) Urban wastewater treatment plants as hotspots for antibiotic resistant bacteria and genes spread into the environment: a review. Sci Total Environ 447:345–360

Rodríguez-Minguela CM, Apajalahti JH, Chai B, Cole JR, Tiedje JM (2009) Worldwide prevalence of class 2 integrases outside the clinical setting is associated with human impact. Appl Environ Microbiol 75:5100–5110

Rysz M, Alvarez PJ (2004) Amplification and attenuation of tetracycline resistance in soil bacteria: aquifer column experiments. Water Res 38:3705–3712

Salyers AA, Amabile-Cuevas CF (1997) Why are antibiotic resistance genes so resistant to elimination? Antimicrob Agents Chemother 41:2321

Seiler C, Berendonk TU (2007) Heavy metal driven co-selection of antibiotic resistance in soil and water bodies impacted by agriculture and aquaculture. Front Microbiol 3:399

Shade A, Peter H, Allison SD, Baho DL, Berga M, Bürgmann H, Huber DH, Langenheder S, Lennon JT, Martiny JB (2012) Fundamentals of microbial community resistance and resilience. Front Microbiol 3:417

Shintani M, Sanchez ZK, Kimbara K (2015) Genomics of microbial plasmids: classification and identification based on replication and transfer systems and host taxonomy. Front Microbiol 6:242

Silliman BR, van de Koppel J, McCoy MW, Diller J, Kasozi GN, Earl K, Adams PN, Zimmerman AR (2012) Degradation and resilience in Louisiana salt marshes after the BP–Deepwater Horizon oil spill. Proc Natl Acad Sci 109:11234–11239

Silliman BR, Dixon PM, Wobus C, He Q, Daleo P, Hughes BB, Rissing M, Willis JM, Hester MW (2016) Thresholds in marsh resilience to the Deepwater Horizon oil spill. Sci Rep 6:32520

Smillie C, Garcillán-Barcia MP, Francia MV, Rocha EP, de la Cruz F (2010) Mobility of plasmids. Microbiol Mol Biol Rev 74:434–452

Sobecky PA, Hazen TH (2009) In: Horizontal gene transfer and mobile genetic elements in marine systems. Horizontal Gene Transfer: Genomes in Flux. Humana Press, pp 435–453

Stalder T, Barraud O, Casellas M, Dagot C, Ploy M-C (2007) Integron involvement in environmental spread of antibiotic resistance. Front Microbiol 3:87–100

Stalder T, Barraud O, Jové T, Casellas M, Gaschet M, Dagot C, Ploy M-C (2014) Quantitative and qualitative impact of hospital effluent on dissemination of the integron pool. ISME J 8:768–777

Stanton TB (2007) Prophage-like gene transfer agents—novel mechanisms of gene exchange for *Methanococcus*, *Desulfovibrio*, *Brachyspira*, and *Rhodobacter* species. Anaerobe 13:43–49

Steiner CF, Stockwell RD, Kalaimani V, Aqel Z (2011) Dispersal promotes compensatory dynamics and stability in forced metacommunities. Am Nat 178:159–170

Stepanauskas R, Glenn TC, Jagoe CH, Tuckfield RC, Lindell AH, King CJ, McArthur J (2006) Coselection for microbial resistance to metals and antibiotics in freshwater microcosms. Environ Microbiol 8:1510–1514

Stokes HW, Gillings MR (2011) Gene flow, mobile genetic elements and the recruitment of antibiotic resistance genes into Gram-negative pathogens. FEMS Microbiol Rev 35:790–819

Stokes HW, Nesbø CL, Holley M, Bahl MI, Gillings MR, Boucher Y (2006) Class 1 integrons potentially predating the association with Tn402-like transposition genes are present in a sediment microbial community. J Bacteriol 188:5722–5730

Stuart RK, Brahamsha B, Busby K, Palenik B (2013) Genomic island genes in a coastal marine *Synechococcus* strain confer enhanced tolerance to copper and oxidative stress. ISME J 7:1139–1149

Swingley WD, Sadekar S, Mastrian SD, Matthies HJ, Hao J, Ramos H, Acharya CR, Conrad AL, Taylor HL, Dejesa LC (2007) The complete genome sequence of *Roseobacter denitrificans* reveals a mixotrophic rather than photosynthetic metabolism. J Bacteriol 189:683–690
Tang H, Yao Y, Wang L, Yu H, Ren Y, Wu G, Xu P (2012) Genomic analysis of *Pseudomonas putida*: genes in a genome island are crucial for nicotine degradation. Sci Rep 2:377
Taylor NG, Verner-Jeffreys DW, Baker-Austin C (2011) Aquatic systems: maintaining, mixing and mobilising antimicrobial resistance? Trends Ecol Evol 26:278–284
Teitzel GM, Parsek MR (2003) Heavy metal resistance of biofilm and planktonic *Pseudomonas aeruginosa*. Appl Environ Microbiol 69:2313–2320
Timoney J, Port J, Giles J, Spanier J (1978) Heavy-metal and antibiotic resistance in the bacterial flora of sediments of New York Bight. Appl Environ Microbiol 36:465–472
Ullrich S, Kube M, Schübbe S, Reinhardt R, Schüler D (2005) A hypervariable 130-kilobase genomic region of *Magnetospirillum gryphiswaldense* comprises a magnetosome island which undergoes frequent rearrangements during stationary growth. J Bacteriol 187:7176–7184
Uyaguari MI, Scott GI, Norman RS (2013) Abundance of class 1–3 integrons in South Carolina estuarine ecosystems under high and low levels of anthropogenic influence. Mar Pollut Bull 76:77–84
van der Meer JR, Sentchilo V (2003) Genomic islands and the evolution of catabolic pathways in bacteria. Curr Opin Biotechnol 14:248–254
van Essen-Zandbergen A, Smith H, Veldman K, Mevius D (2007) Occurrence and characteristics of class 1, 2 and 3 integrons in *Escherichia coli*, *Salmonella* and *Campylobacter* spp. in the Netherlands. J Antimicrob Chemother 59:746–750
Venter JC, Remington K, Heidelberg JF, Halpern AL, Rusch D, Eisen JA, Wu D, Paulsen I, Nelson KE, Nelson W (2004) Environmental genome shotgun sequencing of the Sargasso Sea. Science 304:66–74
von Wintersdorff CJH, Penders J, van Niekerk JM, Mills ND, Majumder S, van Alphen LB, Savelkoul PHM, Wolffs PFG (2016) Dissemination of antimicrobial resistance in microbial ecosystems through horizontal gene transfer. Front Microbiol 7:173
Wagner-Döbler I, Biebl H (2006) Environmental biology of the marine *Roseobacter* lineage. Annu Rev Microbiol 60:255–280
Wang X, Kim Y, Ma Q, Hong SH, Pokusaeva K, Sturino JM, Wood TK (2010) Cryptic prophages help bacteria cope with adverse environments. Nat Commun 1:147
Wang F-H, Qiao M, Lv Z-E, Guo G-X, Jia Y, Su Y-H, Zhu Y-G (2014) Impact of reclaimed water irrigation on antibiotic resistance in public parks, Beijing, China. Environ Pollut 184:247–253
Wright MS, Baker-Austin C, Lindell AH, Stepanauskas R, Stokes HW, McArthur JV (2008) Influence of industrial contamination on mobile genetic elements: class 1 integron abundance and gene cassette structure in aquatic bacterial communities. ISME J 2:417–428
Yang J, Wang C, Shu C, Liu L, Geng J, Hu S, Feng J (2013) Marine sediment bacteria harbor antibiotic resistance genes highly similar to those found in human pathogens. Microb Ecol 65:975–981
Yoshida M, Suzuki S (2014) Heavy oil exposure increases viral production in natural marine bacterial populations. J Oceanogr 70:115–122
Zhang T, Zhang X-X, Ye L (2011) Plasmid metagenome reveals high levels of antibiotic resistance genes and mobile genetic elements in activated sludge. PLoS ONE 6:e26041
Zhaxybayeva O, Swithers KS, Lapierre P, Fournier GP, Bickhart DM, DeBoy RT, Nelson KE, Nesbø CL, Doolittle WF, Gogarten JP (2009) On the chimeric nature, thermophilic origin, and phylogenetic placement of the Thermotogales. Proc Natl Acad Sci 106:5865–5870

Chapter 7
Genetic Adaptations of Bacteria for Metabolism of Polycyclic Aromatic Hydrocarbons

Vidya de Gannes and William J. Hickey

Abstract Polycyclic aromatic compounds (PAH) are a family of chemicals containing two or more fused benzene rings, which present ecotoxicological concerns ranging from acute toxicity in aquatic organisms to carcinogenesis in mammals. In contrast, microbial ecotoxicology of PAH centers on metabolic activities that enable utilization of PAH as growth supporting substrates. This chapter focuses on PAH biodegradation by aerobic bacteria, and examines characteristics that are important in PAH metabolism at three levels: enzyme systems that mediate catabolism and carbon assimilation, regulatory circuitry that controls expression of catabolic enzymes and cell structures and physiological activities that affect PAH uptake. The goal is to present a holistic view of the organisms and their biology that is relevant to PAH biodegradation. Of these areas, catabolism is the most developed and key enzymes and mechanisms have been elucidated. However, comparatively little is known about the regulatory systems that control expression of these enzymes. Uptake from the environment is the single most important factor affecting PAH degradation and while cellular characteristics that affect uptake are known, the process is still largely a "black box" and mechanistic details are lacking, especially regarding molecules that may facilitate PAH access. Mechanisms of regulation and uptake remain areas in need of future research. Future work should also focus on moving beyond studies of individual organisms, and gaining an understanding of PAH biodegradation processes operating within microbial consortia.

Keywords Polycyclic aromatic hydrocarbons · Biodegradation · Genetic adaptations · Catabolic pathways · Exopolymeric substance · Ring hydroxylating dioxygenase · Regulatory elements · Biomarkers

---

V. de Gannes
Department of Food Production, University of the West Indies,
St. Augustine, Trinidad and Tobago

W.J. Hickey (✉)
O.N. Allen Laboratory for Soil Microbiology, Department of Soil Science,
University of Wisconsin, Madison, WI 53706, USA
e-mail: wjhickey@wisc.edu

C. Cravo-Laureau et al. (eds.), *Microbial Ecotoxicology*,
DOI 10.1007/978-3-319-61795-4_7

7.1 Introduction

Polycyclic aromatic hydrocarbons (PAHs) are a family of compounds, members of which contain two or more benzene rings fused in linear, angular or clustered geometries (Fig. 7.1). While benzene rings form the structural core of PAHs, some of these chemicals, termed non-alternate PAH (Fig. 7.1), are composed of benzene rings fused to other non-benzene structures (Harvey 1997). The PAHs of environmental concern are categorized as low-molecular weight (LMW, 2–3 rings) or high-molecular weight (HMW, 4–6 rings) and each group has different toxicological impacts on macrofauna; LMW PAHs may have significant acute toxicity to aquatic organisms, while some of the HMW are potential carcinogens for both wildlife and humans. Based upon demonstrated or suspected potential to cause human cancer, the US EPA identified a subset of 16 PAH as Priority Pollutants (Fig. 7.1) upon which the majority of research has focused.

PAHs are released into the environment by a variety of natural and anthropogenic processes (Harvey 1997). But, point sources from the latter, primarily spills or improper disposal of hydrocarbon liquids containing PAHs, represent the main ecotoxicological concerns, and cases where microbial activity is of greatest interest *vis a vis* bioremediation. Marine oil spills, such as those of the *Deepwater Horizon*, garner much attention and can inflict widespread ecological damage. But, while the

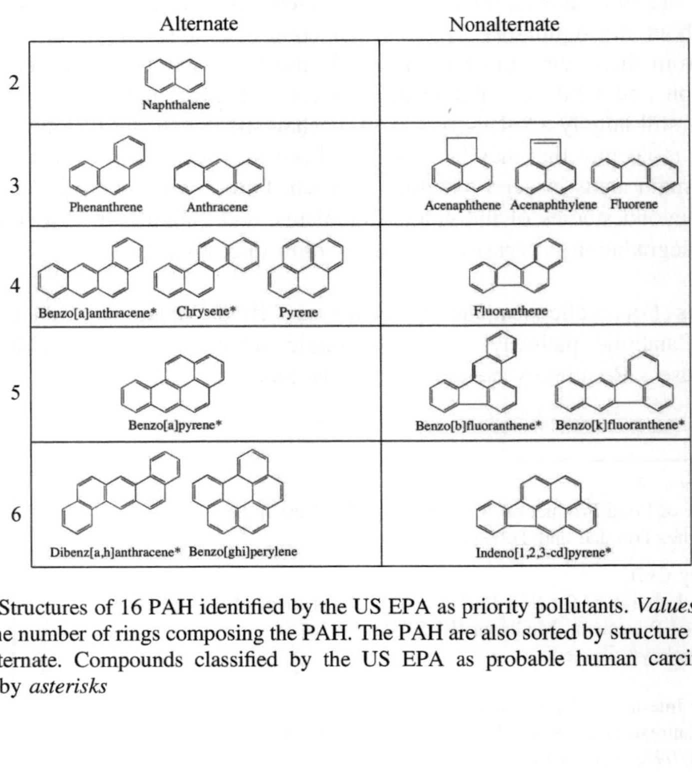

Fig. 7.1 Structures of 16 PAH identified by the US EPA as priority pollutants. *Values on the left* indicate the number of rings composing the PAH. The PAH are also sorted by structure as alternate or non-alternate. Compounds classified by the US EPA as probable human carcinogens are indicated by *asterisks*

fate of PAHs released in crude oil is a concern, the PAH content of oil is relatively small, ranging from 0.2 to 7% (Harvey 1997). Oil-derived fuels (e.g., gasoline, diesel) are wide spread environmental pollutants, but typically have a PAH content that is less than the oil parent material. In contrast, in liquids derived from coal (e.g., coal tar and creosote), PAHs comprise the largest single class of chemicals (>50%) and thus pose significant point sources of PAHs (Birak and Miller 2009; Brown et al. 2006; Wehrer et al. 2011). Coal tar and creosote are primarily threats to terrestrial ecosystems (soil, ground water, lakes, rivers) as they were associated with land-based operations; coal tar is a legacy contaminant at former sites of manufactured coal gasification plants and creosote (a derivative of coal tar) is a legacy contaminant at wood- and lumber-treatment facilities. Thus, PAH ecotoxicology and microbial ecotoxicology is relevant to a wide-range of contaminated environments, particularly at sites impacted by coal tar or creosote.

Traditional ecotoxicology of PAHs focuses on the deleterious effects of these chemicals on humans and wildlife. In contrast, microbial ecotoxicology of PAH includes deleterious impacts on microbes as well as positive interactions of PAH with these organisms. The latter of these two dimensions has been the dominant focus in this area, and centers upon the ability of microbes to consume PAH and obtain carbon and energy to support growth. Microbial biodegradation of PAH is therefore a potentially important ecosystem service and can mitigate exposure of humans and wildlife to these chemicals. The scope of research on microbial ecotoxicology of PAH ranges from single organism studies focused on elucidation of the molecular mechanisms that enable growth on of PAH, to community-scale investigations seeking to determine associations between the diversity, abundance or activity of PAH-degraders and PAH contamination. A variety of microbes and microbial processes are potentially operative in PAH biodegradation, which includes transformations mediated by bacteria and fungi in aerobic environments and bacterial transformations in anaerobic systems (Bouchez-Naitali et al. 2008; Doyle et al. 2008; Fernandez-Luqueno et al. 2011; Jain 2015; Johnston and Johnston 2012; Peng et al. 2008; Seo et al. 2009; Van Hamme et al. 2003). This chapter will focus on PAH biodegradation by aerobic bacteria, which has been the subject of the vast majority of research in the microbial ecotoxicology of these chemicals. Genetic adaptations that are key in enabling bacteria to grow on PAH will be examined at three levels (Fig. 7.2): (1) enzymes that mediate catabolism and carbon assimilation, (2) regulatory circuitry that controls expression of catabolic enzymes and (3) cellular structures that affect PAH uptake. All of these characteristics are endowed by an organism's genotype, and thus as a group represent the genetic adaptations that affect its capability to degrade PAH. A holistic view such as this enables recognition of the range of genetic adaptations that affect bacterial degradation of PAH, and an appreciation of the interactions between adaptations at different levels that collectively result in PAH biodegradation. For community-level studies, elucidation of genetic adaptations yields insights into genes that may be useful as biomarkers of PAH-degrading bacteria. Moreover, an appreciation of the interactions between a biomarker and other components of a cell that contribute to the biodegradation process is equally important to understand the extent to which

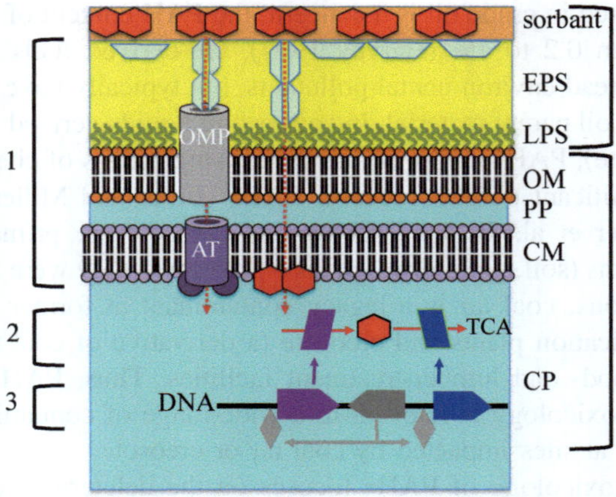

Fig. 7.2 Overview of cellular structures and processes affecting PAH biodegradation. *Numbers* listed along the *left* indicate key cellular components: *1* structures involved in uptake, *2* catabolic enzymes and *3* regulatory systems controlling expression of catabolic enzymes. These components are grouped (*right*) as either extracellular (*XC*) and intracellular (*IC*). The diagram depicts a model *Proteobacterium* possessing exopolymeric substance (*EPS*), lipopolysaccharide (*LPS*), outer membrane (*OM*), periplasm (*PP*), cytoplasmic membrane (*CM*) and cytoplasm (*CP*). The cell is growing on the surface of an environmental sorbent (organic matter or hydrocarbon liquid) and shows the PAH diffusing through the extracellular and intracellular space to reach the catabolic enzymes of the cytoplasm. Structures that may facilitate diffusion are molecules creating hydrophobic domains in the polysaccaride matrix of the EPS/LPS (e.g., biosurfactants, lipids, proteins, shown as green structures in EPS) and OM proteins (*OMP*). The possibility for active transport (*AT*) across the CM is also indicated. Otherwise, PAH diffusion from the sorbent to the cytoplasm is driven solely by concentration gradients and partitioning behavior of the PAH across the different EC and IC regions. The catabolic enzymes are depicted as grouped into the upper pathway (*purple rectangle*) that transforms the PAH to a monoaromatic acid and lower pathway (*blue rectangle*) that completes metabolism. The operons encoding enzymes of the upper and lower pathways are depicted as *purple* and *blue block arrows*, and a regulatory element controlling expression of the operon is depicted in *gray*

inferences can be drawn about interactions between PAH and communities of PAH-degraders based upon the biomarker analysis. Thus, while this chapter primarily focuses on the cellular systems, examples of applications of PAH biomarkers to community analysis will also be examined.

7.2 PAH Catabolic Pathways

Bacterial pathways for PAH catabolism have been examined extensively, and are the subject of a number of reviews to which the reader is referred for details (Bouchez-Naitali et al. 2008; Doyle et al. 2008; Peng et al. 2008; Seo et al. 2009). Our understanding of bacterial genetics and metabolism underlying PAH biodegradation is rooted in the studies of naphthalene done with the archetypical plasmid pNAH7 from *Pseudomonas putida* G7 (Yen and Serdar 1988). A key concept formulated from those studies is the physiological and genetic division of PAH catabolism into "upper" and "lower" segments (Schell et al. 1990). The upper pathways are composed of the initial ring hydroxylating dioxygenases (RHD) and subsequent enzymes that collectively achieve fission of one of the aromatic rings, release of carbon (e.g., pyruvate or other aliphatic acid) and generation of an aromatic acid, which is the starting substrate for the lower pathway (Fig. 7.3). In the NAH7 model, naphthalene is metabolized via the upper pathway to salicylate and an adjacent operon encodes the lower pathway for salicylate catabolism (Fig. 7.3). Both operons are positively regulated by salicylate (Fig. 7.4). While the NAH7 model has been useful to establish many basic principles of bacterial PAH, there are a number of important differences between it and other PAH pathways. For example, instead of salicylate, upper pathways may yield *o*-phthalate, gentisate or protocatechuate, and which of these compounds is produced depends upon the parent PAH, variations in regiospecificity of the RHD and the occurrence of upper pathway enzymes mediating transformations of key intermediates. Differences between upper pathways in the aromatic acid product are significant, because salicylate is thus far the only compound conclusively identified to regulate transcription of the cognate operons. Thus, for pathways lacking salicylate as an intermediate, compounds that serve this critical function are unknown. Also, while upper- and lower-pathway operons on the NAH7 plasmid are in relatively close proximity, the two operons may be chromosomally encoded in many PAH degraders, and in these cases it is not uncommon for the upper- and lower-pathway operons to be encoded at loci distant from each other (Shetty et al. 2015; Tang et al. 2011; Cao et al. 2015; Choi et al. 2015; Dong et al. 2014; Kallimanis et al. 2011; Kim et al. 2008; Kwak et al. 2014; Lai et al. 2012; Li et al. 2016; Maeda et al. 2013; Wang et al. 2014a, b, 2015, 2016; Zhang et al. 2012; Messina et al. 2016).

The cornerstone of PAH degradation pathways are the RHD (Rieske non-heme iron dioxygenase; E.C. 1.14.12), which activate PAH for catabolism via reductive dihydroxylation. RHD consist of an NADH-dependent reductase, a [2Fe-2S] ferredoxin and a terminal oxygenase. The latter is comprised of two separate polypeptides, a large subunit (α-subunit, RHDα) containing [2Fe-2S] and mononuclear Fe(II) at the active site, and small subunit (β-subunit, RHDβ) that appears to serve a structural function (Parales and Ju 2011). The reductase and ferredoxin transport electrons to the α-subunit, which inserts molecular oxygen into the substrate to give *cis*-dihydrodiols, which are subsequently transformed to diols by a dehydrogenase (Fig. 7.5). Most PAH RHD are classified as lateral dioxygenases and add oxygen to two adjacent carbon atoms of an aromatic ring PAH (Fig. 7.5). A smaller group of PAH RHD are categorized as angular dioxygenases, and act on non-alternate PAH, adding oxygen to the carbon atom of an aromatic ring and to an adjacent carbon atom bonded to a non-benzene structure (Nojiri et al. 2001; Schuler et al. 2008). To date, the best characterized angular dioxygenase active in PAH metabolism is the FlnA enzyme in the fluorene-degrader *Sphingomonas* sp. LB126 (Schuler et al. 2008).

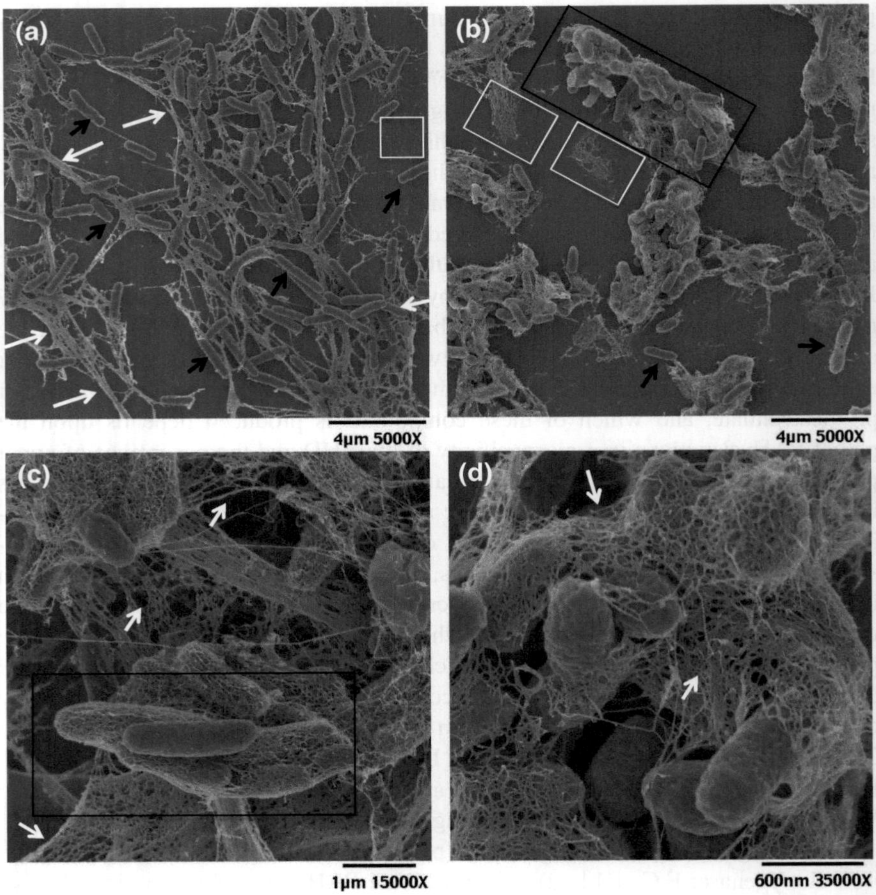

Fig. 7.3 Production of exopolymeric substance (*EPS*) by a phenanthrene-degrading biofilm of *Delftia acidovorans* Cs1-4. Cells of *D. acidovorans* Cs1-4 were collected on 0.22 µm polycarbonate membranes, and then floated on the surface of a mineral salt medium containing phenanthrene crystals (1 mg/mL). The filters were incubated statically, and growth examined by scanning electron microscopy (*SEM*) after 2 d (Panel **a**) or 7 d (Panels **b–d**). Sample preparation for SEM imaging involves dehydration, and thus the originally hydrated EPS appears as an extensive network of web-like strands (*white arrows*) connecting and surrounding cells (*black arrows*) as well as coating the membrane surface (*white boxes*). In the early stage (Panel **a**), cells are distributed primarily as a dispersed monolayer, which progress to form multilayer cell clumps (Panel **b**, *black box*). In later stages, the EPS serves to encase groups of cells (Panel **c**, *black box*)

PAH RHD are broadly distributed across microbial taxa, and comparisons of similarities in PAH-RHDα amino acid sequence have revealed distinct PAH-RHDα families (Habe and Omori 2003; Moser and Stahl 2001), the phylogenies of which generally follow those of the host (Table 7.1). Thus, PAH-RHDα of Proteobacteria and Gram-positive bacteria (*Actinobacteria*) separate into two distinct clades and

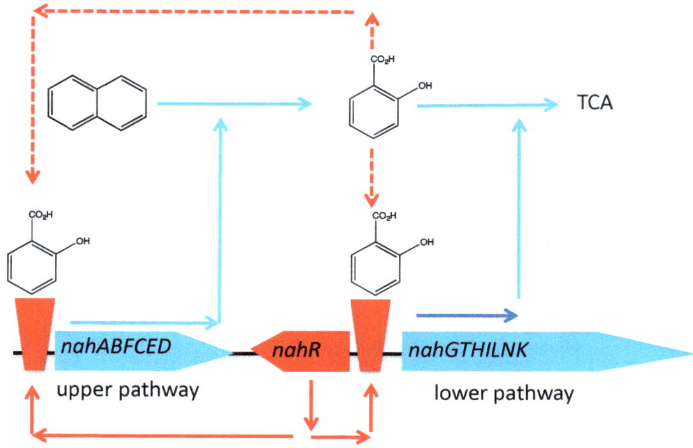

Fig. 7.4 PAH degradation based upon the naphthalene-NAH7 model showing catabolic and regulatory systems in *blue* and *red*, respectively. Two operons encode catabolic enzymes that transform naphthalene to salicylate (*upper pathway*) and salicylate to TCA intermediates (*lower pathway*). The *narR* gene encodes a LysR-type transcriptional repressor-activator that binds to promoters upstream of both the upper- and lower-pathways and represses transcription allowing only a basal level of expression. Binding of salicylate to NahR at the promoters induces a conformational change in DNA-bound NahR that enables transcription to occur

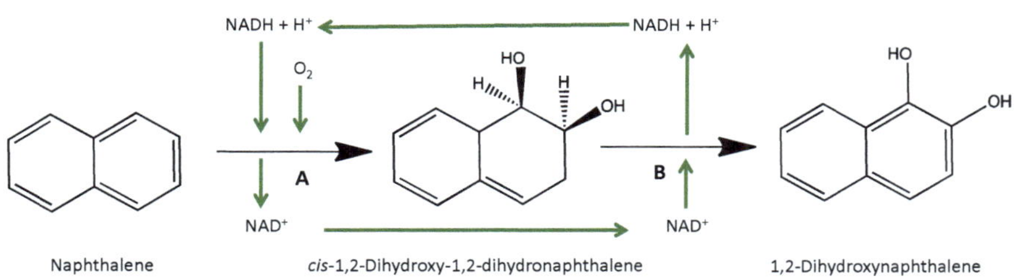

Fig. 7.5 Mechanism of reductive ring dihydroxylation by an example PAH RHD, and coupling of electron flow with subsequent dehydrogenation of the dihydrodiol intermediate. The enzyme catalyzing *Step A* is Naphthalene dioxygenase while *Step B* is mediated by 1,2-Dihydroxy-1,2-dihydronaphthalene dehydrogenase

within each of these broad groups PAH-RHDα subclades are associated with particular bacterial classes. In the Proteobacteria, there are at least five PAH-RHDα families (Table 7.1), the largest of which is the NahAc family that is generally associated with aquatic/terrestrial bacteria of the Gammaproteobacteria, primarily pseudomonads. Notably, the PAH-RHDα of marine gammaproteobacterial PAH-degraders, such as *Cycloclasticus*, is more closely related to the PAH-RHDα of the alphaproteobacterial Sphingomonads then it is to the NahAc group (Kasai et al. 2003). Other proteobacterial PAH-RHDα families are NagAc, $PhnAc_{AFK2}$, $PhnAc_{RP007}$ and AhdA1/BphA1. The NagAc family is closely related to the NahAc

Table 7.1 Families of PAH upper pathway ring hydroxylating dioxygenases

Family	Class or genera	Lower pathway
Proteobacteria		
NahAc	Gamma-/Beta-Proteobacteria	Salicylate
NagAc	Betaproteobacteria	Gentisate
$PhnAc_{AFK2}$	Betaproteobacteria	*o*-Phthalate
$PhnAc_{RP007}$	Betaproteobacteria	Salicylate
AhdA1/BphA1	Alphaproteobacteria (Sphingomonads)	Salicylate or *o*-Phthalate
Actinobacteria		
NidA/PdoA1	*Mycobacterium*	*o*-Phthalate
NidA3/FadA1	*Mycobacterium/Terrabacter*	*o*-Phthalate
NarA	*Rhodococcus*	Gentisate
PdoA2/PhdA	*Mycobacterium/Nocardia*	*o*-Phthalate

group and characteristically produces gentisate, rather than salicylate, as the lower pathway substrate (Nag = *Na*phthalene to *g*entisate (Fuenmayor et al. 1998)). The $PhnAc_{AFK2}$ family is named after bacterium in which it was first identified, *Alcaligenes faecalis* AFK2 (Kiyohara et al. 1982), and the "Phn" designation is used to indicate that PAH-degraders with a $PhnAc_{AFK2}$ genotype are typically limited to phenanthrene as the sole PAH utilized to support growth. The $PhnAc_{AFK2}$–type PAH-RHDα is also characteristically linked to upper pathways producing *o*-phthalate (Kiyohara et al. 1982). The $PhnAc_{RP007}$ family was discovered in *Burkholderia* sp. RP007 (now *Paraburkholderia sartisoli* RP007 (Laurie and Lloyd-Jones 1999)) and despite the similarity in name to $PhnAc_{AFK2}$, $PhnAc_{RP007}$ and $PhnAc_{AFK2}$ genotypes differ in that the former: (1) utilize naphthalene as well as phenanthrene to support growth, and (2) produce salicylate from the upper pathway rather than *o*-phthalate (Laurie and Lloyd-Jones 1999). The PAH-RHDα of Sphingomonands (AhdA1/BphA1) may be associated with upper pathways generating either salicylate or *o*-phthalate (Waigi et al. 2015). Actinobacteria are represented by four groups PAH-RHDα groups, NidA/PdoA1, NidA3/FadA1NarA and PdoA2/PhdA and are generally associated with upper pathways that generate *o*-phthalate (Table 7.1).

A key difference between PAH-RHD are physical characteristics of the active site, which affect the type and range of PAH on which the enzyme will act and funnel to a productive catabolic (lower) pathway. Proteobacterial RHD are generally most active with low molecular weight PAH ($\leq$ three rings) with high molecular weight PAH (HMW PAH, $\geq$ four rings) transformed at comparatively low levels (Baboshin et al. 2014; Jouanneau et al. 2006). In contrast, RHD of the actinobacterial NidA and NidA3 families show higher activity towards HMW PAH, than they do with LMW PAH (Kweon et al. 2010; Krivobok et al. 2003). The preference for LMW- versus HMW-PAH can be explained at least in part by differences between RHD in the substrate-binding pocket, which is larger in NidA/NidA3 enzymes than in proteobacterial RHD (Kweon et al. 2010). Geometry of the active site also effects regiospecificity of dihydroxylation. For example, the NidAB and NidA3B3 homologs from *Mycobacterium vanbaalenii* PYR1 transformed a range of PAH, but displayed the highest regioselectivity with the preferred substrate, pyrene and fluoranthene, respectively (Kweon et al. 2010). Regioselective dihydroxylation by RHD is important in yielding products utilized as substrates for the subsequent enzymes of the upper pathway, and ultimately yield carbon to central metabolism. The RHD reactions consume electrons (Parales and Ju 2011) and thereby impose a negative energy yield, unless coupled with a subsequent dehydrogenase that recover electrons (Fig. 7.4). Thus, coordination between the RHD and dehyrdogenases in the range of substrates utilized by each is important to minimize energy loss. Nevertheless, PAH-RHD exhibit varying degrees of non-selectivity, and will hydroxylate PAH that are not utilized as growth supporting substrates in a process termed cometabolism. For an individual organism, cometabolism is negative as it results in energy loss, and oxidation of PAH that are not growth-supporting may compete with that occurring with PAH that do sustain growth. But, from a community perspective, cometabolism may be beneficial as it could generate metabolites (hydroxylated PAH) that maybe utilized by commensal organisms.

While the activities of RHD and dehydrogenases are important first steps, productive metabolism of a PAH is equally dependent upon the collective activities of the upper pathway enzymes. For bacteria that utilize more than one PAH as a growth substrate, or generate multiple pathways for any given PAH (Seo et al. 2012), the upper pathway would include a combination of core enzymes with broad specificities, which ultimately couple with enzymes producing key aromatic acid intermediates. For example, in *Pseudomomas putida* NCIB 9816, a common set of upper pathway enzymes enables growth on naphthalene, fluorene and phenanthrene (Yang et al. 1994). Likewise, in *Sphingomonas paucimobilis* EPA505 mutational analysis indicated that catabolism of naphthalene, anthracene, and phenanthrene was mediated by a common set of enzymes (Story et al. 2001).

As noted above, the monoaromatic compounds that link the upper and lower pathways are limited to salicylate, gentisate or *o*-phthalate. Pathways for catabolism of these compounds are wide spread in bacteria, and occur in many organisms that are not PAH degraders. Thus, these systems are not unique to PAH degraders, and the reader is referred to reviews for detailed descriptions (Albaiges 2013; Diaz et al. 2013; Ladino-Orjuela et al. 2016; Vamsee-Krishna and Phale 2008). The broad distribution of the lower pathways in microbial communities may present a mechanism for cooperative metabolism, wherein organisms that mediate primary degradation (upper pathway) secrete metabolites that are utilized by commensals. A number of PAH-degrading consortia have been reported, and commensalism is generally proposed as a mechanism supporting the stability of these diverse assemblages although the actual metabolites that might be exchanged are undetermined (Boonchan et al. 2000; Kuppusamy et al. 2016; Lafortune et al. 2009; Stach and Burns 2002; Vinas et al. 2005). Unraveling metabolic interactions occurring within consortia remains a goal of future research.

7.3 Regulation of PAH Gene Expression

Regulation of PAH gene expression is a key aspect in controlling the biodegradation activities of bacteria in the environment and also important in applied technologies such as strain engineering, development of biosensors and devising strategies for bioremediation. Regulation is multi-tiered system: expression of enzymes effecting PAH degradation is controlled at the level of the cognate genes by regulatory proteins (transcription factors) specific for the catabolic operon, which interact with RNA polymerase (RNAP) to activate or de-repress gene expression. These genes may also be subject to regulation by elements that operate globally, and integrate carbon assimilation from PAH into the overall metabolism of the cell. For example, while most PAH upper pathway operons are transcribed by a σ^{70}-RNAP holoenzyme (e.g., bind to σ^{70}-dependent promoter) others are expressed with the σ^{54}-RNAP complex and require interaction with global regulatory elements (Table 7.2).

Genes encoding each of the PAH-RHD families tend to be associated with a specific type of regulatory element from either the LysR, TetR, MarR, GntR or NtrC classes. However, the only regulatory system that has been studied in detail in association with PAH metabolism is that of NahR, a LysR-type element that regulates expression of the NahAc family RHD (and NagAc family, Table 7.2) and the associated genes of the upper and lower pathways (Park et al. 2002). Thus, most of the regulatory systems that control expression of enzymes for PAH metabolism are uncharacterized. This is a key knowledge gap in our understanding PAH biodegradation, and it's unknown how differences in these systems may impact growth of PAH-degraders in the environment. This is significant as PAH-degraders with LysR systems represent only one segment of environmental communities of PAH-degraders (*nah* and *nag* types, Table 7.2), which often do not appear to be the most abundant genotype (see section below on microbial ecology and PAH biomarkers). Thus, information about regulatory elements other than LysR would benefit our understanding of PAH degradation at the cellular and community levels.

In the NahR system, the gene encoding the regulator (*nahR*) is immediately upstream of the of lower pathway operon (Fig. 7.4). NahR is continuously bound to a promoter between *nahR* and the downstream operon (P_{sal}), as well as to a promoter upstream of the upper pathway operon (P_{nah}) and its presence allows only a low level of a constitutive expression. Repression is lifted when the promoter-bound NahR binds salicylate (the effector), inducing a conformational shift that enables RNAP binding and full level transcription of both the upper- and lower-pathway operons (Schell et al. 1990; Huang and Schell 1991; Lonneborg and Brzezinski 2011; Park et al. 2005; Wilkinson and Grove 2006; Maddocks and Oyston 2008; Schell 1985). The Nag RHD family is also associated with an NahR ortholog (NagR), and although the product of Nag upper pathway is gentisate rather than salicylate, only the latter compound induces expression (Jones et al. 2003).

In contrast to the Nah model, regulation of PAH degradation by bacteria possessing a $\mathrm{PhnAc_{AFK2}}$ family RHD appears to be more complex, as the upper

Table 7.2 Summary of regulatory elements associated with families of PAH ring hydroxylating dioxygenases alpha subunits (PAH-RHDα) and upper pathway operons

Class	Action[a]	*Proteobacteria*		*Actinobacteria*		Effector	Promoter	Binding Site[d]	Other factors
		PAH-RHDα[b]	Gene or locus[c]	PAH-RHDα	Gene or locus				
LysR	A/R	NahAc, NagAc, $PhnAc_{RP007}$	*nahR*, *nagR* *phnS*	–	–	Salicylate	σ^{70}	Operator	–
TetR	R	$PhnAc_{AFK2}$	DelCs14_1747	–	–	Unknown	σ^{70}	Operator	–
MarR	R	$PhnAc_{AFK2}$	DelCs14_1753	NidA	*nidR*	Unknown	σ^{70}	Operator	–
GntR	R	–	–	NarA	*narR1*	Unknown	σ^{70}	Operator	–
AraC	A	–	–	PhdA	ORF12	Unknown	various	Various	Various
NtrC	A	AhdA1, $PhnAc_{RP007}$	*ahdR*, *phnR*	NarA	*narR2*	Unknown	σ^{54}	UAS	IHF, ATP

[a]Abbreviations: *A* activator; *R* repressor
[b]See Table 7.1
[c]Name of gene encoding the indicated regulatory element. Locus or ORF is given when the cognate gene is unnamed
[d]Region to which regulatory protein binds

pathway includes both TetR and MarR regulatory elements while the lower pathway genes are associated with two additional regulators. The actinobacterial Nid family also has a gene encoding a MarR regulator (NidR (Kim et al. 2006)) and expression is presumably controlled by that system, although it has not been empirically substantiated. The potential functions of TetR or MarR in controlling *phn* or *nid* expression are unknown, but regulatory activities have been elucidated with other catabolic pathways. For the TetR group, most notably is CymR, which represses expression of *cym* genes encoding degradation of *p*-cymene. CymR represses *cym* expression by binding to a an operator adjoining a σ^{70}-dependent promoter and blocking RNAP; repression is released when the CymR binds the effector, *p*-cumate a metabolite in the cymene pathway (not the parent compound, cymene), and subsequently disassociates from the promoter, allowing RNAP access and transcription. A TetR-type regulator has also recently been shown to negatively regulate testosterone catabolism in *Comamonas testosteroni,* however the effector molecules were not identified (Pan et al. 2015). MarR regulators generally act as transcriptional repressors in a manner similar to TetR, and bind operators nearby σ^{70}-dependent promoters thereby preventing RNAP binding and transcription of cognate operons (Alekshun and Levy 1999). MarR regulation has been demonstrated for operons encoding phenylpropenoid catabolism in Firmicutes (CinR, *Butyrivibrio fibrisolvens* (Dalrymple and Swadling 1997)) and in Actinobacteria (PhdR, *Corynebacterium glutamicum* (Kallscheuer et al. 2016)). The identities of effectors were not conclusively determined, but MarR regulators in general may have an affinity for phenolic compounds (Alekshun and Levy 1999) and, consistent with that concept, PhdR appeared to have an affinity for hydroxylated phenylpropenoids as ligands.

Other actinobacterial PAH genes are associated with GntR, NtrC or AraC regulatory elements (Table 7.2). The NarA family PAH-RHD associated with *Rhodococcus* and related genera is located downstream of two tandemly arranged transcriptional regulatory genes, *narR1* and *narR2*, which are predicted to encode proteins of the GntR and NtrC families, respectively (Kulakov et al. 2005; Larkin et al. 2005). The GntR-type regulators function like the above-described repressors in preventing effective binding of RNAP to a σ^{70}-dependent promoter, which is lifted by effector binding and consequent disassociation of the GntR from the operator. For NarR1 regulated operons, the identities of the effectors are unknown. However, GntR regulation has been established for biphenyl catabolism in *Acidovorax*, and in this case the effector ligand for the GntR repressor (BphS) is a product of biphenyl ring fission, 2-hydroxy-6-oxo-6-phenyl-2,4-dienoic acid (HODPA (Ohtsubo et al. 2001)). Similar metabolites (i.e., aromatic rings with extended aliphatic substituents) occur in PAH degradation pathways (Bouchez-Naitali et al. 2008; Peng et al. 2008, 2011; Seo et al. 2009; Vaidehi and Kulkarni 2012) and could be candidates as potential effectors.

NtrC-type regulators act as transcriptional activators, and are distinct from all of the above-mentioned transcription factors in that they: (1) interact with RNAP at σ^{54}-dependent promoters rather than σ^{70}-dependent promoters, (2) binds ATP as a cofactor in addition to an effector molecule, and (3) require a global regulatory element (Integration Host Factor). NtrC-type factors are representatives of a broad group termed the prokaryotic enhancer binding proteins, which bind DNA in areas called upstream activating sequences (UAS) that are typically located a distance (*ca.* ≥ 100 bp) from the cognate σ^{54} promoter. NtrC activation of transcription is initiated when NtrC dimers in solution bind the effector inducing conformational changes that enable ATP binding, multimerization and subsequent binding to UAS. Integration Host Factor (IHF) binds at a site between the UAS and σ^{54} promoter, which induces bending and allows the DNA-bound NtrC complex to interact with RNAP bound to the σ^{54} promoter. NtrC then activates RNAP (coupled with ATP hydrolysis) by inducing a transition in RNAP conformation from closed to open, thereby enabling transcription.

NtrC regulatory proteins are associated with at least three PAH-RHD families: AhdA1/BphA1 (Khara et al. 2014), $PhnAc_{RP007}$ (Laurie and Lloyd-Jones 1999) and NarA (Kulakov et al. 2005). Details of NtrC regulation have not been developed for operons encoding PAH upper pathways, but the mechanisms have been explored in connection with catabolism of other aromatic compounds including phenol, hydroxybiphenyl, toluene and *m-/p*-xylene (Diaz and Prieto 2000; Tropel and van der Meer 2004). In the pathways for which mechanisms of NtrC control have been defined, the parent compounds as well as some related aromatic compounds act as effectors (Diaz and Prieto 2000; Tropel and van der Meer 2004). The fact that neutral compounds such as toluene and xylenes act as effector ligands for NtrC distinguishes these regulators from all of the others discussed above, for which ligands appear limited to phenolics or amphiphilic compounds (e.g., HOPDA). For PAH operons, an affinity of NtrC for neutral ligands may be important, as it raises the possibility that the PAH parent compounds might act as inducers for their own catabolism. Notably, growth on naphthalene induced the *nar* operon in *Rhodococcus* (Di Gennaro et al. 2010), which is affiliated with an upstream NtrC-like regulatory element. But, it's unclear if naphthalene was an effector, as the cells were able to metabolize naphthalene to salicylate, which also induced the *nar* operon.

In contrast to the Nar family, actinobacterial Phd genes are associated with an AraC-type transcriptional activator (Pagnout et al. 2007). Mechanisms of AraC regulation that are specific to the *phd* operon are unknown, and are difficult to infer because these AraC elements vary widely in fundamental characteristics such as the types of promoters regulated (σ^{70} and others) and their dependence on other regulatory factors for maximal activity (e.g., catabolite repression protein (Gallegos et al. 1997)).

Understanding the mechanisms underlying regulation of genes encoding catabolism of PAH has practical importance in the development of bioremediation strategies. For example, supplementation of contaminated soil with salicylate as a potential inducer of PAH degradation genes has been explored. But the results have been mixed (Song et al. 2015; Carmichael and Pfaender 1997) and are difficult to interpret: a lack of a stimulation could reflect the fact that salicylate is not an effector for the PAH operons at a site, or that salicylate is consumed by non-PAH degraders. Conversely, if stimulation is observed, it's unclear if that is attributable to induction of PAH degradation genes or to activation of other unknown enzymes.

7.4 Cellular Uptake of PAH

The key physicochemical characteristic of PAH affecting uptake by bacteria is their strong hydrophobicity, which increases significantly with PAH size, and is reflected in low aqueous solubility and high octanol-water partition coefficients (K_{ow}, Fig. 7.6). Thus, in the environment, uptake is limited by strong partitioning of PAH to sorbents, primarily natural organic matter and non-aqueous phase liquids (NAPL) of which PAH are constituents. Sorbed PAH are typically regarded as non-bioavailable and inaccessible for uptake, with the transition to a bioavailable state dependent upon relatively slow rates of PAH desorption (Johnsen and Karlson 2004). Thus, while characteristics of enzymes and regulatory elements that mediate catabolism are important, these systems are of little use to the cell if the substrate cannot be acquired. Thus, bioavailability is the single greatest factor limiting bacterial degradation of PAH in the environment (Johnsen et al. 2005; Pouli and Agathos 2011; Juhasz et al. 2014; Mahanty et al. 2011; Rojo-Nieto and Perales-Vargas-Machuca 2012; Simpanen et al. 2016; Yang et al. 2009) and cellular characteristics that may enhance access to sorbed PAH would be crucial to effective degradation of these compounds.

Three pathways have been proposed by which PAH are acquired by bacteria (Fig. 7.7; (Guerin and Boyd 1992; Schippers et al. 2000; Alexander 1994): (1) uptake from aqueous solution, (2) direct uptake from micelles composed of exocellular amphiphiles and (3) direct uptake from sorbents. A wide variety of cellular characteristics and activities can be involved in any of these pathways, but biosurfactant production has a potential role in all three. Biosurfactants are small, amphiphilic molecules possessing hydrophilic and hydrophobic domains (Banat 1995; Lawniczak et al. 2013). There are two main groups of bacterial biosurfactants, glycolipids and peptidolipids (e.g., rhamnolipids and surfactins, respectively), and of these the former have been most extensively studied (Banat 1995; Lawniczak et al. 2013). The term "surfactant" describes the surface-active behavior of these molecules, which preferentially adsorb at interfaces: air-water, liquid-liquid (oil-water) or liquid-solid. The latter two activities are potentially important for PAH uptake as sorption at oil-water boundaries reduces the liquid-liquid interfacial

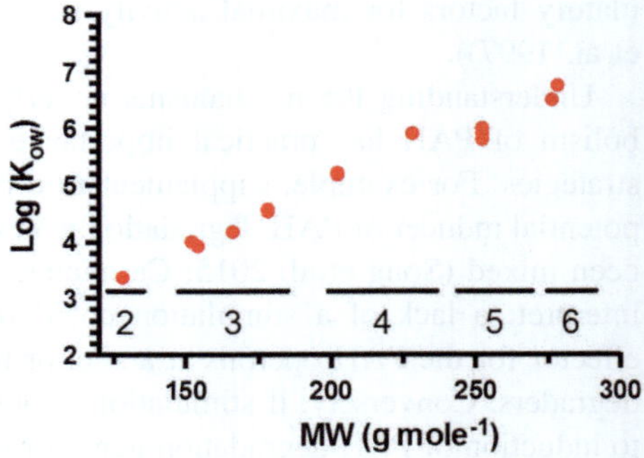

Fig. 7.6 Illustration of the relation between hydrophobicity (K_{ow}) and molecular weight for the 16 priority pollutant PAH. *Values below the horizontal bars* indicate the number of rings in the compounds plotted

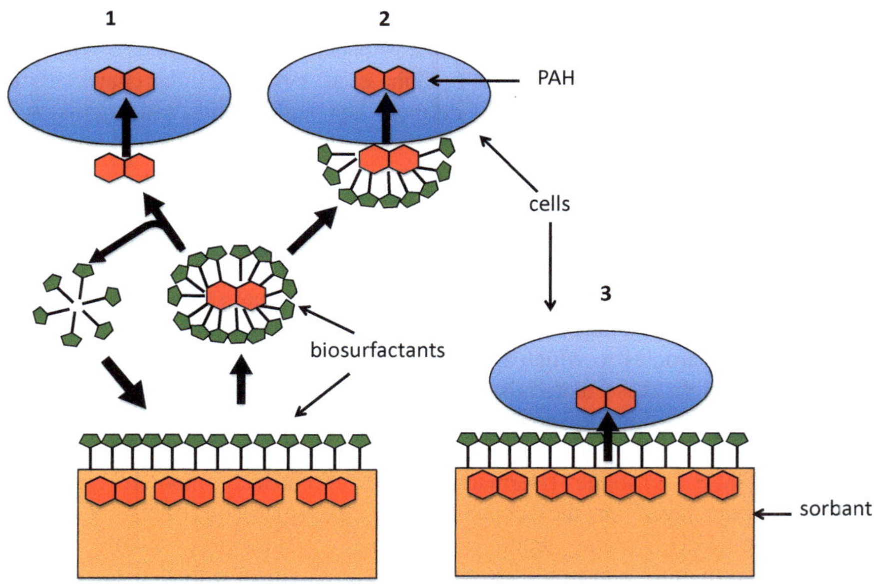

Fig. 7.7 Conceptual pathways of PAH uptake by bacteria and potential roles of biosurfactants. *Pathway 1* represents the conventional pathway with uptake occurring from aqueous solution. In this pathway, biosurfactants potentially enhance mass transfer or extraction of the PAH, but there is no direct transfer from the biosurfactant micelle to the bacterial cell; uptake occurs from the PAH dissolved in the aqueous phase. *Pathway 2* depicts direct uptake of PAH sorbed by a biosurfactant micelle. *Pathway 3* shows direct uptake of PAH from a sorbent

tension, while sorption at the liquid-solid interface affects the substrate "wettability"; both of these activities can facilitate cellular interactions with substrates containing PAH. If levels of surfactant secretion are sufficiently high, surfactant monomers aggregate in solution to form colloidal particles called micelles (Pathways 1 and 2) and the surfactant concentration at which that occurs is termed the critical micelle concentration.

Pathways 1 and 2 are similar in that interfacial activities of biosurfactants enhance mass transfer of PAH from a sorbent to the micellar phase, thereby increasing the apparent aqueous solubility of PAH via pseudo-solubilization or emulsification. However, in Pathway 1, uptake is dependent upon the release of PAH from micelles to the aqueous phase, from which bacterial uptake occurs. In contrast, Pathways 2 and 3 illustrate direct uptake, wherein PAH passes directly from the micelle or sorbant into the cell. Direct uptake has been indicated in cases where mineralization rates of PAH or other hydrophobic organic compounds surpass rates that would be predicted if uptake was dependent upon desorption to the aqueous phase (Calvillo and Alexander 1996; Crocker et al. 1995; Efroymson and Alexander 1991; Guerin and Boyd 1997; Harms and Zehnder 1995; Lahlou and Ortega-Calvo 1999; Laor et al. 1999; Tang et al. 1998). Direct uptake has been indicated for cultures inoculated into soil microcosms, or grown in liquid media containing synthetic surfactants, or humic substances (Guerin and Boyd 1992; Vacca et al. 2005; Guha and Jaffe 1996; Guha et al. 1998). Cellular characteristics that may enable direct uptake are ill defined. But humic acids can interact with cellular membranes (Elayan et al. 2008; Ojwang and Cook 2013), and some bacteria may be able to alter membrane structure in a manner that facilitates direct transfer of PAH to cells when they interact with humic substances containing sorbed PAH (Vacca et al. 2005).

Uptake of PAH via Pathway 3 may have the greatest environmental relevance, as attachment and growth on surfaces as biofilms is the normal state for most bacteria (Dunne 2002; Hall-Stoodley et al. 2004). The surfaces could be solids, such as soil organomineral particles to which PAH sorb, or NAPL that contain PAH, like oil. Direct uptake from these materials is predicated upon adhesion of the cell to the substrate, which is a complex process affected by characteristics of the cell surfaces, the substrate and other physicochemical aspects of environment, like pH and organic and inorganic solutes (Hermansson 1999). Solutes are important in forming a “conditioning layer” on the substrate surface (Hermansson 1999), which may modify its physicochemical parameters in a manner that enables a cell surface to reach close proximity and minimize the effective length of diffusion for a sorbed PAH to reach the intracellular catabolic enzymes (Johnsen et al. 2005). Cell surfaces may also become coated with solutes that affect interaction with the solid sorbents or NAPL. While a variety of organic and inorganic molecules can contribute to the conditioning layer and cellular coatings (Hermansson 1999) biosurfactants are potentially of particular importance because of their surface activity (Feng et al. 2013; Liu et al. 2014; Sotirova et al. 2009).

While biosurfactant secretion could facilitate PAH uptake by any of the three pathways, this activity may not be specifically induced by growth on PAH (Chrzanowski et al. 2012). Production of biosurfactants by bacteria growing on PAH has been demonstrated for some cultures, such as fluorescent pseudomonads growing on naphthalene (Deziel et al. 1996; Dasari et al. 2014). However, other investigators have screened a range of Proteobacteria and Actinobacteria grown on a variety of PAH and found little evidence of significant biosurfactant secretion (Johnsen and Karlson 2004; Willumsen and Karlson 1997). Absence of a strong correlation between biosurfactant production and PAH metabolism probably reflects the fact that global regulatory processes controls the former rather than variations in physicochemical characteristics of the carbon source supporting growth (e.g., aqueous solubility (Antoniou et al. 2015)). For example, rhamnolipids have been the most extensively studied class of bacterial biosurfacants, and their biosynthesis is regulated primarily via quorum-sensing systems (Perfumo et al. 2013; Reis et al. 2011). Quorum sensing regulation is particularly important in biofilms, such as those that may be formed by PAH degraders on the surfaces of NAPL or solid sorbants. For example, in the PAH degrader, *Pseudomonas aeruginosa* N6P6, PAH degradation was enhanced by biofilm formation, which in turn was positively correlated expression of *lasI* (Mangwani et al. 2015). The latter is a gene involved in synthesis of quorum sensing signaling molecules, acyl homoserine lactones. Conversely, inhibiting production of acyl homoserine lactones by *P. aeruginosa* N6P6 diminished both biofilm development and PAH degradation (Mangwani et al. 2015). Thus, quorum sensing can enhance PAH degradation via the promotion of biofilm development, possibly reflecting the beneficial effects of exopolymeric substance (EPS, see below) production and/or biosurfactant secretion. There is as yet no evidence that quorum sensing signaling molecules directly regulate expression of genes encoding enzymes mediating PAH metabolism.

Movement of PAH from a sorbent to the cell cytoplasm necessitates the PAH transverse extra- and intra-cellular regions that differ widely in physico-chemical characteristics (Fig. 7.2). Of particular importance are differences in hydrophobicity/hydrophilicity of these regions, which affect the partitioning behavior of the PAH. As noted above, biofilm formation is likely the natural growth mode for PAH degraders, and a defining characteristic of these structures is the envelopment of cells in a matrix of EPS (Fig. 7.3). Polysaccarides are an essential component of the matrix and provide a foundation for EPS (Bazaka et al. 2011; Fazli et al. 2014; Vu et al. 2009). But, EPS also contain a range of other molecules including lipids and proteins (Dohnalkova et al. 2011; Flemming and Wingender 2010). While the EPS matrix is highly hydrated (>95% water), its heterogeneous composition gives rise to hydrophobic environments on a micro- or nano-scale that may be favorable for entry of PAH, and diffusion to the cell surface (Hobley et al. 2015; Payne and Boles 2016; Zhang et al. 2015).

For most bacteria, the primary cellular boundary is an outer membrane (Sutcliffe 2010; Devos 2014; Sutcliffe et al. 2010), the permeability of which to hydrophobic compounds varies across taxa. In the Proteobacteria, the OM is a barrier for uptake of PAH, a characteristic endowed by the relatively long, hydrophilic carbohydrate chain of the lipopolysaccharide (LPS) dominating the OM outer leaflet. The presence of LPS significantly affects the hydrophobic thickness of the OM, which is ca. 22 A°, compared to ca. 28 A° of the phospholipid bilayer comprising the cytoplasmic membrane (Wu et al. 2014). As noted above, secreted biosurfactants may accumulate on the cell surface offsetting the hydrophilic influence of LPS (Mohanty and Mukherji 2012; Kaczorek et al. 2013; Abbasnezhad et al. 2011), and thereby establish domains conducive to partitioning of hydrophobic compounds. Movement of PAH into and across the OM could thus occur non-specifically, and perhaps be enhanced by biosurfactant modification of cell surface properties (Liu et al. 2014; Li and Zhu 2014).

Transport of PAH across the OM could also be mediated by specific integral OM proteins. The OmpW family of OM proteins possesses hydrophobic channels (Hong et al. 2006; Touw et al. 2010) and has been implicated as important in naphthalene uptake (Neher and Lueking 2009). Another group of potential OM transporters for PAH are FadL-like proteins, which facilitate uptake of hydrophobic compounds by translocation either directly to the periplasm or to the hydrophobic lumen of the OM (Hearn et al. 2008, 2009; van den Berg 2005; van den Berg et al. 2015). Transport by FadL-like proteins is not energy dependent, but instead is driven by conformational changes in the protein following substrate binding (van den Berg 2005). For hydrocarbons, FadL-like transporters have been best characterized with monoaromatic compounds (e.g., tolunene) and include the proteins TodX and TbuX (Kahng et al. 2000; Wang et al. 1995). A FadL-like OM protein had also been associated with uptake of polychlorophenols (Belchik et al. 2010). Whether or not FadL-like proteins have a role in transport of PAH is currently unknown, and as PAH are significantly larger than the monoaromatics, it's possible that the size of PAH may restrict interactions with these transporters.

Two groups of bacteria with OM structure divergent from that of most Proteobacteria are the Sphingomonads of the Alphaproteobacteria (genera *Novosphingobium*, *Sphingobium*, *Sphingopyxis*, and *Sphingomonas*) and *Corynebacterineae* of the *Actinobacteria* (genera *Corynebacterium*, *Mycobacterium*, *Nocardia and Rhodococcus*). The OM of Sphingomonads lacks LPS, and instead contains a unique class of glycolipids, glycosphingolipids, which have a carbohydrate chain smaller and less complex than that of LPS, making the OM surface of sphingomonads less hydrophilic than OM containing LPS (Kawahara et al. 1999). The reduced hydrophilicity of the Sphingomonad OM may enhance access to PAH and facilitate interaction with hydrophobic surfaces (Waigi et al. 2015; Regonne et al. 2013). In the *Corynebacterineae*, the OM is often referred to as the mycomembrane (Bansal-Mutalik and Nikaido 2014; Bayan et al. 2003; Zuber et al. 2008), and a defining feature are mycolic acids, which are a class of β-hydroxy fatty acids unique to the *Corynebacterinea. Mycolic acids* have long carbon chains, ranging from 30 to 90 carbons in length, which makes the mycomembrane characteristically highly hydrophobic (Marrakchi et al. 2014) and growth on PAH can induce a shift in mycolic acid composition that further enhances hydrophobicity of the mycomembrane (Wick et al. 2002). Consequently, the mycomembrane is a structure that may facilitate acquisition of hydrophobic compounds from the environment, and has been cited as factor contributing to the frequent isolation of genera *Mycobacterium*, *Nocardia and Rhodococcus as degraders of PAH, especially HMW PAH* (Song et al. 2011; Uyttebroek et al. 2006; Kanaly and Harayama 2010; Kweon et al. 2011; Schneider et al. 1996), and to the association of these genera with hydrophobic sorbents (Regonne et al. 2013; Bastiaens et al. 2000).

The cytoplasmic membrane is the final barrier for compounds entering cells, and for lipophilic chemicals like PAH, passage through this zone can occur primarily by simple diffusion (Gu et al. 2016). This diffusion process is described by the Overton rule (Missner and Pohl 2009), which holds that the permeation of solutes through lipid membranes is proportional to their oil-water partitioning (K_{ow}). Thus for PAH, permeation decreases with increasing aqueous solubility (e.g., benzo[*a*]anthracene > fluoranthrene, pyrene > anthracene, phenanthrene). Partitioning of PAH into the cytoplasmic membrane is beneficial as subsequent diffusion to cytoplasm ultimately supplies the cell with a carbon and energy source. However, growth benefits of PAH are counterbalanced by potential toxicity, as PAH partitioning into the lipid bilayer can cause structural deformations that degrade membrane integrity and function (Sikkema et al. 1995). The levels to which PAH need to accumulate in cytoplasmic membranes to inflict toxicity are not well defined, and vary as a function of the membrane composition, the type of PAH and other environmental variables (Sikkema et al. 1995). Nevertheless, in model membrane systems, structural alterations can occur when PAH levels reach 10 mol% (Korchowiec et al. 2008) and are generally fluidizing effects directly correlated with PAH size (Liland et al. 2014).

The primary cellular response to mitigate deleterious effects of PAH permeation on the cytoplasmic membrane structure is to modify bilayer fluidity. Fluidity is affected by many factors, but a fundamental aspect is lipid packing, which is determined by structural variations in the fatty acyl chain and head group composition of the constituent lipids (Sikkema et al. 1995; Denich et al. 2003; Murinova and Dercova 2014). While systematic studies of alterations in membrane chemistry that accompany bacterial utilization of PAH are lacking, comparative analysis of phospholipid fatty acid (PLFA) composition in cells grown on PAH versus hydrophilic substrates (sugars, complex media) have demonstrated that growth on PAH induces shifts in PLFA composition that are predicted to result in increased membrane fluidity. Two of the most common PLFA alterations that follow PAH exposure are an increased abundance of odd number cyclopropyl fatty acids and decreased ratios of iso- to anteiso-fatty acids (Vacca et al. 2005; Nam et al. 2002; Wick et al. 2003; Kallimanis et al. 2007). An increase in bilayer fluidity may seem counterintuitive as an adaptation for growth on PAH as it would render the cytoplasmic membrane less permeable to these compounds. However, the shift in fluidity could be a strategy to modulate PAH partitioning into the membrane that balances the benefits of PAH uptake that lead to carbon assimilation versus the drawbacks of membrane damage. The extent to which membrane modification is induced in bacteria exposed to PAH in contaminated environments is largely unknown, but there is some evidence that would be consistent with its occurrence. For example, in soil microcosms spiked with PAH, total PLFA decreased with increasing PAH but the fraction (mol%) of cyclopropyl fatty acid increased (Su and Yang 2009). Also, in a creosote-contaminated soil, cyclopropyl fatty acids were among the most abundant PLFA in locations with the highest PAH levels (Bengtsson et al. 2010).

Uptake of PAH into the cytoplasm likely occurs primarily via passive diffusion, but active transport has also been implicated for some strains. In these cases, energy dependent uptake of PAH (naphthalene or phenanthrene) has been reported for *Mycobacterium* (Miyata et al. 2004), *Rhodococcus* sp. BAP-1, *Arthrobacter* (Kallimanis et al. 2007; Li et al. 2014) and *Pseudomonas* (Whitman et al. 1998). Also, for *Sphingomonas paucimobilis* EPA505, specific transport systems have hypothesized based upon mechanistic interpretations of quantitative structure–activity relationship models, which indicated the rate-limiting step in PAH degradation was a binding and transport process, not interaction of the PAH with the PAH-RHD (Dimitriou-Christidis et al. 2008). Active uptake systems have been identified for other hydrophobic compounds, and include an ABC transporter for hexachlorohexane in *Sphingobium japonicum* UT26 (Endo et al. 2007) and an ATP-dependent permease for styrene in *Pseudomonas putida* CA-3 (Mooney et al.

2006). Similar active transport systems for PAH could be present in cases where energy dependent uptake has been indicated, but these have yet to be identified.

A final point to consider is the possibility that bacteria may enhance access to PAH by sensing these compounds and utilize chemotaxis to locate, and move towards, sources of PAH. Chemotactic responses to naphthalene have been reported for several *Pseudomonas* species (Grimm and Harwood 1997; Ortega-Calvo et al. 2003) and in one of these, *Pseudomonas putida* G3, a gene essential for naphthalene chemotaxis was identified (Grimm and Harwood 1999). This gene was located within the operon encoding naphthalene degradation (*nahY*) and the predicted product was a methyl-accepting chemotaxis protein. While the studies with *Pseudomonas* strains demonstrated the potential for chemotaxis, the significance of chemotaxis to PAH biodegradation in the environment is uncertain. A key issue is the constraints on motility, especially in soil. In this regard, soils represent a partially hydrated environment, and water films in a moist soil (field capacity) are estimated to be only on the order of 10 nm thick, which is only a fraction of a cell's diameter and insufficient to support motility (Ebrahimi and Or 2015; Tecon and Or 2016). The roughness of mineral surfaces is an added constraint (Ebrahimi and Or 2015; Tecon and Or 2016). Models considering hydration level and surface roughness indicate that motility is supported only in narrow window of moisture conditions (Ebrahimi and Or 2015; Tecon and Or 2016).

Biomarkers of PAH degradation and their application in microbial ecology. As PAH-degrading bacteria present a means to mitigate ecotoxicity of PAH, understanding the environmental activities of these organisms is important, and biomarkers of PAH biodegradation and/or PAH degrading bacteria are needed to do so. From the preceding discussion, it's clear that a myriad of cellular systems impact the ability of bacteria to degrade PAH. But, of these, the only genotypic characteristics that are specifically and uniquely affiliated with PAH biodegradation are genes encoding upper pathway enzymes. The PAH-RHD that initiate upper pathway metabolism has been studied in greatest detail and, as noted above, the PAH-RHDα has received particular focus in the development of PAH-RHD families. Thus, genes encoding PAH-RHDα have been targeted as biomarkers for the analysis of PAH-degraders in the environment, primarily via Polymerase Chain Reaction (PCR) methods. As the PAH-RHDα are phylogenetically diverse, genes encoding these polypeptides exhibit relatively low similarity in nucleotide sequence, and degenerate primers are needed in PCR analyses of PAH-RHDα genes to encompass as much as possible of the PAH-degrader community. Consequently, many PCR primer sets have been developed that in principle enable amplification of different segments of the PAH-RHDα family (Cebron et al. 2008; Bordenave et al. 2008).

Ecological investigations employing PAH-RHDα analyses have yielded insights into the distribution of different PAH-RHDα phylotypes in a variety of environments. These studies have provided information about relationships between PAH-degrader community structure and environmental variables that may affect community structure (Bordenave et al. 2008; Chen et al. 2016; Ding et al. 2010; Flocco et al. 2009; Yang et al. 2014, 2015; Fuentes et al. 2015; Gomes et al. 2007;

Li et al. 2015; Sauret et al. 2016; Zhou et al. 2006). PAH-RHDα analyses has also been employed to assess relationships between PAH-degrader abundance and the PAH levels, and these investigations have yielded mixed results. For example, significant positive correlations of PAH-RHDα gene abundance with PAH levels have been reported for coke factory soils (Han et al. 2014), creosote-contaminated soil (Bengtsson et al. 2013), coal tar-contaminated sediments (Dionisi et al. 2004) and roadside soils (Li et al. 2015). Conversely, a lack of correlation between PAH-RHDα gene abundance and PAH concentrations was demonstrated in roadside soils (Johnsen et al. 2014), oil field soils (Yang et al. 2014) and soils from a former coking plant (Cebron et al. 2009). Reasons for the varying results are unknown, and could include technical issues (e.g., PCR bias in RHDα gene amplification, variations in PAH extraction and analysis, etc.) as well as many physicochemical factors that affect bacterial growth in the environment. For example, heterotrophic bacteria that degrade PAH will also grow on a wide range of other organic compounds that may be present, which could alter PAH-RHDα gene abundance irrespective of PAH concentration. There may also be inhibitory or synergistic metabolic interactions between different PAH that uncouple changes in PAH-RHDα gene abundance from changes in PAH concentration (Hennessee and Li 2016). Also, PAH degraders can utilize only bioavailable PAH, which is a small, variable fraction of the total PAH present in soil or sediment. Thus, differences in total PAH may not be ideal to explain differences in PAH-RHDα gene abundance. For example, in a study of Yangtze river water and sediment, PAH-RHDα gene abundance was positively correlated to levels of dissolved PAH (i.e., bioavailable PAH) in water overlying sediments or in sediment pore water (Xia et al. 2015). However, PAH-RHDα gene abundance was not correlated to total PAH levels in sediment (Xia et al. 2015). Thus, while assessments of PAH-RHDα abundance can be informative about community structure, their use as indicators of PAH biodegradation potential is limited by unknowns about the environmental biology of PAH-degraders, as well as by physicochemical complexities of contaminated soils or sediments containing PAH.

7.5 Summary and Future Directions

After several decades of research, much has been learned about the genetic adaptations and processes underlying bacterial metabolism of PAH. The knowledge base is particularly strong in the area of catabolism and characteristics of key enzymes, especially PAH-RHD, which have proven to be a diverse group of proteins. The regulatory systems that control expression of these enzymes have been revealed to be at least equally diverse. Yet, despite the importance of regulation, comparatively little is known about mechanisms by which these systems operate and is a knowledge area in need of significant expansion. Uptake from the environment is the single most important factor affecting PAH degradation and while cellular characteristics that affect uptake are known, the process is still largely a "black box" and there are few mechanistic details, especially regarding molecules that may facilitate PAH access. Elucidating the mechanisms of uptake is another key research need. Lastly, genes encoding PAH-RHDα have served as useful biomarkers to analyze PAH degrader communities in the environment. Yet, interpreting variations in abundance or distribution of different PAH degrader genotypes has been limited, in part by an incomplete understanding of differences between genotypes in other dimensions of cell biology that affect PAH metabolism (i.e., PAH uptake, gene regulation). Filling these knowledge gaps could this help improve the insights that can be gained in ecological studies of PAH biodegradation.

Acknowledgements This work was supported by the O.N. Allen Professorship in Soil Microbiology (to WJH). The authors thank Natalie Kirk for assistance in the development of illustrations.

References

Abbasnezhad H, Foght JM, Gray MR (2011) Adhesion to the hydrocarbon phase increases phenanthrene degradation by *Pseudomonas fluorescens* LP6a. Biodegradation 22(3):485–496

Albaiges J (2013) Organic chemicals in the environment. In: Mechanisms of degradation and transformation, 2nd edn. Int J Environ Anal Chem 93(14):1563

Alekshun MN, Levy SB (1999) The mar regulon: multiple resistance to antibiotics and other toxic chemicals. Trends Microbiol 7(10):410–413

Alexander M (1994) Biodegradation and bioremediation. Academic, San Diego, CA

Antoniou E, Fodelianakis S, Korkakaki E, Kalogerakis N (2015) Biosurfactant production from marine hydrocarbon-degrading consortia and pure bacterial strains using crude oil as carbon source. Front Microbiol 6:274

Baboshin M, Ivashina T, Chernykh A, Golovleva L (2014) Comparison of the substrate specificity of two ring-hydroxylating dioxygenases from Sphingomonas sp. VKM B-2434 to polycyclic aromatic hydrocarbons. Biodegradation 25(5):693–703

Banat IM (1995) Biosurfactants production and possible uses in microbial enhanced oil-recovery and oil pollution remediation—a review. Biores Technol 51(1):1–12

Bansal-Mutalik R, Nikaido H (2014) Mycobacterial outer membrane is a lipid bilayer and the inner membrane is unusually rich in diacyl phosphatidylinositol dimannosides. Proc Natl Acad Sci U S A 111(13):4958–4963

Bastiaens L, Springael D, Wattiau P, Harms H, deWachter R, Verachtert H et al (2000) Isolation of adherent polycyclic aromatic hydrocarbon (PAH)-degrading bacteria using PAH-sorbing carriers. Appl Environ Microbiol 66(5):1834–1843

Bayan N, Houssin C, Chami M, Leblon G (2003) Mycomembrane and S-layer: two important structures of *Corynebacterium glutamicum* cell envelope with promising biotechnology applications. J Biotechnol 104(1–3):55–67

Bazaka K, Crawford RJ, Nazarenko EL, Ivanova EP (2011) Bacterial extracellular polysaccharides. In: Linke D, Goldman A (eds) Bacterial adhesion: chemistry, biology and physics. Adv Exp Med Biol 715:213–226

Belchik SM, Schaeffer SM, Hasenoehrl S, Xun LY (2010) A beta-barrel outer membrane protein facilitates cellular uptake of polychlorophenols in *Cupriavidus necator*. Biodegradation 21(3):431–439

Bengtsson G, Torneman N, Yang X (2010) Spatial uncoupling of biodegradation, soil respiration, and PAH concentration in a creosote contaminated soil. Environ Pollut 158(9):2865–2871

Bengtsson G, Torneman N, De Lipthay JR, Sorensen SJ (2013) Microbial diversity and PAH catabolic genes tracking spatial heterogeneity of PAH concentrations. Microb Ecol 65(1):91–100

Birak PS, Miller CT (2009) Dense non-aqueous phase liquids at former manufactured gas plants: challenges to modeling and remediation. J Contam Hydrol 105(3–4):81–98

Boonchan S, Britz ML, Stanley GA (2000) Degradation and mineralization of high-molecular-weight polycyclic aromatic hydrocarbons by defined fungal-bacterial cocultures. Appl Environ Microbiol 66(3):1007–1019

Bordenave S, Goni-urriza M, Vilette C, Blanchard S, Caumette P, Duran R (2008) Diversity of ring-hydroxylating dioxygenases in pristine and oil contaminated microbial mats at genomic and transcriptomic levels. Environ Microbiol 10(12):3201–3211

Bouchez-Naitali M, Blanchet D, Haeseler F, Vandecasteele JP, VandeCasteele JP (2008) Biodegradation of polycyclic aromatic hydrocarbons (PAHs) 341–411 pp

Brown DG, Gupta L, Kim T-H, Moo-Young HK, Coleman AJ (2006) Comparative assessment of coal tars obtained from 10 former manufactured gas plant sites in the eastern United States. Chemosphere 65(9):1562–1569

Calvillo YM, Alexander M (1996) Mechanism of microbial utilization of biphenyl sorbed to polyacrylic beads. Appl Microbiol Biotechnol 45(3):383–390

Cao J, Lai Q, Yuan J, Shao Z (2015) Genomic and metabolic analysis of fluoranthene degradation pathway in *Celeribacter indicus* P73(T). Sci Rep 5:7741

Carmichael LM, Pfaender FK (1997) The effect of inorganic and organic supplements on the microbial degradation of phenanthrene and pyrene in soils. Biodegradation 8(1):1–13

Cebron A, Norini MP, Beguiristain T, Leyval C (2008) Real-Time PCR quantification of PAH-ring hydroxylating dioxygenase (PAH-RHD alpha) genes from Gram positive and Gram negative bacteria in soil and sediment samples. J Microbiol Methods 73(2):148–159

Cebron A, Beguiristain T, Faure P, Norini MP, Masfaraud JF, Leyval C (2009) Influence of vegetation on the in situ bacterial community and polycyclic aromatic hydrocarbon (PAH) degraders in aged PAH-contaminated or thermal-desorption-treated soil. Appl Environ Microbiol 75(19):6322–6330

Chen SC, Peng JJ, Duan GL (2016) Enrichment of functional microbes and genes during pyrene degradation in two different soils. J Soils Sediments 16(2):417–426

Choi DH, Kwon YM, Kwon KK, Kim S-J (2015) Complete genome sequence of *Novosphingobium pentaromativorans* US6-1(T). Stan Genomic Sci 10:107

Chrzanowski L, Lawniczak L, Czaczyk K (2012) Why do microorganisms produce rhamnolipids? World J Microbiol Biotechnol 28(2):401–419

Crocker FH, Guerin WF, Boyd SA (1995) Bioavailability of naphthalene sorbed to cationic surfactant-modified smectite clay. Environ Sci Technol 29(12):2953–2958

Dalrymple BP, Swadling Y (1997) Expression of a *Butyrivibrio fibrisolvens* E14 gene (cinB) encoding an enzyme with cinnamoyl ester hydrolase activity is negatively regulated by the product of an adjacent gene (cinR). Microbiology-Sgm 143:1203–1210

Dasari S, Subbaiah KCV, Wudayagiri R, Valluru L (2014) Biosurfactant-mediated biodegradation of polycyclic aromatic hydrocarbons-naphthalene. Bioremediat J 18(3):258–265

Denich TJ, Beaudette LA, Lee H, Trevors JT (2003) Effect of selected environmental and physico-chemical factors on bacterial cytoplasmic membranes. J Microbiol Methods 52(2):149–182

Devos DP (2014) PVC bacteria: variation of, but not exception to, the Gram-negative cell plan. Trends Microbiol 22(1):14–20

Deziel E, Paquette G, Villemur R, Lepine F, Bisaillon JG (1996) Biosurfactant production by a soil Pseudomonas strain growing on polycyclic aromatic hydrocarbons. Appl Environ Microbiol 62(6):1908–1912

Di Gennaro P, Terreni P, Masi G, Botti S, De Ferra F, Bestetti G (2010) Identification and characterization of genes involved in naphthalene degradation in *Rhodococcus opacus* R7. Appl Microbiol Biotechnol 87(1):297–308

Diaz E, Prieto MA (2000) Bacterial promoters triggering biodegradation of aromatic pollutants. Curr Opin Biotechnol 11(5):467–475

Diaz E, Jimenez JI, Nogales J (2013) Aerobic degradation of aromatic compounds. Curr Opin Biotechnol 24(3):431–442

Dimitriou-Christidis P, Autenrierh RL, Abraham MH (2008) Quantitative structure-activity relationships for kinetic parameters of polycyclic aromatic hydrocarbon biotransformation. Environ Toxicol Chem 27(7):1496–1504

Ding GC, Heuer H, Zuhlke S, Spiteller M, Pronk GJ, Heister K et al (2010) Soil type-dependent responses to phenanthrene as revealed by determining the diversity and abundance of polycyclic aromatic hydrocarbon ring-hydroxylating dioxygenase genes by using a novel PCR detection system. Appl Environ Microbiol 76(14):4765–4771

Dionisi HM, Chewning CS, Morgan KH, Menn FM, Easter JP, Sayler GS (2004) Abundance of dioxygenase genes similar to Ralstonia sp. strain U2 nagAc is correlated with naphthalene concentrations in coal tar-contaminated freshwater sediments. Appl Environ Microbiol 70(7):3988–3995

Dohnalkova AC, Marshall MJ, Arey BW, Williams KH, Buck EC, Fredrickson JK (2011) Imaging hydrated microbial extracellular polymers: comparative analysis by electron microscopy. Appl Environ Microbiol 77(4):1254–1262

Dong C, Bai X, Lai Q, Xie Y, Chen X, Shao Z (2014) Draft genome sequence of Sphingobium sp. strain C100, a polycyclic aromatic hydrocarbon-degrading bacterium from the deep-sea sediment of the Arctic Ocean. Genome Announcements 2(1):e01210–e01213

Doyle E, Muckian L, Hickey AM, Clipson N (2008) Microbial PAH degradation. In: Laskin AI, Sariaslani S, Gadd GM (eds) Advances in applied microbiology, vol 65. Advances in applied microbiology, pp 27–66

Dunne WM (2002) Bacterial adhesion: seen any good biofilms lately? Clin Microbiol Rev 15(2):155–166

Ebrahimi A, Or D (2015) Hydration and diffusion processes shape microbial community organization and function in model soil aggregates. Water Resour Res 51(12):9804–9827

Efroymson RA, Alexander M (1991) Biodegradation by an arthrobacter species of hydrocarbons partitioned into an organic-solvent. Appl Environ Microbiol 57(5):1441–1447

Elayan NM, Treleaven WD, Cook RL (2008) Monitoring the effect of three humic acids on a model membrane system using P-31 NMR. Environ Sci Technol 42(5):1531–1536

Endo R, Ohtsubo Y, Tsuda M, Nagata Y (2007) Identification and characterization of genes encoding a putative ABC-type transporter essential for utilization of gamma-hexachlorocyclohexane in *Sphingobium japonicum* UT26. J Bacteriol 189(10): 3712–3720

Fazli M, Almblad H, Rybtke ML, Givskov M, Eberl L, Tolker-Nielsen T (2014) Regulation of biofilm formation in Pseudomonas and Burkholderia species. Environ Microbiol 16(7): 1961–1981

Feng WH, Swift S, Singhal N (2013) Effects of surfactants on cell surface tension parameters and hydrophobicity of *Pseudomonas putida* 852 and *Rhodococcus erythropolis* 3586. Colloids Surf B-Biointerfaces 105:43–50

Fernandez-Luqueno F, Valenzuela-Encinas C, Marsch R, Martinez-Suarez C, Vazquez-Nunez E, Dendooven L (2011) Microbial communities to mitigate contamination of PAHs in soil-possibilities and challenges: a review. Environ Sci Pollut Res 18(1):12–30

Flemming HC, Wingender J (2010) The biofilm matrix. Nat Rev Microbiol 8(9):623–633

Flocco CG, Gomes NCM, Mac Cormack W, Smalla K (2009) Occurrence and diversity of naphthalene dioxygenase genes in soil microbial communities from the Maritime Antarctic. Environ Microbiol 11(3):700–714

Fuenmayor SL, Wild M, Boyes AL, Williams PA (1998) A gene cluster encoding steps in conversion of naphthalene to gentisate in Pseudomonas sp. strain U2. J Bacteriol 180(9): 2522–2530

Fuentes S, Ding GC, Cardenas F, Smalla K, Seeger M (2015) Assessing environmental drivers of microbial communities in estuarine soils of the Aconcagua River in Central Chile. Fems Microbiol Ecol 91(10):flv110

Gallegos MT, Schleif R, Bairoch A, Hofmann K, Ramos JL (1997) AraC/XylS family of transcriptional regulators. Microbiol Mol Biol Rev 61(4):393–410

Gomes NCM, Borges LR, Paranhos R, Pinto FN, Krogerrecklenfort E, Mendonca-Hagler LCS et al (2007) Diversity of ndo genes in mangrove sediments exposed to different sources of polycyclic aromatic hydrocarbon pollution. Appl Environ Microbiol 73(22):7392–7399

Grimm AC, Harwood CS (1997) Chemotaxis of Pseudomonas spp. to the polyaromatic hydrocarbon naphthalene. Appl Environ Microbiol 63(10):4111–4115

Grimm AC, Harwood CS (1999) NahY, a catabolic plasmid-encoded receptor required for chemotaxis of *Pseudomonas putida* to the aromatic hydrocarbon naphthalene. J Bacteriol 181(10):3310–3316

Gu HP, Lou J, Wang HZ, Yang Y, Wu LS, Wu JJ et al (2016) Biodegradation, biosorption of phenanthrene and its trans-membrane transport by Massilia sp. WF1 and *Phanerochaete chrysosporium*. Front Microbiol 7

Guerin WF, Boyd SA (1992) Differential bioavailability of soil-sorbed naphthalene to 2 bacterial species. Appl Environ Microbiol 58(4):1142–1152

Guerin WF, Boyd SA (1997) Bioavailability of naphthalene associated with natural and synthetic sorbents. Water Res 31(6):1504–1512

Guha S, Jaffe PR (1996) Bioavailability of hydrophobic compounds partitioned into the micellar phase of nonionic surfactants. Environ Sci Technol 30(4):1382–1391

Guha S, Jaffe PR, Peters CA (1998) Bioavailability of mixtures of PAHs partitioned into the micellar phase of a nonionic surfactant. Environ Sci Technol 32(15):2317–2324

Habe H, Omori T (2003) Genetics of polycyclic aromatic hydrocarbon metabolism in diverse aerobic bacteria. Biosci Biotechnol Biochem 67(2):225–243

Hall-Stoodley L, Costerton JW, Stoodley P (2004) Bacterial biofilms: from the natural environment to infectious diseases. Nat Rev Microbiol 2(2):95–108

Han X-M, Liu Y-R, Zheng Y-M, Zhang X-X, He J-Z (2014) Response of bacterial pdo1, nah, and C12O genes to aged soil PAH pollution in a coke factory area. Environ Sci Pollut Res 21(16):9754–9763

Harms H, Zehnder AJB (1995) Bioavailability of sorbed 3-chlorodibenzofuran. Appl Environ Microbiol 61(1):27–33

Harvey RG (1997) Polycyclic aromatic hydrocarbons. Wiley-VCH, New York

Hearn EM, Patel DR, van den Berg B (2008) Outer-membrane transport of aromatic hydrocarbons as a first step in biodegradation. Proc Natl Acad Sci U S A 105(25):8601–8606

Hearn EM, Patel DR, Lepore BW, Indic M, van den Berg B (2009) Transmembrane passage of hydrophobic compounds through a protein channel wall. Nature 458(7236):367–370

Hennessee CT, Li QX (2016) Effects of polycyclic aromatic hydrocarbon mixtures on degradation, gene expression, and metabolite production in four mycobacterium species. Appl Environ Microbiol 82(11):3357–3369

Hermansson M (1999) The DLVO theory in microbial adhesion. Colloids Surf B-Biointerfaces 14(1–4):105–119

Hobley L, Harkins C, MacPhee CE, Stanley-Wall NR (2015) Giving structure to the biofilm matrix: an overview of individual strategies and emerging common themes. FEMS Microbiol Rev 39(5):649–669

Hong HD, Patel DR, Tamm LK, van den Berg B (2006) The outer membrane protein OmpW forms an eight-stranded beta-barrel with a hydrophobic channel. J Biol Chem 281(11): 7568–7577

Huang JZ, Schell MA (1991) Invivo interactions of the NahR transcriptional activator with its target sequences—inducer-mediated changes resulting in transcription activation. J Biol Chem 266(17):10830–10838
Jain PK (2015) Microbial biodegradation of polycyclic aromatic hydrocarbons. In: Harzevili FD, Chen H (eds), 331–349 pp
Johnsen AR, Karlson U (2004) Evaluation of bacterial strategies to promote the bioavailability of polycyclic aromatic hydrocarbons. Appl Microbiol Biotechnol 63(4):452–459
Johnsen AR, Wick LY, Harms H (2005) Principles of microbial PAH-degradation in soil. Environ Pollut 133(1):71–84
Johnsen AR, Styrishave B, Aamand J (2014) Quantification of small-scale variation in the size and composition of phenanthrene-degrader populations and PAH contaminants in traffic-impacted topsoil. FEMS Microbiol Ecol 88(1):84–93
Johnston CG, Johnston GP (2012) Bioremediation of polycyclic aromatic hydrocarbons. In: Microbial biotechnology: energy and environment, pp 279–296
Jones RM, Britt-Compton B, Williams PA (2003) The naphthalene catabolic (nag) genes of Ralstonia sp. strain U2 are an operon that is regulated by NagR, a LysR-type transcriptional regulator. J Bacteriol 185(19):5847–5853
Jouanneau Y, Meyer C, Jakoncic J, Stojanoff V, Gaillard J (2006) Characterization of a naphthalene dioxygenase endowed with an exceptionally broad substrate specificity toward polycyclic aromatic hydrocarbons. Biochemistry 45(40):12380–12391
Juhasz AL, Aleer S, Adetutu EM (2014) Predicting PAH bioremediation efficacy using bioaccessibility assessment tools: validation of PAH biodegradation-bioaccessibility correlations. Int Biodeterior Biodegradation 95:320–329
Kaczorek E, Salek K, Guzik U, Jesionowski T, Cybulski Z (2013) Biodegradation of alkyl derivatives of aromatic hydrocarbons and cell surface properties of a strain of *Pseudomonas stutzeri*. Chemosphere 90(2):471–478
Kahng HY, Byrne AM, Olsen RH, Kukor JJ (2000) Characterization and role of tbuX in utilization of toluene by *Ralstonia pickettii* PKO1. J Bacteriol 182(5):1232–1242
Kallimanis A, Frillingos S, Drainas C, Koukkou AI (2007) Taxonomic identification, phenanthrene uptake activity, and membrane lipid alterations of the PAH degrading Arthrobacter sp. strain Sphe3. Appl Microbiol Biotechnol 76(3):709–717
Kallimanis A, Karabika E, Mavromatis K, Lapidus A, LaButti KM, Liolios K et al (2011) Complete genome sequence of Mycobacterium sp. strain (Spyr1) and reclassification to *Mycobacterium gilvum* Spyr1. Stan Genomic Sci 5(1):144–153
Kallscheuer N, Vogt M, Kappelmann J, Krumbach K, Noack S, Bott M et al (2016) Identification of the phd gene cluster responsible for phenylpropanoid utilization in *Corynebacterium glutamicum*. Appl Microbiol Biotechnol 100(4):1871–1881
Kanaly RA, Harayama S (2010) Advances in the field of high-molecular-weight polycyclic aromatic hydrocarbon biodegradation by bacteria. Microb Biotechnol 3(2):136–164
Kasai Y, Shindo K, Harayama S, Misawa N (2003) Molecular characterization and substrate preference of a polycyclic aromatic hydrocarbon dioxygenase from Cycloclasticus sp. strain A5. Appl Environ Microbiol 69(11):6688–6697
Kawahara K, Kuraishi H, Zahringer U (1999) Chemical structure and function of glycosphingolipids of Sphingomonas spp. and their distribution among members of the alpha-4 subclass of Proteobacteria. J Ind Microbiol Biotechnol 23(4–5):408–413
Khara P, Roy M, Chakraborty J, Ghosal D, Dutta TK (2014) Functional characterization of diverse ring-hydroxylating oxygenases and induction of complex aromatic catabolic gene clusters in Sphingobium sp. PNB. FEBS Open Bio 4:290–300
Kim SJ, Kweon O, Freeman JP, Jones RC, Adjei MD, Jhoo JW et al (2006) Molecular cloning and expression of genes encoding a novel dioxygenase involved in low- and high-molecular-weight polycyclic aromatic hydrocarbon degradation in *Mycobacterium vanbaalenii* PYR-1. Appl Environ Microbiol 72(2):1045–1054

Kim SJ, Kweon O, Jones RC, Edmondson RD, Cerniglia CE (2008) Genomic analysis of polycyclic aromatic hydrocarbon degradation in *Mycobacterium vanbaalenii* PYR-1. Biodegradation 19(6):859–881

Kiyohara H, Nagao K, Kouno K, Yano K (1982) Phenanthrene-degrading phenotype of alcaligenes-faecalis AFK2. Appl Environ Microbiol 43(2):458–461

Korchowiec B, Corvis Y, Viitala T, Feidt C, Guiavarch Y, Corbier C et al (2008) Interfacial approach to polyaromatic hydrocarbon toxicity: phosphoglyceride and cholesterol monolayer response to phenanthrene, anthracene, pyrene, chrysene, and benzo a pyrene. J Phys Chem B 112(43):13518–13531

Krivobok S, Kuony S, Meyer C, Louwagie M, Willison JC, Jouanneau Y (2003) Identification of pyrene-induced proteins in Mycobacterium sp. strain 6PY1: evidence for two ring-hydroxylating dioxygenases. J Bacteriol 185(13):3828–3841

Kulakov LA, Chen SC, Allen CCR, Larkin MJ (2005) Web-type evolution of Rhodococcus gene clusters associated with utilization of naphthalene. Appl Environ Microbiol 71(4):1754–1764

Kuppusamy S, Thavamani P, Megharaj M, Naidu R (2016) Biodegradation of polycyclic aromatic hydrocarbons (PAHs) by novel bacterial consortia tolerant to diverse physical settings—assessments in liquid- and slurry-phase systems. Int Biodeterior Biodegradation 108:149–157

Kwak Y, Park G-S, Lee S-E, Li QX, Shin J-H (2014) Genome sequence of *Mycobacterium aromaticivorans* JS19b1(T), a novel isolate from Hawaiian soil. J Biotechnol 186:137–138

Kweon O, Kim S-J, Freeman JP, Song J, Baek S, Cerniglia CE (2010) Substrate specificity and structural characteristics of the novel Rieske nonheme iron aromatic ring-hydroxylating oxygenases NidAB and NidA3B3 from *Mycobacterium vanbaalenii* PYR-1. MBio 1(2): e00135-10

Kweon O, Kim S-J, Holland RD, Chen H, Kim D-W, Gao Y et al (2011) Polycyclic aromatic hydrocarbon metabolic network in *Mycobacterium vanbaalenii* PYR-1. J Bacteriol 193 (17):4326–4337

Ladino-Orjuela G, Gomes E, da Silva R, Salt C, Parsons JR (2016) Metabolic pathways for degradation of aromatic hydrocarbons by bacteria. In: DeVoogt P (ed) Reviews of environmental contamination and toxicology, vol 237. Rev Environ Contam Toxicol 237:105–121

Lafortune I, Juteau P, Deziel E, Lepine F, Beaudet R, Villemur R (2009) Bacterial diversity of a consortium degrading high-molecular-weight polycyclic aromatic hydrocarbons in a two-liquid phase biosystem. Microb Ecol 57(3):455–468

Lahlou M, Ortega-Calvo JJ (1999) Bioavailability of labile and desorption-resistant phenanthrene sorbed to montmorillonite clay containing humic fractions. Environ Toxicol Chem 18 (12):2729–2735

Lai Q, Li W, Wang B, Yu Z, Shao Z (2012) Complete genome sequence of the pyrene-degrading bacterium Cycloclasticus sp. strain P1. J Bacteriol 194(23):6677

Laor Y, Strom PF, Farmer WJ (1999) Bioavailability of phenanthrene sorbed to mineral-associated humic acid. Water Res 33(7):1719–1729

Larkin MJ, Kulakov LA, Allen CCR (2005) Biodegradation and Rhodococcus—masters of catabolic versatility. Curr Opin Biotechnol 16(3):282–290

Laurie AD, Lloyd-Jones G (1999) The phn genes of Burkholderia sp. strain RP007 constitute a divergent gene cluster for polycyclic aromatic hydrocarbon catabolism. J Bacteriol 181 (2):531–540

Lawniczak L, Marecik R, Chrzanowski L (2013) Contributions of biosurfactants to natural or induced bioremediation. Appl Microbiol Biotechnol 97(6):2327–2339

Li F, Zhu L (2014) Surfactant-modified fatty acid composition of Citrobacter sp SA01 and its effect on phenanthrene transmembrane transport. Chemosphere 107:58–64

Li Y, Wang H, Hua F, Su M, Zhao Y (2014) Trans-membrane transport of fluoranthene by Rhodococcus sp. BAP-1 and optimization of uptake process. Biores Technol 155:213–219

Li XF, Hou LJ, Liu M, Zheng YL, Li Y, Lin XB (2015) Abundance and diversity of polycyclic aromatic hydrocarbon degradation bacteria in urban roadside soils in Shanghai. Appl Microbiol Biotechnol 99(8):3639–3649

Li Z-Y, Wu Y-H, Huo Y-Y, Cheng H, Wang C-S, Xu X-W (2016) Complete genome sequence of a benzo a pyrene-degrading bacterium *Altererythrobacter epoxidivorans* CGMCC 1.7731(T). Mar Genomics 25:39–41

Liland NS, Simonsen AC, Duelund L, Torstensen BE, Berntssen MHG, Mouritsen OG (2014) Polyaromatic hydrocarbons do not disturb liquid-liquid phase coexistence, but increase the fluidity of model membranes. Chem Phys Lipid 184:18–24

Liu Y, Ma X, Zeng G, Zhong H, Liu Z, Jiang Y et al (2014) Role of low-concentration monorhamnolipid in cell surface hydrophobicity of *Pseudomonas aeruginosa*: adsorption or lipopolysaccharide content variation. Appl Microbiol Biotechnol 98(24):10231–10241

Lonneborg R, Brzezinski P (2011) Factors that influence the response of the LysR type transcriptional regulators to aromatic compounds. BMC Biochemistry 12(1):49

Maddocks SE, Oyston PCF (2008) Structure and function of the LysR-type transcriptional regulator (LTTR) family proteins. Microbiology-Sgm 154:3609–3623

Maeda AH, Nishi S, Ozeki Y, Ohta Y, Hatada Y, Kanaly RA (2013) Draft genome sequence of Sphingobium sp. strain KK22, a high-molecular-weight polycyclic aromatic hydrocarbon-degrading bacterium isolated from cattle pasture soil. Genome Announcements 1(6)

Mahanty B, Pakshirajan K, Dasu VV (2011) Understanding the complexity and strategic evolution in PAH remediation research. Crit Rev Environ Sci Technol 41(19):1697–1746

Mangwani N, Kumari S, Das S (2015) Involvement of quorum sensing genes in biofilm development and degradation of polycyclic aromatic hydrocarbons by a marine bacterium *Pseudomonas aeruginosa* N6P6. Appl Microbiol Biotechnol 99(23):10283–10297

Marrakchi H, Laneelle M-A, Daffe M (2014) Mycolic acids: structures, biosynthesis, and beyond. Chem Biol 21(1):67–85

Messina E, Denaro R, Crisafi F, Smedile F, Cappello S, Genovese M et al (2016) Genome sequence of obligate marine polycyclic aromatic hydrocarbons-degrading bacterium Cycloclasticus sp. 78-ME, isolated from petroleum deposits of the sunken tanker Amoco Milford Haven, Mediterranean Sea. Mar Genomics 25:11–13

Missner A, Pohl P (2009) 110 years of the Meyer-Overton rule: predicting membrane permeability of gases and other small compounds. ChemPhysChem 10(9–10):1405–1414

Miyata N, Iwahori K, Foght JM, Gray MR (2004) Saturable, energy-dependent uptake of phenanthrene in aqueous phase by Mycobacterium sp. strain RJGII-135. Appl Environ Microbiol 70(1):363–369

Mohanty S, Mukherji S (2012) Alteration in cell surface properties of Burkholderia spp. during surfactant-aided biodegradation of petroleum hydrocarbons. Appl Microbiol Biotechnol 94 (1):193–204

Mooney A, O'Leary ND, Dobson ADW (2006) Cloning and functional characterization of the styE gene, involved in styrene transport in *Pseudomonas putida* CA-3. Appl Environ Microbiol 72(2):1302–1309

Moser R, Stahl U (2001) Insights into the genetic diversity of initial dioxygenases from PAH-degrading bacteria. Appl Microbiol Biotechnol 55(5):609–618

Murinova S, Dercova K (2014) Response mechanisms of bacterial degraders to environmental contaminants on the level of cell walls and cytoplasmic membrane. Int J Microbiol 2014

Nam K, Moon HS, Kim JY, Kukor JJ (2002) Linkage between biodegradation of polycyclic aromatic hydrocarbons and phospholipid profiles in soil isolates. J Microbiol Biotechnol 12 (1):77–83

Neher TM, Lueking DR (2009) *Pseudomonas fluorescens* ompW: plasmid localization and requirement for naphthalene uptake. Can J Microbiol 55(5):553–563

Nojiri H, Habe H, Omori T (2001) Bacterial degradation of aromatic compounds via angular dioxygenation. J Gen Appl Microbiol 47(6):279–305

Ohtsubo Y, Delawary M, Kimbara K, Takagi M, Ohta A, Nagata Y (2001) BphS, a key transcriptional regulator of bph genes involved in polychlorinated biphenyl/biphenyl degradation in Pseudomonas sp. KKS102. J Biol Chem 276(39):36146–36154

Ojwang LM, Cook RL (2013) Environmental conditions that influence the ability of humic acids to induce permeability in model biomembranes. Environ Sci Technol 47(15):8280–8287

Ortega-Calvo JJ, Marchenko AI, Vorobyov AV, Borovick RV (2003) Chemotaxis in polycyclic aromatic hydrocarbon-degrading bacteria isolated from coal-tar- and oil-polluted rhizospheres. FEMS Microbiol Ecol 44(3):373–381

Pagnout C, Frache G, Poupin P, Maunit B, Muller J-F, Ferard J-F (2007) Isolation and characterization of a gene cluster involved in PAH degradation in Mycobacterium sp. strain SNP11: expression in *Mycobacterium smegmatis* mc(2)155. Res Microbiol 158(2):175–186

Pan T, Huang P, Xiong G, Maser E (2015) Isolation and identification of a repressor TetR for 3,17 beta-HSD expressional regulation in *Comamonas testosteroni*. Chem Biol Interact 234:205–212

Parales RE, Ju KS (2011) Rieske-type dioxygenases: key enzymes in the degradation of aromatic hydrocarbons. In: Comprehensive biotechnology, vol 6: Environmental biotechnology and safety, 2nd edn, pp 115–134

Park W, Padmanabhan P, Padmanabhan S, Zylstra GJ, Madsen EL (2002) nahR, encoding a LysR-type transcriptional regulator, is highly conserved among naphthalene-degrading bacteria isolated from a coal tar waste-contaminated site and in extracted community DNA. Microbiology-SGM 148:2319–2329

Park HH, Lim WK, Shin HJ (2005) In vitro binding of purified NahR regulatory protein with promoter Psal. Biochim Biophy Acta-General Subjects 1725(2):247–255

Payne DE, Boles BR (2016) Emerging interactions between matrix components during biofilm development. Curr Genet 62(1):137–141

Peng R-H, Xiong A-S, Xue Y, Fu X-Y, Gao F, Zhao W et al (2008) Microbial biodegradation of polyaromatic hydrocarbons. FEMS Microbiol Rev 32(6):927–955

Peng J-J, Wang N, Li H-J, Cai C (2011) Microbial degradation mechanisms of soil high molecular weight PAHs and affecting factors: a review. Shengtaixue Zazhi 30(11):2621–2627

Perfumo A, Rudden M, Smyth TJP, Marchant R, Stevenson PS, Parry NJ et al (2013) Rhamnolipids are conserved biosurfactants molecules: implications for their biotechnological potential. Appl Microbiol Biotechnol 97(16):7297–7306

Pouli M, Agathos SN (2011) Bioremediation of PAH-contaminated sites: from pathways to bioreactors. In: Koukkou AI (ed), 119–147 pp

Regonne RK, Martin F, Mbawala A, Ngassoum MB, Jouanneau Y (2013) Identification of soil bacteria able to degrade phenanthrene bound to a hydrophobic sorbent in situ. Environ Pollut 180:145–151

Reis RS, Pereira AG, Neves BC, Freire DMG (2011) Gene regulation of rhamnolipid production in *Pseudomonas aeruginosa*—a review. Biores Technol 102(11):6377–6384

Rojo-Nieto E, Perales-Vargas-Machuca JA (2012) Microbial degradation of PAHs: organisms and environmental compartments. In: Singh SN (ed) Microbial degradation of xenobiotics. Environ Sci 263–290

Sauret C, Tedetti M, Guigue C, Dumas C, Lami R, Pujo-Pay M et al (2016) Influence of PAHs among other coastal environmental variables on total and PAH-degrading bacterial communities. Environ Sci Pollut Res 23(5):4242–4256

Schell MA (1985) Transcriptional control of the nah and sal hydrocarbon-degradation operons by the nahR gene-product. Gene 36(3):301–309

Schell MA, Brown PH, Raju S (1990) Use of saturation mutagenesis to localize probable functional domains in the NahR protein, a LysR-type transcription activator. J Biol Chem 265 (7):3844–3850

Schippers C, Gessner K, Muller T, Scheper T (2000) Microbial degradation of phenanthrene by addition of a sophorolipid mixture. J Biotechnol 83(3):189–198

Schneider J, Grosser R, Jayasimhulu K, Xue W, Warshawsky D (1996) Degradation of pyrene, benz a anthracene, and benzo a pyrene by Mycobacterium sp. strain RJGII-135, isolated from a former coal gasification site. Appl Environ Microbiol 62(1):13–19

Schuler L, Chadhain SMN, Jouanneau Y, Meyer C, Zylstra GJ, Hols P et al (2008) Characterization of a novel angular dioxygenase from fluorene-degrading Spingomonas sp. strain LB126. Appl Environ Microbiol 74(4):1050–1057

Seo J-S, Keum Y-S, Li QX (2009) Bacterial degradation of aromatic compounds. Int J Environ Res Public Health 6(1):278–309

Seo J-S, Keum Y-S, Li QX (2012) *Mycobacterium aromativorans* JS19b1(T) degrades phenanthrene through C-1,2, C-3,4 and C-9,10 dioxygenation pathways. Int Biodeterior Biodegradation 70:96–103

Shetty AR, de Gannes V, Obi CC, Lucas S, Lapidus A, Cheng JF et al (2015) Complete genome sequence of the phenanthrene-degrading soil bacterium *Delftia acidovorans* Cs 1-4. Stand Genomic Sci 10

Sikkema J, Debont JAM, Poolman B (1995) Mechanisms of membrane toxicity of hydrocarbons. Microbiol Rev 59(2):201–222

Simpanen S, Makela R, Mikola J, Silvennoinen H, Romantschuk M (2016) Bioremediation of creosote contaminated soil in both laboratory and field scale: Investigating the ability of methyl-beta-cyclodextrin to enhance biostimulation. Int Biodeterior Biodegradation 106:117–126

Song X, Xu Y, Li G, Zhang Y, Huang T, Hu Z (2011) Isolation, characterization of Rhodococcus sp P14 capable of degrading high-molecular-weight polycyclic aromatic hydrocarbons and aliphatic hydrocarbons. Mar Pollut Bull 62(10):2122–2128

Song M, Luo CL, Jiang LF, Zhang DY, Wang YJ, Zhang G (2015) Identification of benzo a pyrene-metabolizing bacteria in forest soils by using DNA-based stable-isotope probing. Appl Environ Microbiol 81(21):7368–7376

Sotirova A, Spasova D, Vasileva-Tonkova E, Galabova D (2009) Effects of rhamnolipid-biosurfactant on cell surface of *Pseudomonas aeruginosa*. Microbiol Res 164 (3):297–303

Stach JEM, Burns RG (2002) Enrichment versus biofilm culture: a functional and phylogenetic comparison of polycyclic aromatic hydrocarbon-degrading microbial communities. Environ Microbiol 4(3):169–182

Story SP, Parker SH, Hayasaka SS, Riley MB, Kline EL (2001) Convergent and divergent points in catabolic pathways involved in utilization of fluoranthene, naphthalene, anthracene, and phenanthrene by *Sphingomonas paucimobilis* var. EPA505. J Ind Microbiol Biotechnol 26(6):369–382

Su YH, Yang XY (2009) Interactions between selected PAHs and the microbial community in rhizosphere of a paddy soil. Sci Total Environ 407(3):1027–1034

Sutcliffe IC (2010) A phylum level perspective on bacterial cell envelope architecture. Trends Microbiol 18(10):464–470

Sutcliffe LC, Brown AK, Dover LG (2010) The rhodococcal cell envelope: composition, organisation and biosynthesis. In: Alvarez HM (ed) Biology of Rhodococcus. Microbiology monographs, vol 16, pp 29–71

Tang WC, White JC, Alexander M (1998) Utilization of sorbed compounds by microorganisms specifically isolated for that purpose. Appl Microbiol Biotechnol 49(1):117–121

Tang H, Yu H, Li Q, Wang X, Gai Z, Yin G et al (2011) Genome sequence of *Pseudomonas putida* strain B6-2, a superdegrader of polycyclic aromatic hydrocarbons and dioxin-like compounds. J Bacteriol 193(23):6789–6790

Tecon R, Or D (2016) Bacterial flagellar motility on hydrated rough surfaces controlled by aqueous film thickness and connectedness. Sci Rep 6:19409

Touw DS, Patel DR, van den Berg B (2010) The crystal structure of OprG from *Pseudomonas aeruginosa*, a potential channel for transport of hydrophobic molecules across the outer membrane. PLoS ONE 5(11)

Tropel D, van der Meer JR (2004) Bacterial transcriptional regulators for degradation pathways of aromatic compounds. Microbiol Mol Biol Rev 68(3):474–500

Uyttebroek M, Ortega-Calvo J-J, Breugelmans P, Springael D (2006) Comparison of mineralization of solid-sorbed phenanthrene by polycyclic aromatic hydrocarbon (PAH)-degrading Mycobacterium spp. and Sphingomonas spp. Appl Microbiol Biotechnol 72(4):829–836

Vacca DJ, Bleam WF, Hickey WJ (2005) Isolation of soil bacteria adapted to degrade humic acid-sorbed phenanthrene. Appl Environ Microbiol 71(7):3797–3805

Vaidehi K, Kulkarni SD (2012) Microbial remediation of polycyclic aromatic hydrocarbons: an overview. Res J Chem Environ 16(4):200–212

Vamsee-Krishna C, Phale PS (2008) Bacterial degradation of phthalate isomers and their esters. Indian J Microbiol 48(1):19–34

van den Berg B (2005) The FadL family: unusual transporters for unusual substrates. Curr Opin Struct Biol 15(4):401–407

van den Berg B, Bhamidimarri SP, Winterhalter M (2015) Crystal structure of a COG4313 outer membrane channel. Sci Rep 5

Van Hamme JD, Singh A, Ward OP (2003) Recent advances in petroleum microbiology. Microbiol Mol Biol Rev 67(4):503–549

Vinas M, Sabate J, Guasp C, Lalucat J, Solanas AM (2005) Culture-dependent and -independent approaches establish the complexity of a PAH-degrading microbial consortium. Can J Microbiol 51(11):897–909

Vu B, Chen M, Crawford RJ, Ivanova EP (2009) Bacterial Extracellular polysaccharides involved in biofilm formation. Molecules 14(7):2535–2554

Waigi MG, Kang F, Goikavi C, Ling W, Gao Y (2015) Phenanthrene biodegradation by sphingomonads and its application in the contaminated soils and sediments: a review. Int Biodeterior Biodegradation 104:333–349

Wang Y, Rawlings M, Gibson DT, Labbe D, Bergeron H, Brousseau R et al (1995) Identification of a membrane-protein and a truncated LysR-type regulator associated with the toluene degradation pathway in *Pseudomonas putida* F1. Mol Gen Genet 246(5):570–579

Wang X, Jin D, Zhou L, Wu L, An W, Zhao L (2014a) Draft genome sequence of *Advenella kashmirensis* strain W13003, a polycyclic aromatic hydrocarbon-degrading bacterium. Genome Announcements 2(1):e00003–e00014

Wang X, Jin D, Zhou L, Wu L, Qi L, Li C et al (2014b) Draft genome sequence of halotolerant polycyclic aromatic hydrocarbon-degrading *Pseudomonas bauzanensis* strain W13Z2. Genome Announcements 2(5):e01049–e01114

Wang X, Jin D, Zhou L, Zhang Z (2015) Draft genome sequence of *Ochrobactrum anthropi* strain W13P3, a halotolerant polycyclic aromatic hydrocarbon-degrading bacterium. Genome Announcements 3(4):e00867–e00915

Wang X, Jin D, Zhou L, Zhang Z (2016) Draft genome sequence of *Pannonibacter phragmitetus* strain CGMCC9175, a halotolerant polycyclic aromatic hydrocarbon-degrading bacterium. Genome Announcements 4(1):e01675–e01715

Wehrer M, Rennert T, Mansfeldt T, Totsche KU (2011) Contaminants at former manufactured gas plants: sources, properties, and processes. Crit Rev Environ Sci Technol 41(21):1883–1969

Whitman BE, Lueking DR, Mihelcic JR (1998) Naphthalene uptake by a *Pseudomonas fluorescens* isolate. Can J Microbiol 44(11):1086–1093

Wick LY, Wattiau P, Harms H (2002) Influence of the growth substrate on the mycolic acid profiles of mycobacteria. Environ Microbiol 4(10):612–616

Wick LY, Pelz O, Bernasconi SM, Andersen N, Harms H (2003) Influence of the growth substrate on ester-linked phospho- and glycolipid fatty acids of PAH-degrading Mycobacterium sp. LB501T. Environ Microbiol 5(8):672–680

Wilkinson SP, Grove A (2006) Ligand-responsive transcriptional regulation by members of the MarR family of winged helix proteins. Curr Issues Mol Biol 8:51–62

Willumsen PA, Karlson U (1997) Screening of bacteria, isolated from PAH-contaminated soils, for production of biosurfactants and bioemulsifiers. Biodegradation 7(5):415–423

Wu EL, Fleming PJ, Yeom MS, Widmalm G, Klauda JB, Fleming KG et al (2014) *E. coli* outer membrane and interactions with OmpLA. Biophys J 106(11):2493–2502

Xia XH, Xia N, Lai YJ, Dong JW, Zhao PJ, Zhu BT et al (2015) Response of PAH-degrading genes to PAH bioavailability in the overlying water, suspended sediment, and deposited sediment of the Yangtze River. Chemosphere 128:236–244

Yang YJ, Chen RF, Shiaris MP (1994) Metabolism of naphthalene, fluorene, and phenanthrene - preliminary characterization of a cloned gene-cluster from *Pseudomonas-putida* NCIB-9816. J Bacteriol 176(8):2158–2164

Yang Y, Hunter W, Tao S, Gan J (2009) Microbial availability of different forms of phenanthrene in soils. Environ Sci Technol 43(6):1852–1857

Yang YY, Wang J, Liao JQ, Xie SG, Huang Y (2014) Distribution of naphthalene dioxygenase genes in crude oil-contaminated soils. Microb Ecol 68(4):785–793

Yang Y, Wang J, Liao J, Xie S, Huang Y (2015) Abundance and diversity of soil petroleum hydrocarbon-degrading microbial communities in oil exploring areas. Appl Microbiol Biotechnol 99(4):1935–1946

Yen KM, Serdar CM (1988) Genetics of naphthalene catabolism in pseudomonads. CRC Crit Rev Microbiol 15(3):247–268

Zhang Y, Qin F, Qiao J, Li G, Shen C, Huang T et al (2012) Draft genome sequence of Rhodococcus sp. strain P14, a biodegrader of high-molecular-weight polycyclic aromatic hydrocarbons. J Bacteriol 194(13):3546–3546

Zhang W, Sun J, Ding W, Lin J, Tan R, Lu L et al (2015) Extracellular matrix-associated proteins form an integral and dynamic system during *Pseudomonas aeruginosa* biofilm development. Front Cell Infect Microbiol 5

Zhou HW, Guo CL, Wong YS, Tam NFY (2006) Genetic diversity of dioxygenase genes in polycyclic aromatic hydrocarbon-degrading bacteria isolated from mangrove sediments. FEMS Microbiol Lett 262(2):148–157

Zuber B, Chami M, Houssin C, Dubochet J, Griffiths G, Daffe M (2008) Direct visualization of the outer membrane of mycobacteria and corynebacteria in their native state. J Bacteriol 190 (16):5672–5680

Chapter 8
Microbial Community Responses to Contaminants and the Use of Molecular Techniques

Simone C. Birrer, Katherine A. Dafforn and Emma L. Johnston

Abstract Human activities threaten global ecosystems through the introduction of a range of contaminants. The potential effects of contaminants are commonly tested on organisms, or cell lines, in separate tests: one toxicant and one species at a time. The results have limited predictive capacity in complex ecosystems, and when they are used to calculate guideline toxicity values, multiple safety factors must be applied to mitigate this uncertainty. Community responses can provide a more realistic assessment of contaminant effects but may be more difficult to detect and interpret. In recent years, advances in molecular approaches have significantly improved our ability to investigate community responses at the micro scale. Microbial groups represent a major source of biomass and chemical activity in many ecosystems and they include many of the most chemically sensitive organism groups. Traditional microbial ecotoxicological studies have either measured impacts on diversity or on ecosystem function, with very few attempting to quantify both diversity and function at the same time. Molecular approaches to microbial ecotoxicology have the potential to reveal novel elements of ecosystem change such as the mechanisms behind functional responses to contaminants. This is crucial to understand because structural changes do not necessarily translate to a change in function and vice versa. Increasing our understanding of stress-related structural and functional changes in microbial groups that drive global biogeochemical cycles will enable highly relevant and sensitive predictions of the impact of contaminants on ecosystem health.

Keywords Molecular methods · Whole-community assessment · Omics · Contaminants · Diversity · Function

S.C. Birrer (✉) · K.A. Dafforn · E.L. Johnston
Evolution and Ecology Research Centre, School of Biological, Earth and Environmental Sciences, University of New South Wales, Sydney, NSW 2052, Australia
e-mail: s.birrer@unsw.edu.au

S.C. Birrer · K.A. Dafforn · E.L. Johnston
The Sydney Institute of Marine Science, Mosman, NSW 2088, Australia

C. Cravo-Laureau et al. (eds.), *Microbial Ecotoxicology*,
DOI 10.1007/978-3-319-61795-4_8

8.1 Introduction

Human activities represent a major threat to global ecosystems through the introduction of a range of contaminants. Since the industrial revolution in the early 1800s, industry has been a major source of toxic contaminants such as metals and metalloids (e.g. Birch and Taylor 1999; Nicholson et al. 2003), petroleum based toxicants (Santschi et al. 2001), and complex organic compounds (Birch et al. 1999). These may initially be released at point sources into soils, waterways or the atmosphere, but the final distribution may be widespread. Agricultural practices have also introduced chemical contaminants into much of the world's environment, including fertilisers such as nitrates and phosphates that enrich non-target systems, but also pesticides and herbicides, which are designed for maximum toxicity (e.g. Arias-Estévez et al. 2008; Puckett 1995). Together, the products of agriculture and industry support burgeoning urban populations and transport networks that can introduce additional contaminants, such as metals and polycyclic aromatic hydrocarbons (Zhu et al. 2008; Nguyen et al. 2014). Understanding the ecological effects of these stressors in different systems requires extensive testing and monitoring that uses sensitive indicators of change (Dafforn et al. 2014). Commonly, such ecotoxicological tests are performed on single test species before new chemicals can be approved for either agriculture or industry (Breitholtz et al. 2006). However, as these tests focus on a small group of organisms, the relevance of the generated data to other groups of organisms and ecological processes in general is often unclear (Bourlat et al. 2013).

Past ecosystem monitoring strategies have tended to focus on describing the chemical characteristics of different systems and combining this information with the results of laboratory toxicity tests to predict ecological effects (Scanes et al. 2007). However, modern approaches increasingly recognise the importance of coupling chemical to biological monitoring to increase predictive ability (Dequiedt et al. 2011; Friberg et al. 2011; McKinley et al. 2011; Dafforn et al. 2012; Sun et al. 2012; Clark et al. 2015). Similarly, the need for real-world applicability has driven investment in the development of ecotoxicology and multiple-stressor testing at the level of whole communities in both the laboratory (Edge et al. 2015) and the field (Dafforn et al. 2014; Edge et al. 2014). Community responses to contaminants, rather than single organism toxicity testing, can provide a sensitive assessment of effects against a background of realistic biotic interactions and processes that would naturally be occurring in contaminated systems (Fig. 8.1). Terrestrial model ecosystems have used community-level approaches to test the toxicity of pesticides in a soil community (Scholz-Starke et al. 2011). However, terrestrial model ecosystem tests are based on groups of organisms that can be collected and taxonomically identified by eye, thus excluding all microorganisms associated with the tested community. In this chapter, we highlight the necessity for the incorporation of microbial communities to ecosystem health assessments, and describe modern techniques that enable the inclusion of community-level ecotoxicology tests that incorporate the microbial as well as infaunal organisms.

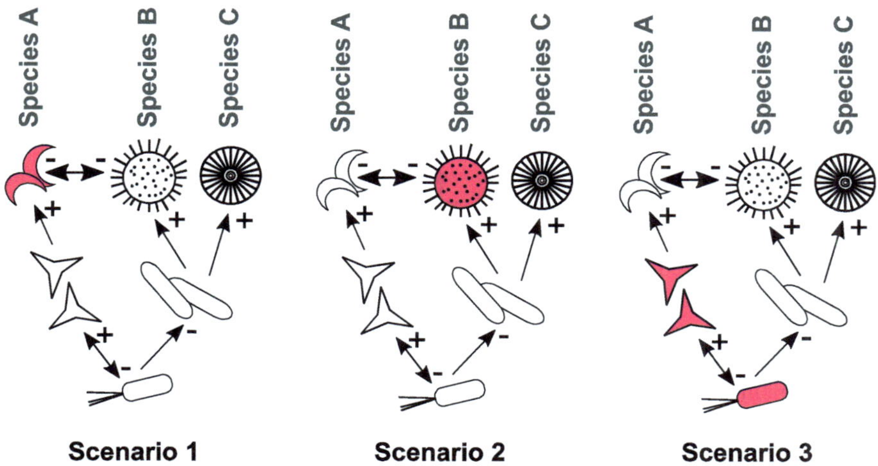

Fig. 8.1 Community response to a stressor provides a more sensitive test than multiple species individually. Three possible scenarios are depicted. Species that change their numbers of activity upon a stressor are highlighted in *pink* and species interactions are represented by *black arrows*. Commonly, multiple species (*species A–C*) are tested individually in toxicity testing. In *scenario 1*, the ecotox tests show that only *species A* responds to the presence of a stressor. However, *species B* will be affected by the response in *species A* in a community setting, due to their direct interaction. In *scenario 2*, *species B* is affected by a stressor in the ecotox test, whereas *species A* and *C* show no response. However, both *species A* and *C* will be affected by the stressor through direct and indirect interactions with *species B*, respectively. In *scenario 3*, none of the tested *species* (*A–C*) respond to the stressor in ecotox tests, however they will all be indirectly affected through their interactions with other organisms in the community, which are sensitive to the stressor. This figure highlights how testing of single species (even if multiple species are tested) can misrepresent the response of those species in a community setting. Therefore, whole-community measurements can provide a much more accurate and detailed representation of species responses to stressors

In recent years, advances in molecular approaches have significantly improved our understanding of community responses to contaminants, particularly at the micro scale (Sun et al. 2012). These approaches have increased the visibility of many parts of ecological communities that were previously a black box, and increased the quantity and resolution of biological measures that can be used to assess contaminant responses (Dafforn et al. 2014; Gibson et al. 2015; Zaiko et al. 2016). Historically, such assessments were restricted to only visible taxa through the use of dissecting or light microscopes and thus ecotoxicological approaches have tended to focus on macrofauna (Chariton et al. 2010; Smith et al. 2011; van der Linden et al. 2016) and a subset of the phytoplankton community (Johnston et al. 2015). More recently, molecular approaches such as next-generation sequencing have opened up all three domains of life (bacteria, archaea and eukarya) and now allow for assessments of community responses to contaminants that incorporate novel diversity that cannot be assessed with microscopy or other methods (Zimmerman et al. 2014).

These relatively invisible microbial communities, which represent a major source of biomass and chemical activity in many systems, include some of the most sensitive groups. Through their direct interactions with the environment, bacterial and archaeal communities have been found to be sensitive to human modification across large spatial scales (Sun et al. 2012). Within the eukarya, some of the most commonly used bioindicators for single-organism toxicity testing include microalgae such as diatoms (Buikema et al. 1982; Araújo et al. 2010), which have also been used to test whole-community responses (Johnston et al. 2015). Sensitivity to altered conditions implies a faster and/or stronger response compared to resistant organisms, which can increase the possibility of early detection or detection at low concentrations of contamination. Thus microbes represent an emerging group of sensitive bioindicators and molecular approaches are aiding the development of these assessment tools.

Microbial ecotoxicology also has the potential to reveal novel elements of ecosystem change such as the mechanisms behind functional responses to contaminants. Heretofore, traditional ecotoxicological studies have primarily measured diversity (Johnston and Mayer-Pinto 2015), with few attempting to quantify processes or function at the same time (Kinsella and Crowe 2016). Yet it is crucial that managers understand both, as increasing evidence highlights that structural change does not necessarily translate to a change in function, if there are high levels of redundancy in a community (Allison and Martiny 2008). Microbes are the most diverse and dominant functional drivers in almost every ecosystem (Falkowski et al. 2008). The potential for 'omic' techniques to fully characterise microbial community change and link these structural changes to altered gene expression opens up options for pursuing diversity-function relationships and how they are impacted by contaminants (Johnston et al. 2015).

Ecotoxicology aims to predict the fate and effect of contaminants in different environments and therefore must take a range of approaches including laboratory testing at multiple levels, and field experiments and monitoring at relevant spatial and temporal scales (Chapman 2002). Microbial indicators can be applied within all components of community assessment. As molecular techniques become faster and cheaper, there will be increasing opportunities for their integration into standard toxicity testing, guidelines and biomonitoring programs.

Various realms have been more or less well studied in relation to microbial ecotoxicology. The majority of progress has centered on terrestrial soils, aquatic sediments, freshwater and marine systems, and this is where we focus our review. These are also the realms that have the highest concentrations of many contaminants due to their inherent binding properties and their use as waste disposal systems for many decades (Birch 1996; Kelly et al. 1996; Holt 2000). As such, these systems have historical legacies of contaminants, and ongoing inputs of a range of chemicals that are potentially toxic to microbes including inorganic (metals, fertilisers i.e. nitrates, phosphates), organic (polyaromatic hydrocarbons, pesticides, herbicides) and emerging contaminants of concern such as microplastics and pharmaceuticals (Kennish 2002; Pal et al. 2010; Browne et al. 2011). In most environments, communities will be exposed to multiple contaminants simultaneously, and microbial communities provide a promising mechanism for greater power to differentiate the impacts of particular chemicals within mixtures due to the specificity of the species level responses.

8.2 How Do Contaminants Change Structural Components of Microbial Communities?

Toxic contaminants such as metals and polycyclic aromatic hydrocarbons have been demonstrated to significantly influence structural components (e.g. diversity and community evenness) of microbial communities in a wide range of environments (Ellis et al. 2001; Gillan et al. 2005). In general, microbes appear to be sensitive indicators of metal stressors. Significant shifts in community composition have been found in marine sediment bacteria in large-scale surveys of contaminated and uncontaminated estuaries (Sun et al. 2012), and in freshwater biofilm bacteria exposed to copper, lead and zinc in mesocosms (Ancion et al. 2010). These differences were detected with DNA fingerprinting (for example ARISA: Automated Ribosomal Intergenic Spacer Analysis, see Box 8.1, Fig. 8.2), which does not directly provide taxonomic information, but nevertheless has proven to be a useful tool for identifying bacterial community change in response to contamination over and above natural variation.

Box 8.1—Overview of molecular methods

Automated Ribosomal Intergenic Spacer Analysis (ARISA). An analysis of the highly variable genetic fragment between the well-conserved 16S and 23S rRNA genes provides information for richness and diversity measures of a community. It is increasingly being replaced by amplicon sequencing or 'omic' techniques as these other tools become cheaper. For an example study using this technique, see Sun et al. (2012).

Denaturing Gradient Gel Electrophoresis (DGGE). A gene of interest is amplified and run through a gel with increasing denaturing conditions. An image of bands on the gel reveals different lengths of the gene and provides information on differences in community structure or diversity of a functional community. For an example study using this technique, see Geets et al. (2006).

Terminal Restriction Fragment Length Polymorphism (TRFLP). A gene of interest is amplified, exposed to restriction enzymes and the resulting fragment lengths are measured. This technique provides information on richness and diversity measures of a community. Using functional genes, the differences in structure of a specific functional part between communities can be determined. For an example study using this technique, see Bissett et al. (2010).

Amplicon sequencing. A gene of interest is amplified and sequenced using a next-generation sequencing platform such as the Roche 454 GS FLX or the Illumina MiSeq or Hiseq. 454 pyrosequencing is becoming less common than Illumina, but still has advantages with respect to sequence length and the assignment of taxonomy to sequences with a reference database (Liu et al. 2012). This taxonomic information exposes community shifts and identifies specific taxa driving these shifts. Community functions can be inferred. Using a functional gene as input, this technique can also be used to assess activity and diversity of a specific functional part of the community. The most commonly used genes for examining community structure are 16S rRNA for bacterial, 18S rRNA for eukaryotic and ITS (nuclear ribosomal internal transcribed spacer) for fungal community analysis. For an example study using this technique, see Sun et al. (2013).

Quantitative Real-Time Polymerase Chain Reaction (qPCR). Genes of interest are amplified using fluorescent primers to determine the quantity of the genes in a community. Using this technique, the functional activity or relative abundance of targeted taxa in a specific sample can be assessed. For an example study using this technique, see Graham et al. (2011).

Microarrays. DNA or RNA from samples of interest is fluorescently labelled and washed over a plate containing fixed probes of the genes of interest, where the sample DNA or RNA binds to according probes. Using the fluorescent signal, relative quantitation of the genes of interest can be determined using the known positions on the microarray plate. Using this technique the relative activity of targeted functional genes or relative abundance of targeted taxa can be assessed. In addition, functional community diversity, evenness and richness can be calculated. For an example study using this technique, see Sun (2016). For information on Geochip and Phytochip, two types of microarrays, see He et al. (2013) and Noyer et al. (2015), respectively.

Metagenomics. Total DNA from an entire community is sequenced using a next-generation sequencing platform. Resulting sequences are assigned to taxonomic/functional genes using a reference database. Information on taxonomy and potential function can be gained from this technique. For an example study using this technique, see Thomas et al. (2010).

Metatranscriptomics. Total RNA (rRNA and mRNA) or only mRNA from an entire community is sequenced on a next-generation sequencing platform. Resulting sequences are assigned to known functional genes. Using the rRNA fraction, taxonomic information can be extracted, and the mRNA sequences provide functional information on the active part of the community. For an example study using this technique, see Stewart et al. (2012).

Recent progress with targeted gene sequencing (e.g. 454 pyrosequencing and Illumina) has removed the need for cloning and gene picking prior to sequencing, which allows for (1) greater sequencing depth with (2) rapid and relatively easy

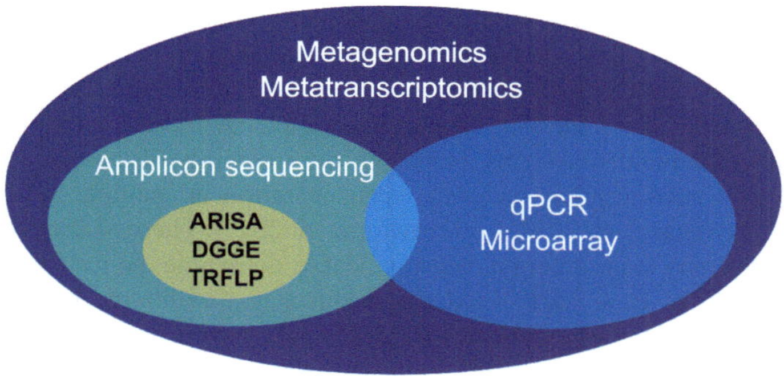

Fig. 8.2 This diagram represents the structural (*greens*) and functional (*light blue*) information that can be gained through different molecular techniques. *Dark blue* highlights the methods that can provide both structural and functional information. *Smaller circles* represent cheaper options with less output, while *larger circles* represent more detailed information and significantly higher costs

taxonomic characterisation of microbial communities at (3) a high level of resolution (Liu et al. 2012). This has advantages over ARISA and other DNA fingerprinting methods because it enables identification of the taxa driving shifts in response to contamination, and provides a link to potential functions. For example, amplicon sequencing of the 16S rRNA gene in sediments from Lake Geneva, Switzerland, revealed differences in composition between contaminated and uncontaminated areas and specific taxa that were related to these differences were members of *Betaproteobacteria* (e.g. *Dechloromonas* sp.), which includes species of contaminant-degrading bacteria, and *Deltaproteobacteria* (e.g. sulphate-reducing bacteria and *Geobacter* sp.), known for reducing Fe(III) (Haller et al. 2011). In another study, 16S rRNA gene sequencing in several estuaries of different disturbance levels (Sun et al. 2013), showed that estuarine sediments are generally dominated by *Deltaproteobacteria*, indicating high levels of sulphur cycling. Gibson et al. (2015) showed that amplicon sequencing can provide more detailed taxonomic information than conventional morphological analyses, thus highlighting the applicability of diversity measures from molecular techniques to monitoring programs.

Molecular approaches to community ecotoxicology provide a powerful mechanism for investigating the impacts of contaminants on microbial communities that are likely to have flow-on effects for macrofauna. In a recent study, Lawes et al. (2016b) investigated the impacts on estuarine microbes of a common agricultural fertiliser mix (Osmocote). Using targeted sequencing of the 16S rRNA gene they found that nutrients had distinct effects on bacterial community structure and connectivity. Responses were able to be assigned to particular bacterial types and are likely to have caused higher level impacts on invertebrates recruiting to the site of the bacterial biofilms (Lawes et al. 2016a).

Other commonly used measures in ecotoxicology to identify contaminant stress include metrics such as species richness, diversity and evenness. These measures have been used most often in monitoring programs that target macrofauna. A recent meta-analysis found that marine species richness decreased by around 40% in response to contaminants (Johnston and Roberts 2009). Johnston and Roberts' systematic review also identified gaps in knowledge about how microbial richness changes in contaminated systems. However, the suitability of diversity metrics for microbial ecotoxicology has received recent attention in multiple systems (Sims et al. 2013; Aylagas et al. 2014; Lejzerowicz et al. 2015). These studies showed that microbial diversity is an accurate measurement to determine the restoration success of wetlands (Sims et al. 2013) and the health status of marine benthic systems (Aylagas et al. 2014; Lejzerowicz et al. 2015).

Community resistance is posited to provide the best buffer against contaminant stressors since overall structure and function remains unchanged in the presence of the stressor (Box 8.2). Many microbial species demonstrate significant resistance to toxic contaminants (Bruins et al. 2000), and because of this they have a wide application in remediation (Antizar-Ladislao 2010; Nikolopoulou et al. 2013). For bacterial species this tolerance and ability to remediate extreme contamination has been strongly linked to horizontal gene transfer where resistance genes may be transferred between organisms other than vertical transmission from parent to offspring (Davison 1999). Another factor contributing to the resistance and resilience of many microbial communities is their ability to rapidly evolve in response to new conditions because of shorter generation times. This has been studied extensively in the laboratory (Reusch and Boyd 2013; Lohbeck et al. 2014), but requires further field validation. Where community resistance is lacking, then resilience can also provide a degree of buffering from contaminant impacts. Microbial community resilience to contaminants tends to be greater than resilience of larger organism groups because of their fast growing rates and hence fast recovery (Shade et al. 2012). Diversity is also an important factor to consider in microbial community ecotoxicology since it can imply functional redundancy. That is, more diverse microbial communities may provide a pool of genes and functions that increases community resilience to disturbance (Bissett et al. 2007). Where a community lacks resistance, resilience and redundancy, then contaminant impacts to structure and function may be significant and irreversible (Fig. 8.3).

Box 8.2—Resistance, resilience and functional redundancy

Microbial communities can be resistant, resilient or functionally redundant to a disturbance (Allison and Martiny 2008). Resistant communities are neither structurally, nor functionally affected by the specific disturbance, whereas resilient communities are initially shifted structurally and functionally, but can recover over time (Fig. 8.3). Functionally redundant communities have a shifted structure (alternative equilibrium) upon a disturbance, but still perform the same functions as the initial community. Only when a community is not resistant, nor resilient or functionally redundant to a disturbance, we see a community that is impacted both structurally and functionally in the long term.

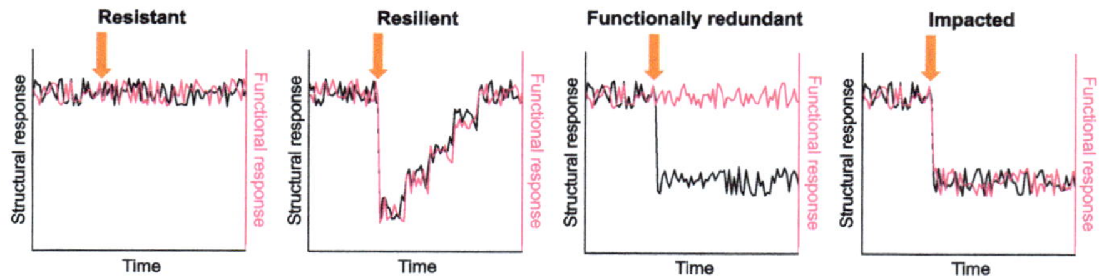

Fig. 8.3 Structural (*black*) and functional (*pink*) responses of resistant, resilient, functionally redundant and impacted communities with time after exposure to a disturbance (*orange arrow*)

8.3 How Do Contaminants Change Functioning in Microbial Communities?

Much like multicellular organisms, microorganisms adapt their behaviour and functions to the environment. They make 'intelligent' decisions in response to specific circumstances based on biological and chemical information (Westerhoff et al. 2014). Responses are not only made by each microbe itself, but are dependent on the responses of others in the same community (Ross-Gillespie and Kümmerli 2014).

In contaminated environments, microbes respond in order to mitigate the impact on their cells and to deal with contaminants directly. Stress-related responses protect the cells through activation of heat-shock proteins (HSPs). HSPs act as molecular chaperones that assist in the folding and unfolding of other proteins and ensure that misfolding due to stress is minimised (Hartl 1996). Affected microorganisms therefore shift their energetic resources to survival-related instead of growth-related mechanisms. This can significantly affect energy and nutrient flows in the ecosystem (Schimel et al. 2007). Hence, contaminants can affect the functioning of an entire ecosystem. However, microbes can also up-regulate functions in response to a drop in organism numbers that are able to perform this specific function, thus exhibiting functional redundancy (Box 8.2). This phenomenon is only possible because microbial communities consist of many more different organisms than any other community and also many of these organisms can perform multiple critical biogeochemical cycling steps at once. This functional redundancy provides microbial communities with the possibility to keep functions stable despite a change in community structure (Bissett et al. 2007).

In addition to these indirect effects, contaminants can alter processes that are directly related to the specific contaminant. For example, an increase in organic material will alter the dynamics of carbon, nitrogen, phosphorous and sulphur cycles, which are responsible for the cycling of nutrients and the removal of excess nutrients from the system (Falkowski et al. 2008). Process rates have commonly been measured using biogeochemical measurements in incubation chambers (Eyre and Ferguson 2005; Kelaher et al. 2013). These chambers seal off the investigated samples, such as soil, water or plants for a pre-determined amount of time. At the beginning and end of each incubation, samples are taken for measurement of different gases and nutrients. A further common technique to directly measure process rates is the isotopic dilution method (Di et al. 2000). In this method either radioactive (^{14}C, ^{33}P, and ^{35}S) or stable (^{13}C and ^{15}N) isotopes are added to a nutrient pool, and changes over time of the amount of labelled nutrients in the pool are monitored. Both of the above techniques provide direct information on functional differences between samples from either different experimental treatments or different environmental conditions through direct measurements of gas production (incubation chambers) or nutrient degradation (isotopic dilution).

Molecular analyses using 'omic' techniques (Box 8.1) provide indirect information on functional changes through the quantification of genetic activity. Molecular techniques are thus able to provide insights into the genetic machinery controlling measurable changes in gas and nutrient fluxes. This approach can inform on all functions in an entire ecosystem at a certain point in time. It generates enormous datasets within a relatively short time frame and is therefore very powerful to further the understanding of contaminant impacts on ecosystem functions . Biogeochemical incubations can provide insight into up- or down-regulation of entire pathways. However, not all genes in a pathway are regulated in the same way; while some genes might be down-regulated or turned off, the activity of others might be increased. Therefore, the alteration of gene expression of functional pathways, can lead to complex results, i.e. accumulation of different intermediate or end products, which can be difficult to decipher. 'Omic' approaches help elucidate the nature of response and the potential interactions of different processes through revealing the underlying genetic mechanisms of observed altered functions.

In anthropogenically impacted ecosystems, we can rarely find the situation where only a single stressor is present. Most often ecosystems are exposed to a variety of stressors at one time (Mayer-Pinto et al. 2015). These stressors can interact and lead to either additive, synergistic or antagonistic responses in biological organisms (Lawes et al. 2016a). Research targeting single stressor impacts on specific ecosystems is crucial for our fundamental understanding of the possible effects of a certain stressor. However, the single-stressor approach is not fully applicable to our natural environments and their current contaminant state. The investigation of multiple stressor impacts is therefore more crucial than ever in providing us with relevant information on real scenario situations and can be applied in modern ecosystem management (Dafforn et al. 2016).

8.4 Linking Diversity and Structure to Function

The functional potential of a community can be extrapolated from structural information, but this requires a high level of expertise on species-traits relationships. Furthermore, a high resolution of taxonomic information is needed, as many phyla and classes contain species that facilitate very distinct reactions. Different aspects of diversity can be important to determine functional traits of a community. On the one hand, the active part of the community (using RNA for analysis) can provide information on functions that are being performed. On the other hand, analysing the DNA provides insight into potential functions. This can help to assess the potential resilience of the community to anthropogenic stressors, as it includes information on dormant organisms which are activated under certain conditions (Jones and Lennon 2010).

However, communities that are structurally different are often likely to be functionally different (Strickland et al. 2009). The diversity of a community is tightly linked to primary productivity, energy flow and a wide variety of ecological processes, such as contaminant remediation and biogeochemical cycles. More diversity provides a higher assembly of functional genes, and emerging niches can be filled quickly from the pool of species already existing in the community (Finlay et al. 1997). Most ecological processes, however, can be performed by a number of different organisms. Thus a higher diversity directly leads to a higher potential for functional resistance/redundancy of a community. For instance, when a community is exposed to a stressor, this stressor will affect sensitive organisms more than others. It might decimate a number of species significantly, but as long as there are other species in the same community that have the potential to perform the same functions with the same efficiency, then this community might very well be resistant to this stressor on a functional level (i.e. functional redundancy, Box 8.2). Such a response is evident when, for example, invasive species replace native species in sessile invertebrate communities exposed to metals (Piola and Johnston 2008).

Pollution Induced Community Tolerance or PICT assessments (Blanck and Wängberg 1988) were introduced to investigate whole community response to contaminant stress. The premise being that different species within a community would demonstrate differential sensitivity to toxicants (Clements and Rohr 2009), and that this could be measured with a combination of short-term bioassays and physiological endpoints such as photosynthetic activity (Blanck and Dahl 1996) or respiration (Tlili et al. 2010). PICT focuses on microbial communities and molecular tools have opened up increasing opportunities to link measured endpoints to structural change (Tlili et al. 2015). However, the application of PICT to real-world contaminated environments requires further field-testing, bioassays that incorporate chemical mixtures, and more relevant endpoints (Tlili et al. 2015).

Community biodiversity and function seem to be unlinked in most cases (Bier et al. 2015). Bier and colleagues observed that functional responses upon stressors generally precede any structural changes to the community. The authors give, among others, the following reasons for a lack of a link between structure and function: dormancy of microorganisms, horizontal gene transfer, phenotypic plasticity, and functional redundancy. Other researchers investigating the link between species richness and carbon cycling in soils, have also found that functional redundancy leads to an unlinked community structure and function (Nielsen et al. 2011). Specifically, in soil systems, which exhibit high levels of functional redundancy, species richness only affected carbon cycling in communities with very poor richness. Functional redundancy further underpins the resilience of ecosystem functions (Oliver et al. 2015).

Stressors can also affect the connectivity of a community as shown by Lawes et al. (2016b) for estuarine biofilm communities exposed to elevated nutrients. This can have an impact on the cohesiveness (i.e. number of species that when removed lead to a disconnected community network, Horvath and Dong 2008) and stability (Proulx et al. 2005) of communities, and thus also impact the functional resistance and community resilience to a stressor (Oliver et al. 2015). Under stressed conditions, a structure dominated by a few resistant organisms takes over from a highly balanced community with high richness and evenness (Johnston and Roberts 2009). However, in these stressed communities, abundant organisms are often most highly affected by stressors, while rare organisms commonly contribute most to community functioning, potentially due to body size and associated trophic level (Radchuk et al. 2016). Therefore, this dynamic can lead to a structurally changed community while functions are being preserved.

8.5 Future Directions for Microbial Community Ecotoxicology

We are currently going through a revolution in our approach to ecosystem health assessment that is being driven by technologies that give us access to big data in the form of molecular information and remotely sensed environments (Dafforn et al. 2014, 2016). Rather than continuing to rely on the combination of single species toxicity testing and chemical surveys, we should now be moving towards the regular assessment of structural changes to entire communities as the routine approach to biomonitoring (Baird and Hajibabaei 2012). Such information can be rapidly coupled to remotely-sensed information to provide close to real-time insight into the macro and micro worlds. Modern molecular techniques have opened up the possibility to rapidly and cost-effectively analyse entire microbial communities at specific time points. Because structural shifts often do not reflect differential function in impacted communities, separate functional measures should be incorporated to accurately understand the whole-community impact of contaminants and make predictions on their consequences for ecosystems.

Sophisticated field experiments using multiple stressors are the basis for our understanding of the complexity of the real world (Box 8.3, Fig. 8.4). They provide the necessary information to understand and distinguish different stressors in complex environments (Lawes et al. 2016a). There has been a dearth of field experimentation that focuses on microbial communities, much of this work being relegated to laboratory culture studies. The new molecular approaches are again critical to the development of field-based ecotoxicological studies of microbial structure and function.

As next-generation sequencing and associated bioinformatic pipelines are becoming increasingly streamlined, the analysis of big data is becoming simpler. These techniques will enable us to better understand the threat that human activities pose to the global ecosystem through their impacts on previously hidden microbial communities. Increasing knowledge of stress-related changes in the microbial groups that drive global biogeochemical cycles will promote the use of microbes as sensitive bioindicators of ecosystem health. A combination of structural and functional measurements will provide a holistic view of microbial contributions to ecosystem function and the sensitivity to microbial identity.

The rapid response of microbial communities to contamination coupled with highly detailed response measurements using modern molecular techniques, increases the potential to improve the accuracy of contaminant detection even at low levels. This can enable the implementation of remediation and management strategies at an early point in contamination and is therefore a crucial step in the advancement of modern biomonitoring.

Box 8.3—Example—Nitrogen metabolism changes in estuarine sediment due to excess organic matter

Birrer et al. in review

Organic matter can enter waterways in solid (e.g. leaves and soils in stormwater run-off, direct sewage input) or dissolved forms (e.g. fertilisers), and is one of the most ubiquitous contaminants in modern-day urbanised areas. Excess organic matter can lead to increased productivity by phytoplankton and eutrophication in the affected systems. Resulting high respiration rates and a lack of oxygen can impact a variety of biogeochemical cycles. Sulphides build up in organically enriched soils and sediments (Meyer-Reil and Köster 2000; Gray et al. 2002), which favour dissimilatory nitrate reduction to ammonia (Kraft et al. 2014), and inhibit the gene involved in nitrous oxide reduction (Brunet and Garcia-Gil 1996). Furthermore, anoxic conditions lead to methane production (Liikanen and Martikainen 2003). Excess organic matter therefore has the potential to increase the production of greenhouse gases (hydrogen sulphide, nitrous oxide and methane) and to accumulate toxic ammonia, which can in turn impact the productivity of the affected community.

We examined changes in gene expression of nitrogen metabolism genes in estuarine sediments in a robust field experiment. Sediments were mixed with 10% dry weight fertiliser to mimic organic enrichment of estuarine systems due to agricultural and urban run-off. Gene expression of control and enriched sediment samples was measured 17 weeks after experiment deployment using metatranscriptomics, and changes in expression of all nitrogen metabolism genes were analysed. Organic enrichment led to an increase in nitrogen fixation and dissimilatory nitrate reduction to ammonium (DNRA). Furthermore, the nitrous oxide reductase gene (*nosZ*) was down-regulated in enriched samples and the first step of denitrification was starting to shut down. These changes can lead to a loss of effectiveness in removing excess N from the system. Specifically, ammonia and nitrous oxide were likely being accumulated in the sediments. Ammonia, on the one hand, is highly toxic and can thus exacerbate the impacts of the fertiliser itself on the sediment community. It can also lead to drastically lower ecosystem productivity levels. This can have an effect on the entire trophic system, which relies on primary producers. Nitrous oxide, on the other hand, is a highly potent greenhouse gas. In fact, it has a warming potential of up to 300 times that of a CO_2 molecule and is therefore of major concern for our global climate.

Hence, eutrophication of highly urbanised waterways can have serious consequences on ecosystem services of affected areas and also potentially impact climate on a global level.

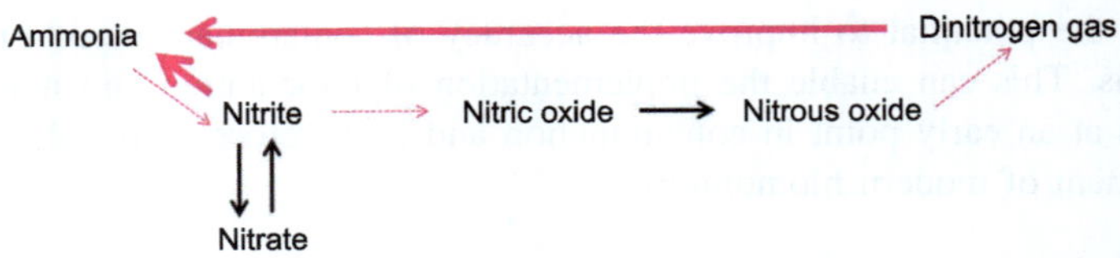

Fig. 8.4 The nitrogen cycling pathway in coastal sediments is driven by microbial activity and was impacted by organic enrichment in a controlled experiment (Birrer et al. *in review*). Altered gene expressions are highlighted in *pink*, whereas *dotted arrows* represent down-regulation and *thicker arrows* represent up-regulation of the genes involved in that step of the pathway

Acknowledgements The authors received funding from the Australian Research Council through LP130100364 awarded to ELJ. We thank two anonymous reviewers for their comments, which helped to improve this chapter.

References

Allison SD, Martiny JBH (2008) Resistance, resilience, and redundancy in microbial communities. Proc Natl Acad Sci 105:11512–11519

Ancion PY, Lear G, Lewis GD (2010) Three common metal contaminants of urban runoff (Zn, Cu & Pb) accumulate in freshwater biofilm and modify embedded bacterial communities. Environ Pollut 158:2738–2745. doi:10.1016/j.envpol.2010.04.013

Antizar-Ladislao B (2010) Bioremediation: working with bacteria. Elements 6:389–394. doi:10.2113/gselements.6.6.389

Araújo CVM, Tornero V, Lubián LM et al (2010) Ring test for whole-sediment toxicity assay with ucode Type="General_Pun">—a—benthic marine diatom. Sci Total Environ 408: 822–828. doi:10.1016/j.scitotenv.2009.10.018

Arias-Estévez M, López-Periago E, Martínez-Carballo E et al (2008) The mobility and degradation of pesticides in soils and the pollution of groundwater resources. Agric Ecosyst Environ 123:247–260. doi:10.1016/j.agee.2007.07.011

Aylagas E, Borja Á, Rodríguez-Ezpeleta N (2014) Environmental status assessment using DNA metabarcoding: towards a genetics based marine biotic index (gAMBI). PLoS ONE 9:e90529. doi:10.1371/journal.pone.0090529

Baird DJ, Hajibabaei M (2012) Biomonitoring 2.0: a new paradigm in ecosystem assessment made possible by next-generation DNA sequencing. Mol Ecol 21:2039–2044

Bier RL, Bernhardt ES, Boot CM et al (2015) Linking microbial community structure and microbial processes: an empirical and conceptual overview. FEMS Microbiol Ecol 91:1–11. doi:10.1093/femsec/fiv113

Birch GF (1996) Sediment-bound metallic contaminants in Sydney's estuaries and adjacent offshore, Australia. Estuar Coast Shelf Sci 42:31–44. doi:10.1006/ecss.1996.0003

Birch GF, Eyre BD, Taylor SE (1999) The distribution of nutrients in bottom sediments of Port Jackson (Sydney Harbour), Australia. Mar Pollut Bull 38:1247–1251

Birch GF, Taylor SE (1999) Source of heavy metals in sediments of the Port Jackson estuary, Australia. Sci Total Environ 227:123–138. doi:10.1016/S0048-9697(99)00007-8

Bissett A, Burke C, Cook PLM, Bowman JP (2007) Bacterial community shifts in organically perturbed sediments. Environ Microbiol 9:46–60. doi:10.1111/j.1462-2920.2006.01110.x

Bissett A, Richardson AE, Baker G et al (2010) Life history determines biogeographical patterns of soil bacterial communities over multiple spatial scales. Mol Ecol 19:4315–4327. doi:10.1111/j.1365-294X.2010.04804.x

Blanck H, Dahl B (1996) Pollution-induced community tolerance (PICT) in marine periphyton in a gradient of tri-n-butyltin (TBT) contamination. Aquat Toxicol 35:59–77. doi:10.1016/0166-445X(96)00007-0

Blanck H, Wängberg S-Å (1988) Induced community tolerance in marine periphyton established under arsenate stress. Can J Fish Aquat Sci 45:1816–1819

Bourlat SJ, Borja Á, Gilbert JA et al (2013) Genomics in marine monitoring: new opportunities for assessing marine health status. Mar Pollut Bull 74:19–31. doi:10.1016/j.marpolbul.2013.05.042

Breitholtz M, Rudén C, Ove Hansson S, Bengtsson BE (2006) Ten challenges for improved ecotoxicological testing in environmental risk assessment. Ecotoxicol Environ Saf 63:324–335. doi:10.1016/j.ecoenv.2005.12.009

Browne MA, Crump P, Niven SJ et al (2011) Accumulation of microplastic on shorelines worldwide: sources and sinks. Environ Sci Technol 45:9175–9179. doi:10.1021/es201811s

Bruins MR, Kapil S, Oehme FW (2000) Microbial resistance to metals in the environment. Ecotoxicol Environ Saf 45:198–207. doi:10.1006/eesa.1999.1860

Brunet RC, Garcia-Gil LJ (1996) Sulfide-induced dissimilatory nitrate reduction to ammonia in anaerobic freshwater sediments. FEMS Microbiol Ecol 21:131–138. doi:10.1111/j.1574-6941.1996.tb00340.x

Buikema AL, Niederlehner BR, Cairns J (1982) Biological monitoring part IV-Toxicity testing. Water Res 16:239–262. doi:10.1016/0043-1354(82)90188-9

Chapman PM (2002) Integrating toxicology and ecology: putting the "eco" into ecotoxicology. Mar Pollut Bull 44:7–15. doi:10.1016/S0025-326X(01)00253-3

Chariton A, Simpson S, Roach A, Batley G (2010) The influence of environmental variable choice on interpreting the spatial patterns of sediment contaminants and their relationships with benthic communities. Mar Freshw Res 61:1109–1122

Clark GF, Kelaher BP, Dafforn KA et al (2015) What does impacted look like? High diversity and abundance of epibiota in modified estuaries. Environ Pollut 196:12–20. doi:10.1016/j.envpol.2014.09.017

Clements WH, Rohr JR (2009) Community responses to contaminants: using basic ecological principles to predict ecotoxicological effects. Environ Toxicol Chem 28:1789. doi:10.1897/09-140.1

Dafforn KA, Baird DJ, Chariton AA et al (2014) Faster, higher and stronger? The pros and cons of molecular faunal data for assessing ecosystem condition. Adv Ecol Res 51:1–40. doi:10.1016/B978-0-08-099970-8.00003-8

Dafforn KA, Johnston EL, Ferguson A et al (2016) Big data opportunities and challenges for assessing multiple stressors across scales in aquatic ecosystems. Mar Freshw Res 67:393–413. doi:10.1071/MF15108

Dafforn KA, Simpson SL, Kelaher BP et al (2012) The challenge of choosing environmental indicators of anthropogenic impacts in estuaries. Environ Pollut 163:207–217. doi:10.1016/j.envpol.2011.12.029

Davison J (1999) Genetic exchange between bacteria in the environment. Plasmid 42:73–91. doi:10.1006/plas.1999.1421

Dequiedt S, Saby NPA, Lelievre M et al (2011) Biogeographical patterns of soil molecular microbial biomass as influenced by soil characteristics and management. Glob Ecol Biogeogr 20:641–652. doi:10.1111/j.1466-8238.2010.00628.x

Di HJ, Cameron KC, McLaren RG (2000) Isotopic dilution methods to determine the gross transformation rates of nitrogen, phosphorus, and sulfur in soil: a review of the theory, methodologies, and limitations. Soil Res 38:213–230. doi:10.1071/SR99005

Edge KJ, Dafforn KA, Simpson SL et al (2015) Resuspended contaminated sediments cause sublethal stress to oysters: a biomarker differentiates total suspended solids and contaminant effects. Environ Toxicol Chem 34:1345–1353. doi:10.1002/etc.2929

Edge KJ, Dafforn KA, Simpson SL et al (2014) A biomarker of contaminant exposure is effective in large scale assessment of ten estuaries. Chemosphere 100:16–26

Ellis RJ, Neish B, Trett MW et al (2001) Comparison of microbial and meiofaunal community analyses for determining impact of heavy metal contamination. J Microbiol Methods 45:171–185. doi:10.1016/S0167-7012(01)00245-7
Eyre BD, Ferguson AJP (2005) Benthic metabolism and nitrogen cycling in a subtropical east Australian estuary (Brunswick): temporal variability and controlling factors. Limnol Oceanogr 50:81–96. doi:10.4319/lo.2005.50.1.0081
Falkowski PG, Fenchel T, Delong EF (2008) The microbial engines that drive earth's biogeochemical cycles. Science (80) 320:1034–1039. doi:10.1126/science.1153213
Finlay BJ, Maberly SC, Cooper JI (1997) Microbial diversity and ecosystem function. Oikos 80:209–213. doi:10.2307/3546587
Friberg N, Bonada N, Bradley DC et al (2011) Biomonitoring of human impacts in freshwater ecosystems: the good, the bad and the ugly. Elsevier Ltd., Amsterdam
Geets J, Borremans B, Diels L et al (2006) DsrB gene-based DGGE for community and diversity surveys of sulfate-reducing bacteria. J Microbiol Methods 66:194–205. doi:10.1016/j.mimet.2005.11.002
Gibson JF, Shokralla S, Curry C et al (2015) Large-scale biomonitoring of remote and threatened ecosystems via high-throughput sequencing. PLoS ONE 10:e0138432. doi:10.5061/dryad.vm72v
Gillan DC, Danis B, Pernet P et al (2005) Structure of sediment-associated microbial communities along a heavy-metal contamination gradient in the marine environment. Appl Environmantal Microbiol 71:679–690. doi:10.1128/AEM.71.2.679
Graham JE, Wantland NB, Campbell M, Klotz MG (2011) Characterizing bacterial gene expression in nitrogen cycle metabolism with RT-qPCR. Methods Enzymol 496:345–372. doi:10.1016/B978-0-12-386489-5.00014-2
Gray JS, Wu RS, Or YY (2002) Effects of hypoxia and organic enrichment on the coastal marine environment. Mar Ecol Prog Ser 238:249–279. doi:10.3354/meps238249
Haller L, Tonolla M, Zopfi J et al (2011) Composition of bacterial and archaeal communities in freshwater sediments with different contamination levels (Lake Geneva, Switzerland). Water Res 45:1213–1228. doi:10.1016/j.watres.2010.11.018
Hartl FU (1996) Molecular chaperones in cellular protein folding. Nature 381:571–579
He Z, Van Nostrand JD, Zhou J (Joe) (2013) GeoChip-based metagenomic technologies for analyzing microbial community functional structure and activities. pp 236–247
Holt MS (2000) Sources of chemical contaminants and routes into the freshwater environment. Food Chem Toxicol 38:S21–S27. doi:10.1016/S0278-6915(99)00136-2
Horvath S, Dong J (2008) Geometric interpretation of gene coexpression network analysis. PLoS Comput Biol 4:e1000117. doi:10.1371/journal.pcbi.1000117
Johnston EL, Mayer-Pinto M (2015) Pollution : effects of chemical contaminants and debris. Mar Ecosyst Hum impacts biodivers Funct Serv 244. doi:10.1017/CBO9781139794763.009
Johnston EL, Mayer-Pinto M, Crowe TP (2015) REVIEW: chemical contaminant effects on marine ecosystem functioning. J Appl Ecol 52:140–149. doi:10.1111/1365-2664.12355
Johnston EL, Roberts DA (2009) Contaminants reduce the richness and evenness of marine communities: a review and meta-analysis. Environ Pollut 157:1745–1752. doi:10.1016/j.envpol.2009.02.017
Jones SE, Lennon JT (2010) Dormancy contributes to the maintenance of microbial diversity. Proc Natl Acad Sci 107:5881–5886. doi:10.1073/pnas.0912765107
Kelaher BP, Bishop MJ, Potts J et al (2013) Detrital diversity influences estuarine ecosystem performance. Glob Chang Biol 19:1909–1918. doi:10.1111/gcb.12162
Kelly J, Thornton I, Simpson PR (1996) Urban geochemistry: a study of the influence of anthropogenic activity on the heavy metal content of soils in traditionally industrial and non-industrial areas of Britain. Appl Geochemistry 11:363–370. doi:10.1016/0883-2927(95)00084-4
Kennish MJ (2002) Environmental threats and environmental future of estuaries. Environ Conserv 29:78–107

Kinsella CM, Crowe TP (2016) Separate and combined effects of copper and freshwater on the biodiversity and functioning of fouling assemblages. Mar Pollut Bull 107:136–143. doi:10.1016/j.marpolbul.2016.04.008

Kraft B, Tegetmeyer HE, Sharma R et al (2014) The environmental controls that govern the end product of bacterial nitrate respiration. Science (80) 345:676–679

Lawes JC, Clark GF, Johnston EL (2016a) Contaminant cocktails: interactive effects of fertiliser and copper paint on marine invertebrate recruitment and mortality. Mar Pollut Bull 102:148–159. doi:10.1016/j.marpolbul.2015.11.040

Lawes JC, Neilan BA, Brown MV et al (2016b) Elevated nutrients change bacterial community composition and connectivity: high throughput sequencing of young marine biofilms. Biofouling 32:57–69. doi:10.1080/08927014.2015.1126581

Lejzerowicz F, Esling P, Pillet LL et al (2015) High-throughput sequencing and morphology perform equally well for benthic monitoring of marine ecosystems. Sci Rep 5:13932. doi:10.1038/srep13932

Liikanen A, Martikainen PJ (2003) Effect of ammonium and oxygen on methane and nitrous oxide fluxes across sediment-water interface in a eutrophic lake. Chemosphere 52:1287–1293. doi:10.1016/S0045-6535(03)00224-8

Liu L, Li Y, Li S et al (2012) Comparison of next-generation sequencing systems. J Biomed Biotechnol. doi:10.1155/2012/251364

Lohbeck KT, Riebesell U, Reusch TBH (2014) Gene expression changes in the coccolithophore *Emiliania huxleyi* after 500 generations of selection to ocean acidification. Proc R Soc B 281:1–7. doi:10.1098/rspb.2014.0003

Mayer-Pinto M, Johnston EL, Hutchings PA et al (2015) Sydney Harbour: a review of anthropogenic impacts on the biodiversity and ecosystem function of one of the world's largest natural harbours. Mar Freshw Res 66:1088. doi:10.1071/MF15157

McKinley AC, Dafforn KA, Taylor MD, Johnston EL (2011) High levels of sediment contamination have little influence on estuarine beach fish communities. PLoS ONE 6: e26353. doi:10.1371/journal.pone.0026353

Meyer-Reil L-A, Köster M (2000) Eutrophication of marine waters: effects on benthic microbial communities. Mar Pollut Bull 41:255–263

Nguyen TC, Loganathan P, Nguyen TV et al (2014) Polycyclic aromatic hydrocarbons in road-deposited sediments, water sediments, and soils in Sydney, Australia: comparisons of concentration distribution, sources and potential toxicity. Ecotoxicol Environ Saf 104:339–348. doi:10.1016/j.ecoenv.2014.03.010

Nicholson FA, Smith SR, Alloway BJ et al (2003) An inventory of heavy metals inputs to agricultural soils in England and Wales. Sci Total Environ 311:205–219. doi:10.1016/S0048-9697(03)00139-6

Nielsen UN, Ayres E, Wall DH, Bardgett RD (2011) Soil biodiversity and carbon cycling: a review and synthesis of studies examining diversity-function relationships. Eur J Soil Sci 62:105–116. doi:10.1111/j.1365-2389.2010.01314.x

Nikolopoulou M, Pasadakis N, Kalogerakis N (2013) Evaluation of autochthonous bioaugmentation and biostimulation during microcosm-simulated oil spills. Mar Pollut Bull 72:165–173. doi:10.1016/j.marpolbul.2013.04.007

Noyer C, Abot A, Trouilh L et al (2015) Phytochip: development of a DNA-microarray for rapid and accurate identification of *Pseudo-nitzschia* spp. and other harmful algal species. J Microbiol Methods 112:55–66. doi:10.1016/j.mimet.2015.03.002

Oliver TH, Heard MS, Isaac NJB et al (2015) Biodiversity and resilience of ecosystem functions. Trends Ecol Evol xx:1–12. doi:10.1016/j.tree.2015.08.009

Pal A, Gin KYH, Lin AYC, Reinhard M (2010) Impacts of emerging organic contaminants on freshwater resources: review of recent occurrences, sources, fate and effects. Sci Total Environ 408:6062–6069. doi:10.1016/j.scitotenv.2010.09.026

Piola RF, Johnston EL (2008) Pollution reduces native diversity and increases invader dominance in marine hard-substrate communities. Divers Distrib 14:329–342. doi:10.1111/j.1472-4642.2007.00430.x

Proulx SR, Promislow DEL, Phillips PC (2005) Network thinking in ecology and evolution. Trends Ecol Evol 20:345–353. doi:10.1016/j.tree.2005.04.004
Puckett LJ (1995) Identifying the major sources of nutrient water pollution. Environ Sci Technol 29:408A–414A. doi:10.1021/es00009a001
Radchuk V, De Laender F, Van den Brink PJ, Grimm V (2016) Biodiversity and ecosystem functioning decoupled: invariant ecosystem functioning despite non-random reductions in consumer diversity. Oikos 125:424–433. doi:10.1111/oik.02220
Reusch TBH, Boyd PW (2013) Experimental evolution meets marine phytoplankton. Evolution (N Y) 67:1849–1859. doi:10.1111/evo.12035
Ross-Gillespie A, Kümmerli R (2014) Collective decision-making in microbes. Front Microbiol 5:54. doi:10.3389/fmicb.2014.00054
Santschi PH, Presley BJ, Wade TL et al (2001) Historical contamination of PAHs, PCBs, DDTs, and heavy metals in Mississippi River Delta, Galveston Bay and Tampa Bay sediment cores. Mar Environ Res 52:51–79. doi:10.1016/S0141-1136(00)00260-9
Scanes P, Coade G, Doherty M, Hill R (2007) Evaluation of the utility of water quality based indicators of estuarine lagoon condition in NSW, Australia. Estuar Coast Shelf Sci 74:306–319. doi:10.1016/j.ecss.2007.04.021
Schimel JP, Balser TC, Wallenstein M (2007) Microbial stress-response physiology and its implications for ecosystem function. Ecology 88:1386–1394. doi:10.1890/06-0219
Scholz-Starke B, Nikolakis A, Leicher T et al (2011) Outdoor terrestrial model ecosystems are suitable to detect pesticide effects on soil fauna: design and method development. Ecotoxicology 20:1932–1948. doi:10.1007/s10646-011-0732-z
Shade A, Peter H, Allison SD et al (2012) Fundamentals of microbial community resistance and resilience. Front Microbiol 3:1–19. doi:10.3389/fmicb.2012.00417
Sims A, Zhang Y, Gajaraj S et al (2013) Toward the development of microbial indicators for wetland assessment. Water Res 47:1711–1725. doi:10.1016/j.watres.2013.01.023
Smith JG, Brandt CC, Christensen SW (2011) Long-term benthic macroinvertebrate community monitoring to assess pollution abatement effectiveness. Environ Manage 47:1077–1095. doi:10.1007/s00267-010-9610-3
Stewart FJ, Ulloa O, Delong EF (2012) Microbial metatranscriptomics in a permanent marine oxygen minimum zone. Environ Microbiol 14:23–40. doi:10.1111/j.1462-2920.2010.02400.x
Strickland MS, Lauber CL, Fierer N, Bradford MA (2009) Testing the functional significance of microbial community composition. Ecology 90:441–451
Sun MY (2016) Impacts of anthropogenic modification on the taxonomic and functional structure of estuarine sediment communities. University of New South Wales, Australia
Sun MY, Dafforn KA, Brown MV, Johnston EL (2012) Bacterial communities are sensitive indicators of contaminant stress. Mar Pollut Bull 64:1029–1038. doi:10.1016/j.marpolbul.2012.01.035
Sun MY, Dafforn KA, Johnston EL, Brown MV (2013) Core sediment bacteria drive community response to anthropogenic contamination over multiple environmental gradients. Environ Microbiol 15:2517–2531. doi:10.1111/1462-2920.12133
Thomas T, Rusch D, DeMaere MZ et al (2010) Functional genomic signatures of sponge bacteria reveal unique and shared features of symbiosis. ISME J 4:1557–1567. doi:10.1038/ismej.2010.74
Tlili A, Berard A, Blanck H et al (2015) Pollution-induced community tolerance (PICT): towards an ecologically relevant risk assessment of chemicals in aquatic systems. Freshw Biol 1–11. doi:10.1111/fwb.12558
Tlili A, Bérard A, Roulier JL et al (2010) PO43-dependence of the tolerance of autotrophic and heterotrophic biofilm communities to copper and diuron. Aquat Toxicol 98:165–177. doi:10.1016/j.aquatox.2010.02.008
van der Linden P, Borja Á, Rodríquez JG et al (2016) Spatial and temporal response of multiple trait-based indices to natural- and anthropogenic seafloor disturbance (effluents). Ecol Indic 69:617–628. doi:10.1016/j.ecolind.2016.05.020

Westerhoff HV, Brooks AN, Simeonidis E et al (2014) Macromolecular networks and intelligence in microorganisms. Front Microbiol 5:379. doi:10.3389/fmicb.2014.00379

Zaiko A, Schimanski K, Pochon X et al (2016) Metabarcoding improves detection of eukaryotes from early biofouling communities: implications for pest monitoring and pathway management. Biofouling 32:671–684. doi:10.1080/08927014.2016.1186165

Zhu W, Bian B, Li L (2008) Heavy metal contamination of road-deposited sediments in a medium size city of China. Environ Monit Assess 147:171–181. doi:10.1007/s10661-007-0108-2

Zimmerman N, Izard J, Klatt C et al (2014) The unseen world: environmental microbial sequencing and identification methods for ecologists. Front Ecol Environ 12:224–231. doi:10.1890/130055

Chapter 9
Bacterial Community Response to Hydrocarbon Contamination in Soils and Marine Sediments: A Critical Review of Case Studies

Elpiniki Vandera and Anna I. Koukkou

Abstract The widespread distribution of petroleum hydrocarbons due to anthropogenic activities is becoming a major problem for both marine and terrestrial environments. Indigenous microbial communities with aliphatic and aromatic hydrocarbons degradation capacities play a collaborative role in the dissipation of hydrocarbon contaminants in such environments. Although the role of the uncultured microbes, which represent the majority of the members of the microbial communities in the environment, remains unknown, it is clear that our knowledge on microbial community's responses has considerably improved over the past two decades and its relevance in field studies has been demonstrated in several instances. This critical review provides an in-depth understanding of the microbial responses to oil hydrocarbons in soils and marine sediments through case studies and sheds light on the bacterial diversity from these environments mainly with the application of molecular approaches and the recent "omics" and meta-"omics" technologies. Finally, the interplay of the various factors affecting the abundance, diversity and functional composition of bacterial communities such as soil type, type and concentration of contaminants, chronic contamination, temperature, salinity, pH and biostimulation through nutrient and oxygen supplementation are reviewed.

Keywords Crude oil · Petroleum · Hydrocarbons · Alkanes · PAHs · Aged-PAHs · AlkB · PAH-RHD · Chronic pollution · Biodegradation · Bioremediation · Bacterial community structure · Bacterial community response to hydrocarbons · Soil environments · Marine environments · Bacterial biodiversity · Metagenomics · Metatrancriptomics · Metaproteomics

E. Vandera · A.I. Koukkou (✉)
Sector of Organic Chemistry and Biochemistry, Department of Chemistry,
University of Ioannina, Ioannina, Greece
e-mail: akukku@cc.uoi.gr

E. Vandera
e-mail: evandera@cc.uoi.gr

C. Cravo-Laureau et al. (eds.), *Microbial Ecotoxicology*,
DOI 10.1007/978-3-319-61795-4_9

9.1 Introduction

The onset of the industrial exploitation of petroleum in the middle of the 19th century greatly affected our civilization by promoting mass production and thus launching the growth of world economy by leaps and bounds. Yet the effects on the environment would only be seen clearly years later. The extraction of crude oil from oil reservoirs and various oil industrial activities associated with oil transportation, storage, use of petroleum-derived fuels and accidental release have resulted in the widespread distribution of petroleum hydrocarbons (Yang et al. 2015). Hydrocarbon contamination has essentially affected all environments on Earth including soil and sediments, groundwater, freshwater and oceans which are all exposed to various degrees of contamination (Greer et al. 2010).

Complex petroleum hydrocarbon mixtures, including crude oil, diesel fuel, and creosote, consist of various concentrations of *n*- and branched alkanes, cycloalkanes, phenolics, aromatics, polycyclic aromatic hydrocarbons (PAHs), asphaltenes and resins. The low molecular weight (LMW) compounds are the first to be removed by volatilization or degradation, however branched aliphatic, high molecular weight (HMW)-PAHs and heterocycles, as more recalcitrant chemical compounds, persist in the environment longer (Leahy and Colwell 1990).

The fate of hydrocarbons in soil or marine ecosystems can be determined by a combination of physico-chemical and biological processes with biodegradation, mediated by indigenous microbial communities, being the ultimate fate of a substantial amount of oil-derived hydrocarbons upon entering these habitats (Atlas 1991; Leahy and Colwell 1990; Alexander 1999; Harayama et al. 1999). Hydrocarbon degrading microorganisms are present in most environments and even though in pristine ecosystems they comprise 0.1% of the soil microbiota (Greenwood et al. 2009), or less than 5% of the total number of saprophytic bacteria in non-contaminated intertidal sediments (Delille and Delille 2000) they yet become prevalent in oil-polluted ecosystems.

Although all microorganisms (bacteria, archaea, fungi, algae, protozoa and viruses), are abundant in all ecosystems, bacteria exist in enormous numbers (about 5×10^{30} alone worldwide) and they are probably the most diverse Kingdom (Fuhrman 2009). Given the fact that one gram of soil may contain more than 10^{10} bacteria that may be constituted by 10^4–10^5 different species, or a cubic meter of sea water may contain trillions of microorganisms, including bacteria, one can understand the immense bacterial diversity in soil or marine habitats, as well as, the paucity of knowledge on its correlation to the processes mediated by these complex indigenous microbial communities, which are influenced both by the interactions between taxa and the prevailing environmental conditions (Torsvik and Øvreås 2002; Chikere et al. 2011). Taking into consideration the enormous diversity of bacteria on the one hand, and the fact that alkanes and PAHs are the most commonly detected petroleum hydrocarbon contaminants (Liang et al. 2012) on the other, this review focuses mainly on case studies regarding the bacterial community response to these contaminants in soil and marine ecosystems and their role in pollutant degradation.

Alkanes are saturated hydrocarbons varying in sizes and structures and can constitute up to 50% of crude oil, depending on the oil source. They can be linear (*n*-alkanes), cyclic (*cyclo*-alkanes) or branched (*iso*-alkanes). Due to lack of functional groups, as well as, their very low water solubility, aliphatic hydrocarbons exhibit both, low chemical reactivity and low bioavailability for microorganisms (Rojo 2009). They are also present at low concentrations in diverse non-contaminated sites because many living organisms such as plants, green algae, bacteria and animals produce them (biogenic sources) as chemo-attractants or as protecting agents against water loss, contributing to the background hydrocarbon sources in the impacted area. The chemical fingerprinting of petroleum hydrocarbons, including the distinguishing of biogenic and pyrogenic hydrocarbons (combustion derived) from petrogenic hydrocarbons (petroleum derived) has been well addressed by Wang and Brown (2009). Alkane degradation capability is a widespread function in nature (Rojo 2009; Wang and Shao 2013). Some microorganisms possess the metabolic capacity to use these compounds as carbon and energy sources for their growth (Singh et al. 2012) and numerous bacterial strains able to degrade alkanes have been isolated and characterized. Although most of them belong to the *Alpha-*, *Beta-* and *Gamma-Proteobacteria*, some members of the high G+C Gram-positive *Actinobacteria* have also been reported (Koukkou and Vandera 2011). Alkane monooxygenase AlkB, is a key enzyme in bacterial alkane aerobic degradation acting in a chain reaction with electron carriers to reduce the alkane to an alcohol which subsequently enters bacterial β-oxidation pathway (Rojo 2009; Wang and Shao 2013) (Fig. 9.1).

PAHs are amongst the most important components of crude oil, creosote, asphalt and coal tar, constituting a diverse class of aromatic compounds with two or more fused benzene rings in linear, angular or cluster structural arrangements. Their molecular stability, hydrophobicity and low water solubility are some of the main factors contributing to their persistence in the environment (Cerniglia 1992; Peng et al. 2008; Kanaly and Harayama 2010; Lu et al. 2011). There are more than 100 diverse PAH compounds and their solubility decreases and hydrophobicity increases with an increase in the number of fused benzene rings. They can also be formed as products of incomplete natural combustion sources, such as volcanic eruptions and forest fires and have been thoroughly studied due to their toxicity, mutagenicity and carcinogenicity (Haritash and Kaushik 2009). Biodegradation of PAHs is the most frequently studied degradation process described in the literature as it is considered the most appropriate way for their removal from contaminated environments (Alexander 1981; Cerniglia 1993; Daane et al. 2001). While many bacterial strains with the ability to degrade LMW-PAHs have been isolated and characterized, much less are the isolates with the ability to degrade HMW-PAHs (5 or more rings). Most of them belong to the genera *Acinetobacter*, *Burkholderia*, *Gordona*, *Mycobacterium*, *Pseudomonas*, *Rhodococcus* and *Sphingomonas*, *Stenotrophomonas* (Juhasz and Naidu 2000; Koukkou and Vandera 2011). The

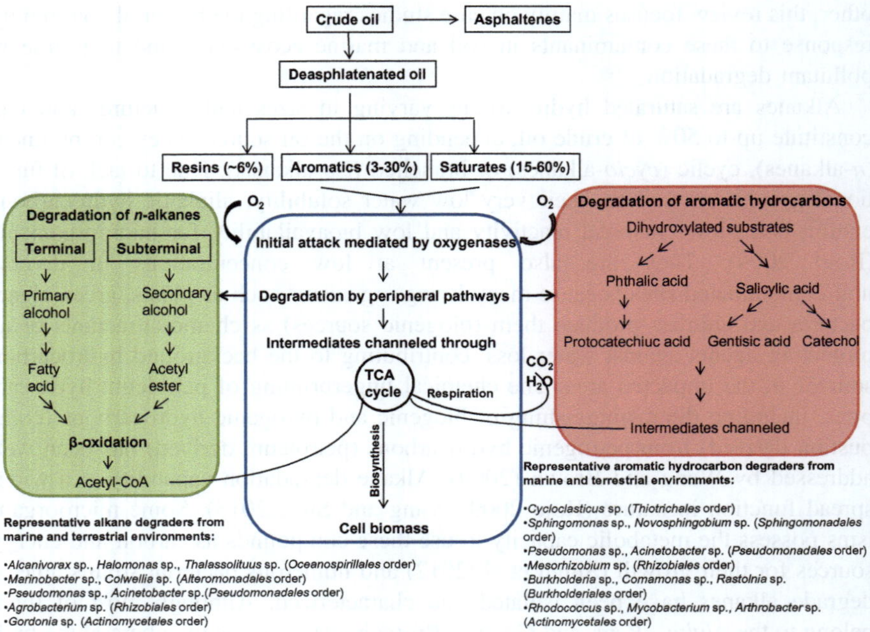

Fig. 9.1 Main principle of aerobic degradation of aromatic hydrocarbons and *n*-alkanes by representative microorganisms. *Single arrows* indicate one-step reactions, whereas *double arrows* indicate two-or more steps reactions

aerobic aromatic hydrocarbon degradation is initiated by the incorporation of both oxygen atoms of an O_2 molecule into the substrate mediated by the aromatic Ring-Hydroxylating Dioxygenases (RHDs) (Cerniglia 1984; Heitkamp et al. 1988) Subsequently, a key step is the opening of the hydroxylated aromatic ring, which is catalyzed by aromatic-ring-cleavage dioxygenases (see review by Peng et al. 2008 for complete pathway) (Fig. 9.1).

The ubiquitous distribution of hydrocarbons from both anthropogenic and natural sources including biogenic sources throughout the environment, has led to the widespread capacity of microbial communities to degrade hydrocarbons even at uncontaminated sites (Couling et al. 2010). Generally, alkanes are more easily degraded in the environment than PAHs. The susceptibility of hydrocarbons to microbial attack however, differs significantly-depending on their size and structure, and hydrocarbons have been ranked in an order of decreasing susceptibility, as follows: *n*-alkanes > branched alkanes > low molecular weight aromatics > cyclic alkanes > high molecular weight aromatics (Leahy and Colwell 1990). The metabolic pathways for the degradation of hydrocarbons can be either aerobic (oxygen is utilised as the primary electron acceptor), or anaerobic (alternative electron acceptors are utilised, such as nitrate and sulphate). Nevertheless, aerobic degradation proceeds more rapidly and is considered to be more effective (Das and

Chadran 2011; Olajire and Essien 2014) therefore in Nature, the biodegradation of hydrocarbons mainly occurs via aerobic processes. The description of the biodegradation pathways of aliphatic and aromatic hydrocarbons is beyond the scope of this chapter and therefore readers are referred to a number of reviews on this topic such as those by Peng et al. (2008), Haritash and Kaushik (2009), Rojo (2009), Obayori and Salam (2010), Cuny et al. (2011) and Singh et al. (2012).

The objective of the present chapter is an approach towards a better understanding of how bacterial communities respond to hydrocarbon contamination along with the factors that influence these responses in order to elucidate their role in biodegradation processes. Knowledge about community functioning is essential to understand the performance management during remediation of soils or marine sediments. In order to determine which microbes play the crucial role in the hydrocarbon biodegradation in these environments, the results from culture-dependent methods should be combined with the data from culture-independent methods, which indicate the qualitative and quantitative importance of particular organisms in situ (Head et al. 2006).

It is now generally accepted that no single species will completely degrade any complex class of hydrocarbons and that indigenous microbial communities with overall broad enzymatic capacities play a collaborative role in the dissipation of hydrocarbon contaminants (McGenity et al. 2012; Fuentes et al. 2014). Traditionally, characterization of microbial community composition in contaminated environments has been limited to the 1% cultivable microorganisms, because more than 99% of these cannot be cultured with readily available technologies. Strains that are fast growing and easily adaptable to particular conditions grow preferentially to less adaptable ones and therefore an accurate representation of the actual microbial community structure is not feasible (Rappé and Giovannoni 2003). On the other hand, although the only way to study the physiology and the actual bacterial degradation pathways is the culturing of isolated strains, intermicrobial competition appears to play an important role in shaping microbial activity and abundance (Stefani et al. 2015) and therefore, the isolated strains might not display the same characteristics under laboratory conditions as they do within their natural environment (Siggins et al. 2012). To investigate the diverse roles played by microbes, for their utilization in bioremediation processes, the study of mixed microbial communities is needed. Many culture-independent approaches for in situ monitoring of microbial dynamics and composition during hydrocarbon bioremediation have been reported, including PLFA, DGGE, ARDRA, T-RFLP, FISH (for review see Malik et al. 2008; Fuentes et al. 2014). Many of these approaches are based on known gene sequences which have been identified through culturing isolated strains, thus limiting our view of the extent of potential community function. Recent advancements in high-throughput 'omics' methods such as metagenomics, metatranscriptomics, metaproteomics, have viewed mixed microbial communities as one meta-organism, allowing the in-depth characterisation of the diversity and the function of microbial communities (for review see Röling et al. 2010; Liu et al. 2013). In addition, the recent integration of stable isotope probing techniques (SIP) with metagenomics has enabled a more comprehensive understanding of the functional community dynamics of entire microbial systems (Uhlik et al. 2013). However, all the above approaches have their limitations, as with any approach (Table 9.1). It is important to note that cultivation-based techniques are still a crucial means to verify and investigate the physiology and genetics of individual contaminant-degrading microorganisms and the challenge now is to integrate the multi-omic data with a strategy which begins at the experimental design phase, by incorporating our knowledge from uncultured and cultured genome sequences (Uhlik et al. 2013; Cravo-Laureau and Duran 2014; Garza and Dutilh 2015; Franzosa et al. 2015).

9.2 Case Studies of Bacterial Community Responses to Hydrocarbon Perturbation in Soil and Marine Ecosystems

Hydrocarbon pollution has been shown to be the main driving force for changes in microbial community structure as it can lead to an increase or a decrease of microbial biomass if the hydrocarbons are used as carbon sources or not, respectively (Wentzel et al. 2007). The response of the microbial communities to the presence of hydrocarbons in association with their biodegradation ability is difficult to be clarified, because our knowledge on the mechanisms involved is still limited. Bacterial communities with similar composition showed different rates of oil degradation, while communities with highly different composition showed similar rates and extents of degradation (Röling et al. 2004). The biodegradation ability of a microbial community may be the result of a distribution between several functional groups of microorganisms and not only of the metabolism of microorganisms that degrade the pollutants (Head et al. 2006). In the environment, the distribution of components in a petroleum hydrocarbon mixture is subject to change due to weathering effects such as volatilization, degradation, oxidation, thus, resulting in changes in the bioavailability of the contaminants, which trigger a series of changes in the bacterial community structure (Acosta-González et al. 2015). However, after hydrocarbon pollution, bacteria belonging to the phyla of *Actinobacteria*, *Proteobacteria* and *Firmicutes* and genera including *Gordonia, Rhodococcus* and *Nocardia* (*Actinobacteria*); *Sphingomonas*, *Acinetobacter*, *Alcanivorax*, *Pseudomonas*, *Caulobacter*, *Cycloclasticus*, *Marinobacter*, *Rhodospirillum*, *Comamonas*, *Burkholderia*, *Neptunomonas*, *Oleiphilus*, *Oleispira*, *Thalassolituus* and *Xanthomonas* (*Proteobacteria*); *Bacillus* and *Exiguobacterium* (*Firmicutes*) constitute the major hydrocarbon degraders and are usually enriched (Kasai et al. 2002; Harayama et al. 2004; Viñas et al. 2005; Brooijmans et al. 2009; McGenity et al. 2012; Yergeau et al. 2012; Bell et al. 2013; Fuentes et al. 2014; Yang et al. 2014c).

In general, aerobic hydrocarbon biodegradation in soils or marine environments depends on the indigenous microbial community structure that is influenced by several factors including the soil type, the type and concentration of contaminant,

Table 9.1 "OMICS" approaches for monitoring microbial community in the environment (mixed microbial communities are viewed as one meta-organism)

Approach	Advantages	Disadvantages	References
Metagenomics	Identification of the functional potential and taxonomic identification of all members of the community present in a environmental sample, without isolating organisms	• Environmental samples: co-extraction of inhibitors, degradation of DNA during procedures may influence the microbial community structure • Include numerous genes of unknown function • Difficult to connect specific microbial phyla to specific functions • No identification of actual active members of community • Unable to differentiate between genes that are actively expressed and genetic potential • Computational and/or Bioinformatic problems	de Menezes et al. (2012), Simon and Daniel (2011) and Sharpton (2014)
Metatranscriptomics	Allows functional insight into microbial populations, including non-cultivable members	• Environmental samples: difficult to extract undegraded mRNA • Lack of correlation between mRNA levels and protein levels • Short time life of mRNA/difficult detection in rapid responses to environmental changes • Insufficient transcriptome databases • Crucial to enrich for mRNA by depleting rRNA before metatranscriptomic sequencing	Franzosa et al. (2015), Gilbert et al. (2008), de Menezes et al. (2012), Singh and Nagaraj (2006) and Siggins et al. (2012)

(continued)

Table 9.1 (continued)

Approach	Advantages	Disadvantages	References
Metaproteomics	Identification of translated proteins at a given time, direct measure of the functional activity of the community without isolating organisms	• Soil samples: lack of effective methods compatible with proteomic techniques for protein extraction, co-extraction with compounds that interfere with proteins • Marine context: problematic sample collection and recovery of the targeted fraction • Improved bioinformatics tools are needed for identification of proteins from unsequenced microorganisms and/or microbial communities • Cannot be viewed as an isolated method, it benefits from genome/metagenome sequencing for protein identification • Molecular and biochemical studies are needed to validate the protein functions	Bastida et al. (2016), Franzosa et al. (2015), Zarraonaindia et al. (2013), Kim et al. (2009), Lacerda and Reardon (2009) and Siggins et al. (2012)

the chronic contamination and long-term effects of contaminants, temperature, salinity, pH and availability of nutrients and oxygen, that will be discussed below in detail.

Impact of soil type Soil physical heterogeneity such as textures, ranging from loam to clay, and different organic matter appear to be the main soil properties which influence hydrocarbon availability, microbial community response to hydrocarbon contamination and consequently affect the biodegradation abilities of the community.

Soil texture affects the permeability and bulk density of soil (Sihag et al. 2014), as well as microbial localization. Indeed, soil fractionation studies have shown that microbial biomass is strongly associated with clays (Taylor et al. 2002). For instance, it has been reported that the remaining PAHs (low and high molecular weight) are much more concentrated in the fine soil fractions (fine silt and clay), which house most PAH degraders and this is where the greatest PAH mineralization in soil occurs (Amellal et al. 2001; Bengtsson et al. 2010). Hamamura et al. (2006) examined different soil types from loam to clay from geographically distinct sites and demonstrated that the bacterial diversity and the structure of the community were more influenced by soil type than by geographical distance. A soil type-dependent response of PAH ring-hydroxylating dioxygenase (PAH-RHDα, catalyses the first step of aerobic degradation of PAHs)-carrying soil microorganisms to phenanthrene amendment, has also been observed. Two unpolluted soils with differences in texture but with similar organic matter, Cambisol (clay loam) and Luvisol (silt loam) were spiked with phenanthrene and incubated for a period of 63 days. The diversity of PAH-RHDα genes in Luvisol was low, with amplicons sharing the highest identity with *phnAc* gene from *Burkholderia* sp. strain. On the contrary, PAH-RHDα genes in Cambisol showed a much higher diversity, with amplicons deriving from Gram-positive *Mycobacteria* and *Rhodococcus* spp., *Alpha-Proteobacteria* (predominantly of the genus *Agrobacterium* and *Novosphingobium*) and *Beta-Proteobacteria* (mainly of the genus *Burkholderia* and *Acidovorax*) (Ding et al. 2010). Furthermore, Ding et al. (2012a) investigated how the bacterial communities change after phenanthrene spiking by analyzing 16S rRNA gene fragments amplified from total community DNA using DGGE. The authors stressed that despite the significant differences in bacterial community structure in both soil types initially, similar genera (especially *Sphingomonas* and *Polaromonas*) increased in relative abundance after spiking. Members of these genera have previously been reported for their PAH degradation capability (Sun et al. 2010; Jones et al. 2011). Indeed, many more functional genes (158 out of the 519 detected) involved in the degradation of aromatic compounds were detected by GeoChip in phenanthrene spiked Luvisol compared to the control, in contrast to the only eight functional genes with significantly higher signal intensity in spiked Cambisol (Ding et al. 2012b). The authors attributed this to the lower clay content of Luvisol that may make the exchange of O_2 into the soil matrix easier, favoring the aerobic populations. Nevertheless, the functional gene structure in both soils converged after spiking and the higher number of functional genes detected by GeoChip in both soils was attributed to the selection of certain populations and their subsequent improved detection (Ding et al. 2012b).

In contrast, Liang et al. (2011) comparing the functional gene diversity in different oil-contaminated fields from five distinct geographic regions in China using GeoChip, reported that even if the overall functional gene patterns differed between all soils, they shared similar organic remediation gene patterns. Aliphatic hydrocarbon degradation genes from *Burkholderia cepacia* and *Rhodococcus* sp. were reported, as well as, aromatic degradation genes from *Xanthomonas campestris*, *Pseudomonas* sp., *Comamonas testosteroni*, *Thauera aromatica*, *Rhodococcus* sp., *Nocardioides* sp., *Mycobacterium* sp., *Burkholderia* sp. and *Mesorhizobium* sp. (Liang et al. 2011). The effect of soil physical heterogeneity on aerobic biodegradation of petroleum vapors was studied on slurry experiments by Kristensen et al. (2010) indicating that the biodegradation potential in different textured soil samples was related to soil type in the order: sandy loam > fine sand > limestone. The air-filled porosity was demonstrated to be a key factor for the intrinsic biodegradation potential in the field. Texture influences the porosity (sandy soils contain mostly large pores) and the degree of soil compaction which in turn, influences the movement and availability of water in the soil (Gebre et al. 2015). So the decreasing of the biodegradation in slurries of the sandy loam was associated with the increasing in situ water saturation (Kristensen et al. 2010). The efficiency of petroleum hydrocarbon bioremediation in different soils from 100% sandy soil to 40% clay was also affected significantly by the soil type with the lowest removal appearing in the clay soil (Haghollahi et al. 2016).

Although microbial communities may locally be highly diverse, Ramette and Tiedje (2007), studying the effects of spatial distance and environmental heterogeneity in microbial communities variations, demonstrated that these variations are most likely related to more contemporary environmental conditions. The association of the distribution of bacterial community composition with local factors, such as soil type, than to more global factors, such as climate or geomorphologic characteristics, has also been reported by others (Dequiedt et al. 2009), whereas Tiedje et al. (2001) demonstrated that soil microbes are basically endemic (geographically unique).

The effect of organic carbon content in different soil types from three distinct areas of the USA was examined by Hamamura et al. (2013), when other parameters such as the water potential, nutrients, temperature and aeration were held constant, and was found to be an important determinant of community responses to hydrocarbon perturbation. In the soil with the lowest organic carbon content, lower biodegradation activities appeared with diesel and kerosene amendments, possibly due to their compounds being less prone to partition in the solid phase, therefore increasing the bioavailable hydrocarbons-especially short-chain *n*-alkanes with higher aqueous solubility, and thus resulting in greater toxic effects to microbiota (Hamamura et al. 2013). Organic matter content is not only an important factor affecting the sorption of nonpolar organic chemicals, but it has also been shown to be fundamental to the development of microbial diversity in soil (Girvan et al. 2005). In general, *Actinobacteria* dominate in soils with organic matter <10%, whereas *Proteobacteria* dominate higher organic matter soils (Bell et al. 2013). Moreover, it has been found that the activity of mycobacterial PAH degraders coincides with the distribution of total organic carbon in soils (Bengtsson et al. 2010). A soil with a diverse organo-mineral composition is associated with a greater microbial resistance and resilience to perturbation stresses (Girvan et al. 2005). Similar results reported by Labud et al. (2007) support this hypothesis. They studied the toxicity through bioassays, microbiological and biochemical parameters of two soils differing in their clay and organic matter content, contaminated by three types of hydrocarbons differing in their volatilization and toxicity (gasoline, diesel oil, crude petroleum). Once again, higher organic matter and clay content of the soil resulted in a lower inhibition of microbial population size. The contaminants seem to adsorb to the clay and organic matter, decreasing their concentration in the aqueous phase and resulting in a decrease in their toxic effect on soil microorganisms (Labud et al. 2007).

Impact of concentration and type of contaminants Hydrocarbons are mainly present as complex mixtures both in terrestrial and marine environments and the microbial communities may be influenced by the complexity, concentration, as well as, the time of exposure to the contaminant mixtures present. Microbial communities within contaminated ecosystems tend to be dominated by the organisms capable of consuming and/or tolerating toxic organic compounds (Yakimov et al. 2005). Alkanes of shorter chain length (C_{10}–C_{20}) have been reported as more toxic than those of larger chain length (C_{20}–C_{40}), as well as HMW PAHs which were found difficult to degrade due to their low aqueous solubility and bioavailability (Fuentes et al. 2014; Guermouche M'rassi et al. 2015). In general, culture-dependent and culture independent methods have indicated that three bacterial groups are mainly responsive to oil hydrocarbon soil pollution *Proteobacteria*, *Actinobacteria* and *Firmicutes* (Fuentes et al. 2014; Morais et al. 2016), while a general trend for an increase of *Proteobacteria*, *Firmicutes*, *Actinobacteria* and *Bacteroidetes* has been reported for marine environments after oil spills (Acosta-González et al. 2015; Acosta-González and Marques 2016).

Hamamura and co-authors examined different petroleum mixtures (crude oil, kerosene, diesel, diesel plus PCP) and soil types from three distinct areas, spiked with ^{14}C-hexadecane or ^{14}C-tridecane. The changes in soil bacterial populations following contamination with hydrocarbons mixtures were monitored using 16S rRNA gene targeted DGGE. Different hydrocarbon mixtures selected both unique and common bacterial populations across all three soils, indicating that petroleum mixture type influenced hydrocarbon degradation and therefore bacterial population selection. In specific, a known alkane-degrading actinobacterium, *Rhodococcus erythropolis* was observed in all mixture-type amendments, while, the alkane-degrading *Burkholderia* populations were not found in crude oil and *Pseudomonas* populations were only observed in diesel amendment, indicating a shift to more Gram-negative populations in shorter chain-length alkane mixtures (diesel and kerosene) compared to the emergence of Gram-positive populations in crude oil amendments (Hamamura et al. 2013). Viggor et al. (2013) also studied the impact of different types of oil (crude oil, shale oil or diesel fuel) on microbial species composition in Baltic Sea coastal seawater microcosms. Analysis of these microcosms was performed using alkane monooxygenase (*alkB*) and 16S rRNA marker genes in PCR-DGGE experiments. All the tested oil products and their concentrations had a substantial impact on microbial community composition as well as on the diversity of *alkB* genes, mainly during the early stage of the experiment (after 3 weeks). *Gamma-Proteobacteria* (particularly the genus *Pseudomonas*) and to a lesser extent *Alpha-Proteobacteria* were dominant in all microcosms treated with oils, while *Alkanivorax*-like sequences were detected at the end of the experiment. *Beta-Proteobacteria* become abundant only in diesel fuel spiked microcosms, whereas they have not been reported as the dominant microorganisms in the original seawater community (Viggor et al. 2013). Microcosms with shale oil that contain more recalcitrant substrates (alkene content 7%, high-polar compounds >10%), were dominated by a decreased number of bacterial phylogenetic types (phylotypes) as compared to other microcosms and were dependent on the shale oil concentration. These observations are in accordance with other studies where a reduction of marine microbial community by oil spills was also reported, due to strong selection for a limited number of hydrocarbon-degrading species (Röling et al. 2002; McKew et al. 2007a; Vila et al. 2010).

Mono- and dioxygenase encoding genes (e.g. *alkB*, RHDα), which are responsible for hydrocarbon degradation, because of their phylogenenetic distribution and high sequence divergence, have also been used in many studies as a molecular marker in hydrocarbon biodegradation, where it has been shown that changes in substrate availability and quality, as well as other environmental conditions, could enrich specific members resulting in distinct spatial patterns (Fuentes et al. 2014; Smith et al. 2013). Powell et al. (2006) demonstrated that the proportion of *alkB* containing microorganisms was positively correlated to the level of *n*-alkanes and the presence of alkane monooxygenases with low similarity compared to those previously described for Gram-positive bacteria such as *Mycobacterium*, *Gordonia*, *Rhodococcus* and *Aeromicrobium* was also reported (Jurelevicius et al. 2012b). Total petroleum hydrocarbons (TPH) levels influenced the spatial distribution of the alkane-degrading bacteria, probably because the TPH content could be toxic for some alkane-degrading bacteria, whereas others could use it as carbon source. In contrast, *Gamma-Proteobacterial* AlkB was found to be the dominant phylotype in sediments of Timor Sea, a region where natural seeps are sources of widespread petroleum hydrocarbons, while *alkB* sequence diversity did not appear to be influenced by the level of measured alkanes. Moreover, AlkB diversity varied substantially depending on the depth of sediment (greater diversity in depth <100 m) (Wasmund et al. 2009). *AlkB* genes were found to be the most prevalently expressed at high alkane concentrations in comparison to genes involved in the degradation of aromatic hydrocarbons (Mason et al. 2012; Beazley et al. 2012; Rivers et al. 2013). Gamma-Proteobacterial AlkB (*Alcanivorax* and *Marinobacter*) has also been reported as the dominant phylotype in bacterioplankton of the northern Gulf of Mexico (Smith et al. 2013), suggesting their abundance in many natural marine environments.

Several studies have shown that an increase in PAH concentration or the addition of PAHs result in the abundance of the PAH-RHD functional genes and the modification of the PAH-degrading bacterial diversity (Bengtsson et al. 2013;

Zhang et al. 2013). Enrichment of a coal-tar contaminated soil with different PAHs (naphthalene, phenanthrene, pyrene) resulted in the increase of the relative abundance of a variety of PAH dioxygenase genes (*nidA*, *pdo*, *nidB3*, *ndoB*, *phdB* and *nahB*) which were related to phylotypes of the *Alpha-*, *Epsilon-* and *Gamma-Proteobacteria*, *Acidobacteria*, *Actinobacteria*, *Firmicutes*, *Gemmatimonadetes* and *Deinococci* (Kumar and Khanna 2010). In diesel oil contaminated soil the presence of PAH-RHDα coding genes from both Gram-positive and Gram-negative bacteria was observed (Jurelevicius et al. 2012a), whereas, in phenanthrene and fluoranthene amended soil, an increase in the abundance of Gram negative PAH-RHDα genes was observed during degradation of contaminants (Sawulski et al. 2014). Because PAH degraders may possess functional genes not detected by existing primers, metatranscriptomic and metaproteomic approaches could provide a better correlation between soil microbial diversity and function. Recently, a metatranscriptomic microcosm study monitors the microbial community responses in soil from a former timber treatment facility amended with phenanthrene. The authors observed a rise in actinobacterial transcript numbers in phenanthrene-amended soil, largely represented by the suborders of *Corynebacterineae* and *Micrococcieneae* and particularly by members of the genera *Mycobacterium* and *Arthrobacter*. Transcript numbers associated with aromatic metabolism, such as dioxygenases which are involved in the initial steps of phenanthrene degradation, and genes associated with stress responses and detoxification, increased on exposure to phenanthrene, whereas transcripts associated with general metabolism were much less affected by phenanthrene amendment. Most dioxygenase transcripts belonged to the *Actinobacteria*, key Gram-positive organisms in soil, whereas only few of the dioxygenases transcripts identified belonged to the *Beta-* and *Gamma-Proteobacteria* (de Menezes et al. 2012). Phenanthrene has been used as a model PAH compound in several studies monitoring the soil and marine bacterial community structure (Table 9.2), but depending on the PAH compounds the response of the microbial community differ.

Enrichment of the bacterial population in the presence of model PAHs consisting of two, three and four rings revealed changes in both the microbial community structure and the dioxygenase gene profile, suggesting the selection of PAH-specific bacterial communities (Ni Chadhain et al. 2006). The soil bacterial community was the most diverse with phenanthrene followed in order by pyrene and naphthalene. The higher diversity in the presence of phenanthrene was possibly due to the prior adaptation of the bacterial community because the original soil contained relatively high phenanthrene concentrations. The enrichment of specific populations in PAH-treated soils has also been reported by others (Cheung and Kinkle 2005; Leys et al. 2005). The presence of low molecular weight PAHs was not reported as a significant driver of overall microbial community change comparing to the more toxic and recalcitrant PAHs (Muckian et al. 2009). Nevertheless, phenanthrene- and fluoranthene-affected communities responded rapidly to the contaminant, in contrast to the high molecular weight PAH-affected communities which responded later, reflecting the pattern of degradation of the different PAHs; phenanthrene and fluoranthene were degraded rapidly compared to HMW-PAHs (Sawulski et al. 2014;

Table 9.2 Phenanthrene used as a model PAH compound in indicative studies monitoring soils and marine bacterial communities

Habitat/geographic location	Soil or marine ecosystem	Specific treatment	Initial contamination	Duration of treatment	Techniques to monitor community shifts	Dominant members in contaminated microbial community	References
Ultuna, Sweden or Farm in Scheyern, Munich, Germany	Cambisol or Luvisol	Phenanthrene spiking 20 mg/g soil	None	0, 21, 63 days	PAH-RHDa gene amplicons sequences	Cambisol soil: Low diversity of PAH-RHDα genes/ *Burkholderia* Luvisol soil: *Actinobacteria* (*Mycobacterium, Rhodococcus*), *Beta-Proteobacteria* (*Burkholderia, Acidovorax*), *Alpha-Proteobacteria* (*Agrobacterium, Novosphingobium*)	Ding et al. (2010, 2012a, b)
		Phenanthrene spiking 2 mg/g soil			DGGE band sequences	After 21 days *Beta-Proteobacteria* (*Sphingomonas*) were detectable in both soils After 63 days there were significant responses of *Alpha-Proteobacteria* (*Polaromonas*) and *Actinobacteria* in both soils	
					GeoChip	Cambisol soil: One-ring aromatic compound genes *Rhodococcus* sp., *Streptomyces* sp., *Pseudomonas* sp. Luvisol soil: Dioxygenases genes *Pseudomonas* sp., *Ralstonia* sp., *Neptunomonas* sp., *Mycobacterium* sp.	

(continued)

Table 9.2 (continued)

Habitat/geographic location	Soil or marine ecosystem	Specific treatment	Initial contamination	Duration of treatment	Techniques to monitor community shifts	Dominant members in contaminated microbial community	References
Microcosms Soil (0–15 cm) from former timber treatment facility (Ireland)		Phenanthrene spiking 725 mg/kg soil	Low level PAHs (4.7 mg/kg) (no detectable phenanthrene)	17 days	Metatranscriptomics	Rise in actinobacterial (*Micrococceae* and *Corynebacterineae*) transcript numbers/*Mycobacterium, Arthrobacter* Reduction in deltaproteobacterial (*Myxococcales*) transcript numbers	de Menezes et al. (2012)
Two different sites	Grassland soil Composting site soil	Phenanthrene spiking 100 µg/g soil	None	30 days	TRFLP and 16S rRNA gene library	Major shift in: *Bacteroidetes*, *Alpha-*, *Beta-*, *Gamma-* and *Delta-Proteobacteria* *Beta-Ptoteobacteria*, *Acidobacteria*	Zhang et al. (2011)
	Long term matured artificial soils with different clay minerals	Phenanthrene spiking 2 mg/g soil	None	70 days	DGGE band sequences	Montmorillonite: *Beta-Proteobacteria* Ferrihydrite, aluminum hydroxide: *Actinobacteria* Montmorillonite + charcoal: *A. polychromogenes* Illite: *P. dioxxanivorans*	Babin et al. (2013)
Former timber preservation facility, Ireland	Surface soil	Phenanthrene spiking 200 mg/kg soil	Low level PAHs (4.7 mg/ kg)	20 days	ARISA and TRFLP fingerprinting, 16S rRNA gene library	Enrichment of *Acidobacteria, Gamma-Proteobacteria* Decrease of *Actinobacteria*	Sawulski et al. (2014)

(continued)

Table 9.2 (continued)

Habitat/geographic location	Soil or marine ecosystem	Specific treatment	Initial contamination	Duration of treatment	Techniques to monitor community shifts	Dominant members in contaminated microbial community	References
PAH-impacted site, Somerset, NJ	Medium loam soil	Phenanthrene spiking 5 mg/g soil	PAHs	85 h	DGGE, 16s RNA gene library, ARDRA	Enriched by *Gamma-Proteobacteria* (*Nitrosococcus*, *Stenotrophomonas*), *Beta-Proteobacteria* (*Massilia*)	Ni Chadhain et al. (2006)
Yvetot, Upper Normandy, France	Silty soil	Phenanthrene spiking 300 mg/kg	low level PAHs (<1 mg/kg)	90 days	A-RISA fingerprinting, qPCR on 16S rRNA, PAH-RHDα	After 60 days in soils spiked with phenanthrene there was an abundance of *Actinobacteria* (*Mycobacterium* genus), *Firmicutes*, *Beta-Proteobacteria*, *Gamma-Proteobacteria* (*Burkholderia* genus)	Niepceron et al. (2014)
Lagoon Etang de Berre, France	Mats	Phenanthrene spiking 100 mg/l	Oil refinery waste	30 days	DGGE	*Sphingomonas, Paracoccus, Xanthomonas*	Hernandez-Raquet et al. (2006)
Baltic sea, Sweden	Sediments	Phenanthrene spiking 10 μg/ml	PAHs	60 days	T-RFLP	Dominated: *Exiguobacterium, Schewanella, Methylomonas, Pseudomonas, Bacteroides,* an uncultured Delta-Proteobacterium and a Gamma-Proteobacterium	Edlund and Jansson (2008)
Carnota, Spain	Oiled sand (OS) Oiled rock (OR)	Phenanthrene 0.1%	Prestige oil (>52% PAHs)	15 days	Culture depended methods	OS: *Gamma-Proteobacteria* 67% *Alpha-Proteobacteria* 33% OR: *Gamma-Proteobacteria* 18% *Alpha-Proteobacteria* 82%	Alonso-Gutiérrez et al. (2009)

(continued)

Table 9.2 (continued)

Habitat/geographic location	Soil or marine ecosystem	Specific treatment	Initial contamination	Duration of treatment	Techniques to monitor community shifts	Dominant members in contaminated microbial community	References
Scharlab, Barcelona, Spain	Artificial seawater	Phenanthrene 0.2 g/l	Prestige oil (>52% PAHs)	15 days	DGGE, 16S rRNA analysis	Mainly *Alpha-Proteobacteria* *Marinobacter* (Gamma-Proteobacterium)	Vila et al. (2010)
DWH, Gulf of Mexico	Sea water	Phenanthrene agar plate sprayed	Oil contaminated	15 days	SIP, 16s rRNA gene library,	*Cycloclasticus* 66% *Colwellia* 13%	Gutierrez et al. (2013)

Muckian et al. 2009). Moreover, Muckian et al. (2007) found by TRFLP and sophisticated multivariate statistical analysis that PAH ring structure rather than total PAH concentration played the major role in bacterial community response in soil samples. Similarly, Vila et al. (2010) demonstrated distinctive bacterial communities composition in seawater cultures inoculated with a marine microbial consortium from a beach affected by oil spill, using different PAHs (phenanthrene, fluoranthene, benzo[a]anthracene, pyrene) as carbon sources.

Microbial community composition has a heterogeneous spatial distribution, mostly depending on the spatial distribution of PAHs, which were spatially autocorrelated and appeared in several hot spots (Törneman et al. 2008; Bengtsson and Törneman 2009). The difference in spatial distributions of individual PAHs may depend on their lipophilicity (Bengtsson and Törneman 2009). Pyrene exhibits stronger sorption to the soil organic matter (higher K_{ow}) and lower diffusion rates (lower solubility) compared to phenanthrene, consequently, the flux of pyrene to pyrene-degraders is limited, leading to fewer cells and slower growth and degradation rates (Johnsen and Karlson 2005). A PAH such as pyrene, is degraded by specialized degraders which show low diversity and usually belong to the *Actinobacteria* with *Mycobacterium* spp. being by far the most commonly reported HMW-PAH degraders (Karabika et al. 2009; Kanaly and Harayama 2000). This specialization may lead to increased spatial variation (Hybholt et al. 2011), although it has been reported that both PAH degraders and PAHs were omnipresent at the mm-scale in diffusely polluted urban soil (Johnsen et al. 2014).

Bacterial community structure of marine sediments is also characterized by high dissimilarity between sites, although studies comparing the global diversity in these environments are limited (Acosta-González and Marques 2016). Newton et al. (2013) examined the range of temporal and spatial variation in correlation with bacterial community stability or disturbance impacts, in response to oil perturbation of marine environments. Despite the lesser spatial and larger temporal variation of microbial community in water samples, community similarity of sandy samples decreased as horizontal distance increased from one to hundreds of meters. In contrast, several relevant phylotypes in the communities from two distinct, highly distant sites, were found to be identical or closely related, suggesting that the observed response was likely to be a common pattern in the extended polluted area (Jiménez et al. 2007; Alonso-Gutiérrez et al. 2009). Because the chemical nature of the oil hydrocarbons differs between the spills due to weathering effects that are variable depending on where the contaminants are deposited, the responses of the bacterial communities were depended on the bioavailability of the individual components of the spill and they were shifted over time, space or depth [extensive and comprehensive reviews of the oil spills impacts on microbial community composition and on microbial activity across ecosystems over time have been provided by Joye et al. (2014), Kimes et al. (2014), King et al. (2015), Acosta-González et al. (2015) and Acosta-González and Marques (2016)]. In general, the major trend of the communities changes is an initial increase of *Gamma-Proteobacteria* (mainly *Alcanivorax*, *Marinobacter*, *Oleispira*-known hydrocarbonoclastic groups) when more reactive components such as *n*-alkanes abound (Kostka et al. 2011; Mason et al. 2012; Kappell et al. 2014), but are subsequently succeeded by *Alpha-Proteobacteria* and *Actinobacteria* when recalcitrant oil hydrocarbons predominate (Jiménez et al. 2007; Alonso-Gutiérrez et al. 2009; Kostka et al. 2011; Acosta-González et al. 2013).

Impact of fresh versus long-term contamination Studies on microbial community responses focus on relatively short time periods (Margesin et al. 2003b; Viñas et al. 2005; Lors et al. 2010; Ding et al. 2012a) and on soils in which PAHs are still available, however, the long-term effects of PAH contamination are less studied. Hydrophobic contaminants, such as PAHs, are largely adsorbed to soil contributing to their persistence (Leahy and Colwell 1990). The resistance of PAHs to desorption in soils and sediments is increased dramatically with time (aged PAHs), thereby, becoming less extractable and bioavailable to microorganisms (Alexander 2000; Huesemann et al. 2004).

Microbial ccommunities exposed to petroleum hydrocarbon contamination undergo selective enrichment and genetic changes which result in increased proportions of hydrocarbon-degrading bacteria, therefore being able to respond to the presence of pollutants in a short period of time (Margesin et al. 2003b; Head et al. 2006; Chikere et al. 2011). A number of studies observed a reduction in bacterial richness and diversity resulting from short-term effects of oil contamination in microbial communities (Viñas et al. 2005; Ding et al. 2012a; Guo et al. 2012; Ribeiro et al. 2013), whereas other studies determined an increase in microbial diversity (Juck et al. 2000; Kaplan and Kits 2004; dos Santos et al. 2011).

On the other hand, communities previously exposed to hydrocarbon contamination exhibit higher biodegradation rates than communities with no history of contamination, indicating that the environmental impact of hydrocarbon contamination depends on the adaptive status of the microbial community in these sites (Leahy and Colwell 1990; Al-Wasify and Hamed 2014). Cébron et al. (2014) studied the bacterial community structure in a short-time experiment (6 days, phenanthrene spiking sand) using as inoculum pre-adapted bacteria from an aged PAH and heavy metal-contaminated soil. The dominant strains found were of the genera *Pseudomonas* and *Paenibacillus* that are fast-growing bacteria, known as *r*-strategists, whereas the phenanthrene degradation activity decreased due to the toxic concentration of phenanthrene used. The short-term effect of phenanthrene on microbial community pre-adapted to the pollutant, affected neither the density nor the identity of dominant bacteria but had an impact on the metabolic and functional properties of bacteria, as detected by different proteomic approaches. Similarly, microbial communities pre-adapted to oil spill, as a result of chronic pollution [such as the Deepwater Horizon (DWH) oil spill into the Gulf of Mexico] responded very fast and the substantial changes which occurred, were in correlation with the hydrocarbon contamination level (Acosta-González and Marques 2016). Rodriguez-R et al. (2015) examined the consecutive patterns of functional and taxonomic diversity over the course of one year, after the DWH oil was deposited on Pensacola Beach sands, using metagenomic and 16S rRNA gene amplicon techniques. The taxonomic diversity decreased the first four months after the deposits of DWH oil, with a significant recovery one year later; in contrast, the functional diversity increased the first three months and significantly decreased after one year when oil concentration became undetectable. Members of the *Alcanivorax*, *Borrelia*, *Spirochaeta*, *Micavibrio* and *Bacteroides* genera responded rapidly, reaching high abundances the first months and then decreased over time, while members of the genera *Hyphomonas*, *Treponema*, *Sphingopyxis* and *Hirschia* showed a peak in abundance after six months and appeared to significantly drop after one year (recovered samples). These results indicated that the early responders to oil contamination, likely degrading aliphatic hydrocarbons, were replaced by populations capable of aromatic hydrocarbon degradation. Similar results were obtained by other researchers who also studied the microbial diversity in samples over time (before, during and for at least one year after DWH incident) (Beazley et al. 2012; Yang et al. 2014b; Kleindienst et al. 2015). It is estimated that functional diversity, rather than phylogenetic diversity, plays a significant role in ecosystems functions. Genotypic diversity confers an advantage in the event of changing environment and endows the ecosystem with greater capacity to perform a particular function (Cravo-Laureau et al. 2011).

In areas chronically polluted by hydrocarbons, bacteria have acquired the metabolic versatility in order to be able to adjust to their contaminated habitat by using petroleum products as a sole carbon and energy, or tolerate pollutants by detoxification mechanisms (Taketani et al. 2010; Peng et al. 2008; Bargiela et al. 2015). Cébron et al. (2008) showed that genes encoding PAH-ring-hydroxylating dioxygenases appeared in higher levels in aged PAH-contaminated soils and sediments, indicating that the selection pressures, exerted by aged PAH contamination, would enrich PAH-degrading bacteria. A long-term adaptation of a microbial community towards a diversified and pollutant-resistant community has also been showed by others (Bourceret et al. 2016). In a recent study conducted by Liao et al. (2015) it was demonstrated that the changes of microbial communities caused by long-term oil contamination were similar in the two different soils examined, despite the fact that in the uncontaminated soils the pristine communities differed significantly in structure, metabolic activity and function. Their results, were similar to others reported (dos Santos et al. 2011; Ji et al. 2013; Peng et al. 2015; Bourceret et al. 2016), showing that oil contamination resulted in the higher abundance, bacterial richness diversity and larger fraction of hydrocarbon-degrading groups (Liao et al. 2015), and contrary to other studies where it was reported that total diversity of the aged soil had been reduced (Cunliffe and Kertesz 2006; Sutton et al. 2013).

Hydrocarbons and heavy metals often co-exist in long-term contaminated soils and could have strong and complex impacts on the microbial community structure (Thavamani et al. 2012b). The presence of heavy metals in PAH-contaminated soils was found to reduce the diversity of the microbial population and lead to an increase in the abundance of few distinctive species by exerting selective pressure in soils (Thavamani et al. 2012a). A reduction of *n*-alkane degrading bacterial community was also found in a site chronically co-contaminated with mineral oil hydrocarbons and metals using the *alkB* gene as a biomarker (Perez-de-Mora et al. 2011). However, it has been reported that the *alkB* gene presence, diversity or expression were not specific enough for studying the microbial community response to chronically oil-contaminated environments, because algae, plants, non-biogenic resources and the whole carbon contents of the environment contribute to the alkane input in natural marine and terrestrial environments (Paisse et al. 2011).

The bacterial community spatial structure was found to be altered in aged creosote-contaminated site by Mukherjee et al. (2014). The authors demonstrated that *Proteobacteria* (include well-known alkane and PAH degraders) were dominant in the zones with high pollution, whereas *Actinobacteria* were dramatically decreased. It was suggested that among both groups of robust PAH degraders, *Actinobacteria* were outcompeted by *Proteobacteria* in this aged creosote-contaminated site. However, the total microbial activity was found to be increased in areas with higher pollution, in contrast to the bacterial diversity, probably due to the adaptation or enrichment of specific indigenous microorganisms that use hydrocarbons as carbon source (Mukherjee et al. 2014; Cheema et al. 2015). Long-term oil-contamination significantly affects soil microbial community spatial structure decreasing microbial alpha-(gene number, richness) and beta-diversity, acting as an environmental filter to decrease the regional differences distinguishing soil microbial communities (Liang et al. 2015). The contradiction in the results of several studies on the microbial community response in aged-hydrocarbons contaminated soil could be explained by several reasons, such as the time of hydrocarbon contamination, competition for nutrient due to imbalanced C:N:P ratio, soil type, pollutant bioavailability, as well as, the fact that contamination could be toxic to many microbial populations reducing microbial diversity as has been found by many studies, or it could be a source of carbon substrates for other populations facilitating the development of rather complex microbial communities (Morais et al. 2016; Patel et al. 2016). Reduced diversity could be caused by the selective ecological pressures of TPH contamination, including the introduction of toxic concentrations of hydrocarbons and biodegradation products (Sutton et al. 2013). On the contrary, increased diversity could be detected under the long-term selective ecological pressure of contamination where soil microbial communities might have already adapted to this condition (Liao et al. 2015).

Impact of temperature In an attempt of a continental-scale description of soil bacterial communities and the environmental factors that influence their biodiversity, it was shown that bacterial community composition is independent of geographic distance and does not relate with latitude, site temperature and other variables that serve as predicting factors for plant and animal diversity (Fierer and Jackson 2006). However in recent studies, temperature has been demonstrated to contribute to marine and soil microbial diversity evolution (Garcia-Pichel et al. 2013; Bargiela et al. 2015). Particularly, in hydrocarbon contaminated environments temperature plays a significant role in controlling the nature and extent of microbial hydrocarbon metabolism and directly affects both the rate of biodegradation, and the physico-chemical behavior of hydrocarbons (Coulon et al. 2007). Diffusion rates, volatilization, solubility and therefore bioavailability of hydrocarbons decrease in lower temperatures, resulting in a slow-down of microbial metabolism (Coulon et al. 2007; Haritash and Kaushik 2009). Although hydrocarbon biodegradation can occur over a wide range of temperatures, the rate of biodegradation generally decreases with the decreasing temperature. Highest degradation rates generally occur in the range 30–40 °C in soil environments, 20–30 °C in some freshwater environments and 15–20 °C in marine environments, but it can continue even in colder environments (<5 °C) (Eriksson et al. 2001; Okoh 2006; Das and Chadran 2011; Jain et al. 2011).

Even if low temperature environments are known to be inhospitable, however, they are colonized by cold-adapted psychrophilic and psychrotolerant microorganisms. A wide variety of cold-adapted, indigenous microorganisms with the ability to degrade aliphatic and aromatic hydrocarbons at low temperatures, thus playing a significant role in the in situ biodegradation process in Polar and Alpine soils, as well as, Arctic and Antarctic marine environments (including seawater, sea-ice, and sediments), have been studied (Lo Guidice et al. 2010). The biodiversity, community response to hydrocarbon perturbation and bioremediation potential of microorganisms in extremely cold environments has been extensively reviewed (Aislabie and Foght 2010; Atlas 2010; Shivaji and Reddy 2010). Representatives of the genera *Pseudomonas*, *Pedobacter*, *Brevundimonas*, *Rhodococcus, Arthrobacter* and *Bacillus* have been found in contaminated alpine soils (Margesin and Schinner 2001; Margesin et al. 2003a, b; Zhang et al. 2012). A shift in bacterial community composition towards a psychrotolerant and cold-adapted proteobacterial strains following hydrocarbon contamination has been observed in alpine and arctic soils, as well as, in the upper permafrost especially in the early stages of hydrocarbon degradation (Yang et al. 2014a; Yergeau et al. 2012; Zhang et al. 2012; Labbé et al. 2007). Among *Proteobacteria*, members of *Sphingomonadaceae* (esp. *Novosphingobium* and *Sphingobium*) and *Caulobacter* have been reported to drastically dominate in oil-contaminated upper permafrost samples. *Sphingomonas* species are able to degrade aromatic compounds at low temperatures due to the ability to alter membrane fluidity, increasing their tolerance in a broad temperature range, whereas, some species of *Caulobacter* have been reported to be resistant to freezing (Yang et al. 2014a). Strains of the genera *Acinetobacter* and *Pseudomonas* have previously been reported as cold-tolerant (Margesin et al. 2003a; Aislabie et al. 2006; Ma et al. 2006). Specifically, *Pseudomonas* species have been correlated with high degradation rates, with their abundance significantly decreased when residual hydrocarbons are almost depleted. In decreased degradation rates *Rhodococcus* species became abundant (Yergeau et al. 2012).

In a study monitoring the influence temperature exerts on biodegradation rates, two microbial consortia extracted from heavily contaminated with PAHs and unspoiled soils were reported to degrade low- (naphthalene, phenanthrene, anthracene) and high-molecular weight PAHs (pyrene, perylene) at both high (15–25 °C) and low (5–15 °C) temperature ranges. Strains of the genera *Acinetobacter* and *Pseudomonas* were present in both microcosms and they were able to degrade HMW-PAHs at a low temperature range (5–15 °C), but at lower rates than at higher temperature (Simarro et al. 2013). In PAH-contaminated soil samples collected near a solid waste combustion facility in Station Nord, Greenland, benzoic acid was shown to be mineralized at −5 °C to a small extent. Increasing temperature to 0 °C resulted in an increased rate and extent of mineralization of benzoic acid and revealed the presence of phenanthrene degraders belonging to *Sphingomonas* spp. and *Pseudomonas* spp. (Sørensen et al. 2010). Through a multi-omics approach (metagenome- and metabolome-wide scan) temperature was found to be the core environmental parameter regulating microbial population and catabolic activities in seven major chronically oil-polluted and geographically separated marine sites along the coastlines of the Mediterranean Sea and in one from the Red Sea. Lower temperature resulted in an increase in total biodiversity, with a marked negative effect on catabolic (i.e., pollutant degradation) diversity, regardless of the geographic location or other environmental constraints (Bargiela et al. 2015).

The microbial community response to the *Deepwater Horizon* oil spill was distinct from that observed in previous spills or mesocosm studies because of the depth of the spill. Some oil components remained in the deep ocean plumes where the temperature was just 4–6 °C, much colder than surface water. Deep water plume communities were characterized by low diversity and were dominated by just three groups of *Gamma-Proteobacteria*, namely *Oceanospirillales*, *Colwellia* and *Cycloclasticus* none of which were abundant in surface oil samples (Hazen et al. 2010; Valentine et al. 2010; Redmond and Valentine 2012). Members of *Oceanospirillales* are known psychrophilic *n*-alkane and cycloalkane degraders (Mason et al. 2012), whereas *Cycloclasticus* is an obligate PAH degrader (Kleindienst et al. 2015). Finally, members of *Colwellia* are known for their ability to degrade gaseous alkanes, aromatics and PAHs (Valentine et al. 2010; Redmond and Valentine 2012; Gutierrez et al. 2013).

Impact of salinity Bacterial hydrocarbon degradation in hypersaline environments, is deemed a difficult process, since an increase in salinity results in a reduction of the solubility of both hydrocarbons and oxygen. Non-extremophilic microorganisms are unable to efficiently degrade hydrocarbon contaminants at high salt concentrations. Halophilic microorganisms though, are metabolically different and are adapted to extreme salinity (Le Borgne et al. 2008); moreover, halophilic hydrocarbon degraders have to cope with the conflicting demands of salt and oil, both having a direct impact on cell envelope hydrophobicity (higher for hydrocarbon uptake and lower for salt tolerance) (McGenity 2010). Microbes in environments with moderate salinities (2.5–10%) can degrade hydrocarbons at salinities higher than the salinity of their habitat. In soil from an oilfield of 10% salinity, Zhao et al. (2009) reported that biodegradation of phenanthrene occurred at 5, 10 and 15% salinity, whereas no biodegradation took place at 0.1 and 20% salinity.

Salinity, rather than extremes of temperature, pH, or other physical and chemical factors, has been reported to be the major environmental determinant of microbial community composition in diverse physical environments (Lozupone and Knight 2007). After analyzing the composition of hydrocarbon-degrading bacterial communities from freshwater, marine, and hypersaline aquatic ecosystems with salinities corresponding to 0.2, 4 and 5%, respectively, enriched with different hydrocarbons (heptadecane, naphthalene, crude oil) the authors report that bacterial selection depended more on the environmental properties than on the different hydrocarbons used in the microcosms (Jurelevicius et al. 2013). Soil salinity was reported to have a suppressive effect on most bacterial populations without changing their dominance and some species were strengthened with the aggravation of salinization. *Actinobacteria*, *Gamma-Proteobacteria*, *Firmicutes* and *Deinococcus thermus* were the dominant bacteria in dual stresses of salinization and oil contamination (Gao et al. 2015).

Among microbial taxa the most common inhabitants of high salinity environments with the potential to degrade various hydrocarbons are members of the genus *Halomonas*, *Marinobacter,* and *Alcanivorax* (Fathepure 2014). A mixed culture consisting of *Halomonas* sp. and *Marinobacter* sp. from hydrocarbon-contaminated saline soil from five different regions in Iran was obtained and the microorganisms were also reported to degrade several PAHs as sole carbon sources (Dastgheib et al. 2012). Halomonads have been identified in halophilic oil degrading consortia. Some strains can degrade some aliphatic hydrocarbons present in crude oil (Mnif et al. 2009) but the majority of Halomonads in oily hypersaline environments degrade metabolites of hydrocarbon degradation (Calvo et al. 2002). *Alcanivorax* and *Marinobacter* species are found whenever oil is present in saline conditions (McGenity 2010); *Alcanivorax* strains (*Oceanospirillales* order), typically known as marine hydrocarbonoclastic bacteria, are primary aliphatic hydrocarbon degraders (McKew et al. 2007a) and have been applied in saline soil bioremediation resulting in the degradation of alkanes (Dastgheib et al. 2011). *Marinobacter* strains are more commonly associated with hypersaline environments (Jurelevicius et al. 2013) and two halophilic *Marinobacter* strains have been reported to successfully mineralize crude oil in nutrient media as well as in hypersaline soil and water microcosms (Al-Mailem et al. 2013).

Impact of pH In contrast to most aquatic ecosystems, soil pH can be highly variable, ranging from 2.5 in mine spoils to 11.0 in alkaline deserts (Leahy and Colwell 1990). Soil pH has been shown to exert a strong influence on the composition and diversity of microbial bacterial communities. It can affect both the overall bacterial community composition and the composition of individual bacterial groups within the community (Rousk et al. 2010). Extremes in pH would be expected to have a negative influence on the ability of microbial populations to degrade hydrocarbons (Leahy and Colwell 1990). Generally, microbial diversity is lower in acidic, when compared to neutral soils, at a continental scale and the microbial community compositions are affected by changes in the abundance of *Actinobacteria*, *Acidobacteria* and *Bacteroidetes* (Fierer and Jackson 2006; Lauber et al. 2009).

Proteobacterial classes that dominated in high pollution levels in an aged creosote-polluted site were shown to respond to the pH gradient. *Burkholderiaceae* and *Acetobacteraceae* prevailed in lower pH, whereas *Pseudomonadaeceae* and *Sphingomonadaceae* were shown to prefer higher pH (Mukherjee et al. 2014). The contrasting tendencies of pH preference displayed by different classes of *Acidobacteria* in this site were in agreement with a study conducted in a railway diesel-contaminated site (Sutton et al. 2013). *Acidobacteria* are among the most abundant groups of soil bacteria and generally thrive in low pH (Mukherjee et al. 2014). Functional genes involved in organic contaminant degradation were higher in abundance in slightly acidic soils, with a distinct representation of the genera *Acidiphilium*, *Acidobacteria* and *Acidothermus*, than in saline-alkali oil-contaminated soils from exploration sites in North and South China (Liang et al. 2014). This could be attributed to the fact that high salinities may limit both the availability of oxygen as well as the microbial access to hydrocarbons (Al-Mailem et al. 2013).

However, it is not clear whether pH itself is the factor shaping microbial bacterial communities (Rousk et al. 2010). A number of soil characteristics (e.g., nutrient availability, cationic metal solubility, organic C characteristics, soil moisture regimen, and salinity) are often directly or indirectly related, and often co-vary with changes in soil pH, also contributing to the observed changes in community composition (Lauber et al. 2009).

Impact of nutrient addition and oxygen Biostimulation is defined as addition of various forms of rate-limiting nutrients and electron acceptors such as phosphorus, nitrogen, oxygen, or carbon in order to increase the population or activity of naturally occurring microorganisms available for bioremediation (Louati et al. 2013). It is of utmost importance to understand the role of microorganisms and their responses towards bioremediation approaches in order to obtain efficient tools to assist in mitigation of hydrocarbon contaminants.

Nutrient Amendment

Contamination with petroleum, which is a mixture of carbon and hydrogen compounds, basically provokes an imbalance in the carbon-nitrogen ratio at the spill site, limiting the availability of inorganic nutrients essential for bacterial growth, and in particular nitrogen (N) and phosphorus (P) (Olajire and Essien 2014; McKew et al. 2007b). Consequently, growth of hydrocarbon-degrading bacteria and hydrocarbon degradation can be strongly enhanced by fertilization with inorganic N and P and it is well documented that degradation of oil is often significantly enhanced by this method (Röling et al. 2002; McKew et al. 2007b; Nikolopoulou and Kalogerakis 2010).

All proteobacterial classes, except for *Epsilon-Proteobacteria*, have been reported to respond to the supplementation of nutrients, as well as, to the input of hydrocarbon substrates (Greer et al. 2010). In nutrient amended (N, P) soils a prevalence of *Beta-Proteobacteria* has been observed in cases of diesel contamination (Bell et al. 2013), or a significant increase in their relative abundance in cases of PAH contamination as reported by Guazzaroni et al. (2013) through metagenomic and metaproteomic approaches (Table 9.3). This relative abundance of *Beta-Proteobacteria* was strongly and positively related to degradation, in contrast to a negative correlation of their abundance to degradation in unamended contaminated soils. This strongly suggests that their effectiveness as degraders may be dependent on sufficient availability of nutrients (Bell et al. 2013). The fact that *Beta-Proteobacteria* are enriched in nutrient-amended soils, can be attributed to the fact

Table 9.3 Key studies describing the impact of nutrient and oxygen on soil and marine bacterial communities

Habitat/geographic location	Type of sample and contamination type	Nutrient amendment	Techniques to monitor community shifts	Dominant members (microbial genera) in initial microbial community	Dominant members in initial contaminated microbial community	Dominant members (microbial genera) in contaminated/amended microbial community	References
N and P							
18 disparate Arctic and sub-Arctic locations	Soil treated with diesel fuel followed by 1 week of weathering	Monoammonium phosphate-MAP	Multiplexed 16S rRNA gene sequencing	*Proteobacteria* (*Alpha-*, *Beta-*, *Gamma-*, *Delta-*) *Actinobacteria*, *Acidobacteria*, *Firmicutes*, *Bacteroidetes*	*Actinobacteria* (low-organic matter soils) *Proteobacteria* (high-organic matter soils)	*Actinobacteria* (*Nocardioidaceae*) *Beta-Proteobacteria* (*Burkholderiaceae*)	Bell et al. (2013)
Chemical plant, Oviedo, Spain	Loamy clay soil treated with PAHs	Biostimulation with calcium ammonia nitrate, NH_4NO_3 and KH_2PO_4,	Metaproteomics metagenomics		*Alpha-Proteobacteria* (*Rhizobium, Devosia, Novosphingobium, Sphingomonas, Brevundimonas*) *Gamma-Proteobacteria* (*Pseudomonas, Stenotrophomonas, Pseudoxanthomonas*)	*Beta-Proteobacteria* (*Acidovorax, Tetrathiobacter*) *Gamma-Proteobacteria* (*Pseudomonas*) *Bacteroidetes*	Guazzaroni et al. (2013)
Stanford le Hope saltmarsh near BP oil refineries, Thames Estuary, UK	Seawater treated with weathered crude oil	N (as NH_4NO_3), P (as KH_2PO_4)	qPCR amplification of functional genes	*Thalassolituus*, *Alcanivorax*, *Cycloclasticus* were not detected in the original seawater	*Cycloclasticus* *Thalassolituus*	*Alcanivorax* *Cycloclasticus* *Thalassolituus*	McKew et al. (2007b)
Upper part of the intertidal zone of	Sediment treated with an emulsion of	N (as sodium nitrate) and P (as potassium	DGGE band sequences		*Gamma-Prtoebacteria* (*Alcanivorax*	*Alpha-Proteobacteria* (*Erythrobacter longus*, *Erythrobacter citreus*)	Röling et al. (2002)

(continued)

Table 9.3 (continued)

Habitat/geographic location	Type of sample and contamination type	Nutrient amendment	Techniques to monitor community shifts	Dominant members (microbial genera) in initial microbial community	Dominant members in initial contaminated microbial community	Dominant members (microbial genera) in contaminated/amended microbial community	References
Stert Flats, Somerset, UK	weathering crude oil	dihydrogen phosphate)			*borkumensis*, *Fundibacter jadensis*)	strong selection for alkane-degrading *Gamma-Proteobacteria* (*Alcanivorax borkumensis*, *Fundibacter jadensis*)	
Ishikawa beach, Japan Sea	Sea water, polluted gravel, heavy oil pastes	P (Linstar 30) and N (Super IB) fertilizers	DGGE band sequences			10 months after the *Nakhodka* oil spill predominance of *Alcanivorax*	Kasai et al. (2002)
Compost							
Industrial zone in Celje, Slovenia (Technosol)	Loamy sand soil highly contaminated with hydrocarbons	C1: shred shrubs and trees (yard waste) C2: food residues, grass clippings and other woody material, vegetable and flower residues	qPCR of the *alkB* gene		*Alpha-Proteobacteria* (*Acetobacter* sp.), *Gamma-Proteobacteria* (*Pseudomonas* sp., *Thalassolituus* sp., *Acinetobacter* sp.) *Firmicutes* (*Bacillus* sp., *Geobacillus* sp.)	*Actinobacteria* (*Mycobacterium* sp. *Gordonia* sp., *Rhodococcus* sp.) *Gamma-Proteobacteria* (*Shewanella* spp. *Hydrocarboniphaga* spp.) *Alpha-Proteobacteria* (*Agrobacterium* sp.)	Wallisch et al. (2014)

(continued)

Table 9.3 (continued)

Habitat/geographic location	Type of sample and contamination type	Nutrient amendment	Techniques to monitor community shifts	Dominant members (microbial genera) in initial microbial community	Dominant members in initial contaminated microbial community	Dominant members (microbial genera) in contaminated/amended microbial community	References
Abandoned agricultural area, Santomera, Spain	Haplic calcisol soil treated with crude oil	Garden and crop prunings	PFLAs and metaproteomics	*Alpha-Proteobacteria* (*Rhizobiales*) *Beta-Proteobacteria* (*Burkholderiales*) *Gamma-Proteobacteria* (*Pseudomonadales*, *Enterobacterales*)	*Alpha-Proteobacteria* (*Caulobacterales*) *Beta-Proteobacteria* (*Burkholderiales*) *Gamma-Proteobacteria* (*Pseudomonadales*) *Actinobacteria* (*Actinomycetales*)	*Alpha-Proteobacteria* (*Sphingomonadales*) *Beta-Proteobacteria* (*Burkholderiales*) *Gamma-Proteobacteria* (*Pseudomonadales*) *Actinobacteria* (*Actinomycetales*)	Bastida et al. (2016)
Northern Italy: Diesel oil storage site Coal-tar processing facility	Soil contaminated by weathered diesel oil Soil contaminated by PAHs	Mixture of municipal and green wastes	ARDRA/16S rRNA sequences		*Alpha-Proteobacteria*, *Actinobacteria* *Alpha-Proteobacteria*, *Gamma-Proteobacteria*	*Beta-Proteobacteria*, *Acidobacteria* *Bacteroidetes*	Gandolfi et al. (2010)
	Artificial soils spiked with phenanthrene	Dried plant litter (maize and potato leaves)	16S gene rRNA sequences		*Stenotrophomonas maltophilia* *Arthrobacter* sp.	*Beta-Proteobacteria* *Alpha-Proteobacteria* *Pseudomonas* sp.	Babin et al. (2014)

that members of this bacterial class grow rapidly in nutrient-rich environments (*r* strategists), therefore predominating where nutrient sources are abundant (Labbé et al. 2007).

The predominant growth of *Alcanivorax* after nutritional supplementation has also been proved using culture-independent methods in oil-contaminated seawater samples (Röling et al. 2002; Syutsubo et al. 2001; Head et al. 2006). When heavy oil paste from the *Nakhodka* oil spill in the Japan Sea was used supplied with phosphorus and nitrogen, the predominance of *Alcanivorax* bacteria was demonstrated after one week, while the authors stressed that they were not the predominant population without the nutrient addition (Kasai et al. 2002). Others report that the predominance of *Alcanivorax* occurred only between 1 and 6 weeks after the addition of fertilizers (McKew et al. 2007b) (Table 9.3), whereas it has also been reported that different genotypes of *Alcanivorax* are adapted to different concentrations of nutrients (Head et al. 2006). The success of *Alcanivorax* spp. could be attributed to their ability to use branched-chain alkanes more effectively than other hydrocarbon-degrading bacteria, giving these species a selective advantage (Hara et al. 2003). Marine bacteria of the genera *Cycloclasticus* and *Thalassolituus* have been reported to increase their relative abundance in the presence of nutrients (McKew et al. 2007b) and have been assigned to the degradation of PAHs (naphthalene and phenanthrene) and *n*-alkanes (C_{12}–C_{32}) (Kasai et al. 2003; McKew et al. 2007a).

Compost Amendment

Among ex situ bioremediation techniques, the composting treatment of soils was proved to be effective in the degradation of PAHs, reaching a satisfactory percentage of removal (>90% in some cases) (Moretto et al. 2005). Compost can act as a stabilizer of soil structure, improving oxygen diffusion and water availability, and it can also act as a source of nutrients for degrading microorganisms (Gandolfi et al. 2010). Moreover, compost contains high numbers of microbes that are capable to degrade hydrocarbons (Xu and Lu 2010; Wallisch et al. 2014). The biostimulation capacity for hydrocarbon removal caused by compost amendment is linked with profound changes in the bacterial community in terms of composition and biomass (Table 9.3). However, it is not clear if the positive effects of compost addition are related to the introduced microbes and their biodegradative potential or to a general shift of microbial community structure in soil as a result of extra nutrients provided by the compost (Wallisch et al. 2014; Gandolfi et al. 2010).

Oxygen Supply

Large concentrations of biodegradable organics in the top layer soil of contaminated sites deplete oxygen reserves in the soil and therefore, the rates of oxygen diffusion into deep layers, slows down (Olajire and Essien 2014). Oxygen can act as an electron acceptor under aerobic metabolism and enhance bacterial growth and oil emulsification (Gao et al. 2013). In soils an applicable in situ technology for biostimulation with the use of oxygen is aeration, whereas in marine environments, where oxygen delivery seems very difficult, this is not applicable, with the exception of the aeration of the top layers of sediments in the coastal zones (Cuny et al. 2011). Aeration was found to stimulate microbial community activities, including those involved in hydrocarbon degradation and/or may facilitate the interactions between the pollutants and the naturally occurring bacteria, inducing a constant degradation of the pollutants. This is probably due to a combination of physical actions and biotic actions by the biostimulation of the microbial communities (Yang et al. 2005; Militon et al. 2010).

Even though proteobacterial phylotypes dominate through the aeration treatments there are contrasting findings regarding their diversity. In a study by Militon et al. (2010), even though they are dominant after 12 months of soil aeration treatment the active phylotypes appear to be less diversified. On the contrary, throughout a 2.5-year aeration treated soil the bacterial community under adaptation, concomitantly with an observed degradation of contaminants in situ, a dynamic shift of Gram-negative (initially restricted to *Pseudomonadales*) developing an increased diversity comprising new proteobacterial classes and Gram-positive bacteria, was observed (Kabelitz et al. 2009). The microbial diversity was shown to decrease in a combined bioremediation approach on a creosote-contaminated soil, using both biostimulation with aeration and the addition of nutrients (Viñas et al. 2005). However, in a similar combined effect approach to simulate in situ bioremediation of PAHs, Singleton et al. (2013) observed that the longest exposure to oxygen and nutrients led to an increase in bacterial community biodiversity. Penetration of oxygen to previously anoxic regions had a marked effect on the increase of bacterial biodiversity resulting in the abundance of PAH degrading uncharacterized *Gamma-Proteobacteria* and *Acidovorax* (Singleton et al. 2013). The genus *Acidovorax* has been reported to predominate in enrichments from soils with PAHs and use a number of them as carbon sources (Eriksson et al. 2003; Singleton et al. 2009).

The increase in bacterial diversity can be attributed to the gradual increase of potential degraders and the reduction of the concentration of toxic organic compounds. This reduction allows the reproduction of bacteria- not necessarily able to degrade aromatics and sensitive to higher concentrations of organic compounds- that are capable of growing on the cross-feeding metabolites excreted from the initial biomass of degraders (Uhlik et al. 2013). The predominance and resilience of *Gamma-Proteobacteria* is marked (Viñas et al. 2005; Kabelitz et al. 2009; Militon et al. 2010; Singleton et al. 2013), however, when comparisons are made on the taxonomical scale order strong shifts in composition inside the order are observed (Kabelitz et al. 2009; Militon et al. 2010). This "gamma-shift" has been reported to occur upon contamination of soil, when the contaminant acts as a source of nutrient for the degraders or under conditions of nutrient oversupply (Popp et al. 2006). In the long-term of aeration treatments a predominance of *Actinobacteria* was reported (Kabelitz et al. 2009; Militon et al. 2010).

Environments such as soils and coastal marine sediments are often submitted to oxic/anoxic switches and therefore constitute particular functional systems that impact bacterial diversity and its role on the degradation of organic compounds. The frequency and duration of oscillations determines the adaptation of bacterial communities (Cuny et al. 2011). Previous studies examining the impact of oscillating conditions on hydrocarbon removal, have demonstrated that a higher efficiency of hydrocarbon removal was achieved in conditions of intermittent aeration (Vieira et al. 2009) or found no difference between oscillating and oxic conditions (Löser et al. 1998; Cravo-Laureau et al. 2011). Apart from addressing the fate of hydrocarbon degradation, it is important to focus on the behavior of bacterial communities under such conditions. In bioreactors subjected to anoxic/oxic switches or continuously oxic conditions it was found that the efficiency of hydrocarbon removal was similar under both conditions, demonstrating that indigenous bacteria presented a strong capacity for adaptation to changes in their environment (aeration vs. anoxia). The PAH-degrading bacterial communities were found to be stable under the oscillating anoxic/oxic mode according to the presence of RHD transcripts, whereas a shift in the structure of the metabolically active communities was observed with 16S rRNA analyses (Vitte et al. 2011, 2013). Indigenous bacteria presented a strong capacity for adaptation to changes in their environment (aeration vs. anoxia). *Gamma-Proteobacteria* affiliated to *Pseudomonadales* were abundant in samples before treatment and remained abundant after a 26-day-treatment under changing anoxic/oxic conditions (Vitte et al. 2011). *Pseudomonads* are considered as pioneer bacteria in the degradation of PAHs since they have been reported to oxidize readily bioavailable compounds (Heiss-Blanquet et al. 2005; Popp et al. 2006), and in addition have the ability to produce biosurfactants (Ron and Rosenberg 2002).

9.3 Conclusion and Perspectives

Because hydrocarbons are mainly present as complex mixtures in the environment, the challenge of understanding the roles of hydrocarbon-degrading bacteria in metabolizing the plethora of petroleum substrates in contaminated sites is clearly substantial for bioremediation approaches. Biodegradation of hydrocarbons in both terrestrial and marine environments is directly dependent on the bacterial community composition which shifts temporally and spatially in response to changing environmental conditions. It has become clear, that in spite of the many studies on the effects of soil or sediment hydrocarbon contamination on microbial community, our understanding needs to be improved. The numerous factors affecting the responses of communities to the contaminants, such as toxic concentrations of contaminant, complexity of soil organic matter, reduced nutrient availability, imbalances in C:N:P, influence of PAH aging, horizontal gene transfer, different approaches for the communities study, as well as the fact that the pollutant disturbance alters the environment and the organisms adapt over time to the changed conditions (Smalla et al. 1998; Kaplan and Kitts 2004; Top and Springael 2003). In many cases, taxonomic diversity of a microbial community could be negatively impacted by hydrocarbon perturbation, but the functional diversity could be increased as an effect of the disturbance. In an excellent review by van Hamme et al

(2003), taking together evidences by culture-dependent and culture-independent approaches, it has been reported that the oil resulted in diminished microbial population diversity and in the selection for metabolic generalists even after extended exposure times. The diversity and functional composition of the microbial community after hydrocarbon perturbation seems to follow the "disturbance specialization hypothesis", according to which the generalists are more resilient to disturbances that alter the niches whereas the most specialist taxa are selected (Rodriguez-R et al. 2015).

The current status is that the majority of bacterial species from different environmental niches remains unknown and uncultured. In PAH polluted sites it was demonstrated that the dominant PAH degraders differed from the bacterial species previously isolated and studied in standard laboratory conditions. Therefore, along with the advancements in culture-independent approaches, it would be necessary to develop new isolation procedures and cultivation conditions for the study of uncultured strains and the identification of their PAH catabolic enzymes and whole catabolic pathways. In situ studies combined with the "omics" and the most recent advances in "meta-omics" technologies, in association with culture depending methods and stable isotope probing will enable scientists to link phylogeny with functionalities in single cells and microbial communities. Moreover, the above approaches will provide further association of the complex microbial interactions and the effect of the prevailing environmental conditions (microbial ecology, physicochemical interactions and factors) with more efficient in situ bioremediation approaches in contaminated sites.

Acknowledgements The authors would like to thank Dr. Hatziloukas Efstathios, Associate Professor in the Department of Biological Applications and Technologies, University of Ioannina, Ioannina, Greece for comments that improved the manuscript.

References

Acosta-González A, Marques S (2016) Bacterial diversity in oil-polluted marine coastal sediments. Curr Opin Biotechnol 38:24–32

Acosta-González A, Rosselló-Móra R, Marqués S et al (2013) Characterization of the anaerobic microbial community in oil-polluted subtidal sediments: aromatic biodegradation potential after the *Prestige* oil spill. Environ Microbiol 15:77–92

Acosta-González A, Martirani-von Abercron S-M, Rosselló-Móra R et al (2015) The effect of oil spills on the bacterial diversity and catabolic function in coastal sediments: a case study on the *Prestige* oil spill. Environ Sci Pollut Res 22:15200–15214

Aislabie J, Foght JM (2010) Response of polar soil bacterial communities to fuel spills. In: Bej AK, Aislabie J, Atlas RM (eds) Polar microbiology: the ecology, biodiversity, and bioremediation potential of microorganisms in extremely cold environments. CRC Press, Boca Raton, FL, pp 215–230

Aislabie J, Saul DJ, Foght JM (2006) Bioremediation of hydrocarbon contaminated polar soils. Extremophiles 10:171–179

Alexander M (1981) Biodegradation of chemicals of environmental concern. Science 211:132–138

Alexander M (1999) Biodegradation and bioremediation, 2nd edn. Academic Press, London, England

Alexander M (2000) Aging, bioavailability and overestimation of risk from environmental pollutants. Environ Sci Technol 34:4259–4265

Al-Mailem DM, Eliyas M, Radwan SS (2013) Oil-bioremediation potential of two hydrocarbon-oclastic, diazotrophic Marinobacter strains from hypersaline areas along the Arabian Gulf coasts. Extremophiles 17:463–470

Alonso-Gutiérrez J, Figueras A, Albaigés J et al (2009) Bacterial communities from Shoreline environments (Costa da Morte, Northwestern Spain) affected by the Prestige oil spill. Appl Environ Microbiol 75:3407–3418

Al-Wasify RS, Hamed SR (2014) Bacterial biodegradation of crude oil using local isolates. Int J Bacteriol 2014:1–8

Amellal N, Portal J, Vogel T et al (2001) Distribution and location of polycyclic aromatic hydrocarbons (PAHs) and PAH-degrading bacteria within polluted soil aggregates. Biodegradation 12:49–57

Atlas RM (1991) Microbial hydrocarbon degradation, bioremediation of oil spills. J Chem Technol Biotechnol 52:149–156

Atlas RM (2010) Microbial bioremediation in polar environments: current status and future directions. In: Bej AK, Aislabie J, Atlas RM (eds) Polar microbiology: the ecology, biodiversity, and bioremediation potential of microorganisms in extremely cold environments. CRC Press, Boca Raton, FL, pp 373–391

Babin D, Ding GC, Pronk GJ et al (2013) Metal oxides, clay minerals and charcoal determine the comparison of microbial communities in matured artificial soils and their response to phenanthrene. FEMS Microbiol Ecol 86:3–14

Babin D, Vogel C, Zuhlke S et al (2014) Soil mineral composition matters: response of microbial communities to phenanthrene and plant litter addition in long-term matured artificial soils. PLoS ONE 9:e106865

Bargiela R, Mapelli F, Rojo D et al (2015) Bacterial population and biodegradation potential in chronically crude oil-contamnated marine sediments are strongly linked to temperature. Sci Rep 5:11651

Bastida F, Jehmlich N, Lima K et al (2016) The ecological and physiological responses of the microbial community from a semiarid soil to hydrocarbon contamination and its bioremediation using compost amendment. J Prot 135:162–169

Beazley MJ, Martinez RJ, Rajan S et al (2012) Microbial community analysis of a coastal saltmarsh affected by the Deepwater Horizon oil spill. PLoS ONE 7:e41305

Bell TH, Yergeau E, Maynard C et al (2013) Predictable bacterial composition and hydrocarbon degradation in Arctic soils following diesel and nutrient disturbance. ISME J 7:1200–1210

Bengtsson G, Törneman N (2009) A spatial approach to environmental risk assessment of PAH contamination. Risk Anal 29:48–61

Bengtsson G, Törneman N, Yang X (2010) Spatial uncoupling of biodegradation, soil respiration, and PAH concentration in PAH contaminated soil. Environ Pollut 158:2865–2871

Bengtsson G, Törneman N, De Lipthay JG et al (2013) Microbial diversity and PAH catabolic genes tracking spatial heterogeneity of PAH concentrations. Microb Ecol 65:91–100

Bourceret A, Cébron A, Tisserant E et al (2016) The bacterial and fungal diversity of an aged PAH- and heavy metal-contaminated soil is affected by plant cover and edaphic parameters. Microb Ecol 71:711–724

Brooijmans RJW, Pastink MI, Siezen RJ (2009) Hydrocarbon-degrading bacteria: the oil-spill clean-up crew. Microb Biotechnol 2:587–594

Calvo C, Martinez-Checa F, Toledo FL et al (2002) Characteristics of bioemulsifiers synthesized in crude oil media by *Halomonas eurihalina* and their effectiveness in the isolation of bacteria able to grow in the presence of hydrocarbons. Lett Appl Microbiol 11:260–263

Cébron A, Norini M-P, Beguiristain T et al (2008) Real-time PCR quantification of PAH-ring hydroxylating dioxygenase (PAH-RHDα) genes from Gram positive and Gram negative bacteria in soil and sediment samples. J Microbiol Methods 73:148–159

Cébron A, Arsène-Ploetze F, Bauda P et al (2014) Rapid impact of phenanthrene and arsenic on bacterial community structure and activities in sand batches. Microb Ecol 67:129–144
Cerniglia CE (1984) Microbial metabolism of polycyclic aromatic hydrocarbons. Adv Appl Microbiol 30:31–71
Cerniglia CE (1992) Biodegradation of polycyclic aromatic hydrocarbons. Biodegradation 3: 351–368
Cerniglia CE (1993) Biodegradation of polycyclic aromatic hydrocarbons. Curr Opin Biotechnol 4:331–338
Cheema S, Lavania M, Lal B (2015) Impact of petroleum hydrocarbon contamination on the indigenous soil microbial community. Ann Microbiol 65:359–369
Cheung PY, Kinkle BK (2005) Changes in Mycobacterium spp. population structure and pyrene mineralization in polycyclic aromatic hydrocarbon amended soils. Soil Biol Biochem 37:1929–1937
Chikere CB, Okpokwasili GC, Chikere BO (2011) Monitoring of microbial hydrocarbon remediation in the soil. Biotech 1:117–138
Couling NR, Towel MG, Semple KT (2010) Biodegradation of PAH in soil: influence of chemical structure, concentration and multiple amendments. Environ Pollut 158:3411–3420
Coulon F, McKew BA, Osborn AM et al (2007) Effects of temperature and biostimulation on oil-degrading microbial communities in temperate estuarine waters. Environ Microbiol 9:177–186
Cravo-Laureau C, Duran R (2014) Marine coastal sediments microbial hydrocarbon degradation processes: contribution of experimental ecology in the omics' era. Front Microbiol 5:39
Cravo-Laureau C, Hernandez-Raquet G, Vitte I et al (2011) Role of environmental fluctuations and microbial diversity in degradation of hydrocarbons in contaminated sludge. Res Microbiol 162:888–895
Cunliffe M, Kertesz MA (2006) Effects of *Sphingobium yanoikuyae* B1 inoculation on bacterial community dynamics and polycyclic aromatic hydrocarbon degradation in aged and freshly PAH-contaminated soil. Environm Pollut 144:228–237
Cuny P, Cravo-Laureau C, Grossi V et al (2011) Biodegradation of hydrocarbons in bioturbated marine sediments. In: Koukkou AI (ed) Microbial bioremediation of non-metals. Caister Academic Press, Norfolk, UK, pp 55–92
Daane LL, Harjono I, Zylstra GJ et al (2001) Isolation and characterization of polycyclic aromatic hydrocarbon-degrading bacteria associated with rhizosphere of salt marsh plant. Appl Environ Microbiol 67:2683–2691
Das N, Chadran P (2011) Microbial degradation of petroleum hydrocarbon contaminants: an overview. Biotechnol Res Int 2011:1–13
Dastgheib SM, Amoozegar MA, Khajeh K et al (2011) A halotolerant *Alcanivorax* sp. strain with potential application in saline soil remediation. Appl Microbiol Biotechnol 90:305–312
Dastgheib SMM, Amoozegar MA, Khajeh K et al (2012) Biodegradation of polycyclic aromatic hydrocarbons by a halophilic microbial consortium. Appl Microbiol Biotechnol 95:789–798
de Menezes A, Clipson N, Doyle E (2012) Comparative metatranscriptomcs reveals widespread community responses during phenanthrene degradation in soil. Environ Microbiol 14:2577–2588
Delille D, Delille B (2000) Field observations on the variability of crude oil impact on indigenous hydrocarbon degrading bacteria from sub-Antarctic intertidal sediments. Mar Environ Res 49:403–417
Dequiedt S, Thioulouse J, Jolivet C et al (2009) Biogeographical patterns of soil bacterial communities. Environ Microbiol Rep 1:251–255
Ding G-C, Heuer H, Zühlke S et al (2010) Soil type-dependent responses to phenanthrene as revealed by determining the diversity and abundance of polycyclic aromatic hydrocarbon ring-hydroxylating dioxygenase genes by using a novel PCR detection system. Appl Environ Microbiol 76:4765–4771
Ding G-C, Heuer H, Smalla K (2012a) Dynamics of bacterial communities in two polluted soils after spiking with phenanthrene: soil type specific and common responders. Front Microbiol 3:290

Ding GD, Heuer H, He Z et al (2012b) More functional genes and convergent overall functional patterns detected by GeoChip in phenanthrene-spiked soils. FEMS Microbiol Ecol 82:148–156
dos Santos HF, Cury JC, do Carmo FL et al (2011) Mangrove bacterial diversity and the impact of oil contamination revealed by pyrosequencing: bacterial proxies for oil pollution. PLoS ONE 6:e16943
Edlund A, Jansson JK (2008) Use of bromodeoxyuridine immunocapture to identify psychrotolerant phenanthrene-degrading bacteria in phenanthrene-enriched polluted Baltic Sea sediments. FEMS Microbiol Ecol 65:513–525
Eriksson M, Jong-Ok K, Mohn WW (2001) Effects of low temperature and freeze-thaw cycles on hydrocarbon biodegradation in Arctic Tundra soil. Appl Environ Microbiol 67:5107–5112
Eriksson M, Sodersten E, Yu Z et al (2003) Degradation of polycyclic aromatic hydrocarbons at low temperature under aerobic nitrate-reducing conditions in enrichment cultures from northern soils. Appl Environ Microbiol 69:275–284
Fathepure BZ (2014) Recent studies in microbial degradation of petroleum hydrocarbons in hypersaline environments. Front Microbiol 5:173
Fierer N, Jackson RB (2006) The diversity and biogeography of soil bacterial communities. PNAS 103:626–631
Franzosa EA, Hsu T, Sirota-Madi A et al (2015) Sequencing and beyond: integrating molecular 'omics for microbial community profiling. Nat Rev Microbiol 13:360–372
Fuentes S, Méndez V, Aguila P et al (2014) Bioremediation of petroleum hydrocarbons: catabolic genes, microbial communities and applications. Appl Microbiol Biotechnol 98:4781–4794
Fuhrman JA (2009) Micobial community structure and its functional implications. Nature 459:193–199
Gandolfi I, Sicolo M, Franzetti A et al (2010) Influence of compost amendment on microbial community and ecotoxicity of hydrocarbon-contaminated soils. Biores Technol 101:568–575
Gao P, Li G, Dai X et al (2013) Nutrients and oxygen alter reservoir biochemical characters and enhance oil recovery during biostimulation. World J Microbiol Biotechnol 29:2045–2054
Gao Y, Wang J, Guo S et al (2015) Effects of salinization and crude oil contamination on soil bacterial community structure in the Yellow River Delta region, China. Appl Soil Ecol 86:165–173
Garcia-Pichel F, Loza V, Marusenko Y et al (2013) Temperature drives the continental-scale distribution of key microbes in top soil communities. Science 340:1574–1577
Garza DR, Dutilh BE (2015) From cultured to uncultured genome sequences: metagenomics and modeling microbial ecosystems. Cell Mol Life Sci 72:4287–4308
Gebre T, Kibru T, Tesfaye S et al (2015) Analysis of watershed attributes for water resources management using GIS: the case of Chelekot micro-watershed, Tigray, Ethiopia. J Geogr Inf Syst 7:177–190
Gilbert JA, Field D, Huang Y et al (2008) Detection of large numbers of novel sequences in the metatranscriptomes of complex marine microbial communities. PLoS ONE 3:e3042
Girvan MS, Campbell CD, Killham K et al (2005) Bacterial diversity promotes community stability and functional resilience after perturbation. Environ Microbiol 7:301–313
Greenwood PF, Wibrow S, George SJ et al (2009) Hydrocarbon biodegradation and soil microbial community response to repeated oil exposure. Org Geochem 40:293–300
Greer CW, Whyte LG, Niederberger TD (2010) Microbial communities in hydrocarbon-contaminated temperate, tropical, alpine and polar soils. In: Timmis KN (ed) Handbook of hydrocarbon and lipid microbiology. Springer, Berlin, pp 2013–2328
Guazzaroni ME, Herbst FA, Lores I et al (2013) Metaproteomic insights beyond bacterial response to naphthalene exposure and bio-stimulation. ISME J 7:122–136
Guermouche M'rassi A, Bensalah F, Gury J et al (2015) Isolation and characterization of different bacterial strains for bioremediation of *n*-alkanes and polycyclic aromatic hydrocarbons. Environ Sci Pollut Res Int 22:15332–15346
Guo H, Yao J, Cai M et al (2012) Effects of petroleum contamination on soil microbial numbers, metabolic activity and urease activity. Chemosphere 87:1273–1280

Gutierrez T, Singleton DR, Berry D et al (2013) Hydrocarbon-degrading bacteria enriched by the Deepwater Horizon oil spill identified by cultivation and DNA-SIP. ISME J 7:2091–2104
Haghollahi A, Fazaelipoor MH, Schaffie M (2016) The effect of soil type on the bioremediation of petroleum contaminated soils. J Environ Manag 180:197–201
Hamamura N, Olson SH, Ward DM et al (2006) Microbial population dynamics associated with crude-oil biodegradation in diverse soils. Appl Environ Microbiol 72:6316–6324
Hamamura N, Ward DM, Inskeep WP (2013) Effects of petroleum mixture types on soil bacterial population dynamics associated with the biodegradation of hydrocarbons in soil environments. FEMS Microbiol Ecol 85:168–178
Hara A, Syutsubo K, Harayama S (2003) Alcanivorax which prevails in oil-contaminated seawater exhibits broad substrate specificity for alkane degradation. Environ Microbiol 5:746–753
Harayama S, Kishira H, Kasai Y et al (1999) Petroleum biodegradation in marine environments. J Mol Microbiol Biotechnol 1:63–70
Harayama S, Kasai Y, Hara A (2004) Microbial communities in oil-contaminated seawater. Curr Opin Biotechnol 15:205–214
Haritash AK, Kaushik CP (2009) Biodegradation aspects of polycyclic aromatic hydrocarbons PAHs: a review. J Hazard Mater 169:1–15
Hazen TC, Dubinsky EA, DeSantis TZ et al (2010) Deep-sea oil plume enriches indigenous oil-degrading bacteria. Science 330:204–208
Head IM, Jones DM, Röling WFM (2006) Marine microorganisms make a meal of oil. Nat Rev Microbiol 4:173–182
Heiss-Blanquet S, Benoit Y, Marechaux C et al (2005) Assessing the role of alkane hydroxylase genotypes in environmental samples by competitive PCR. J Appl Microbiol 99:1392–1403
Heitkamp MA, Freeman JP, Miller DW et al (1988) Pyrene degradation by a *Mycobacterium* sp.: identification of ring oxidation and ring fission products. Appl Environ Microbiol 54:2556–2565
Hernandez-Raquet G, Budzinski H, Caumette P et al (2006) Molecular diversity studies of bacterial communities of oil polluted microbial mats from the Etang de Berre (France). FEMS Microbiol Ecol 58:550–562
Huesemann MH, Hausmann TS, Fortman TJ (2004) Does bioavailability limit biodegradation? A comparison of hydrocarbon biodegradation and desorption rates in aged soils. Biodegradation 15:261–274
Hybholt TK, Aamand J, Johnsen AR (2011) Quantification of centimeter-scale spatial variation in PAH, glucose and benzoic acid mineralization and soil organic matter in road-side soil. Environ Pollut 159:1085–1091
Jain PK, Gupta VK, Gaur RK et al (2011) Bioremediation of petroleum oil contaminated soil and water. Res J Environ Toxicol 5:1–26
Ji X, Ripp SA, Layton AC et al (2013) Assessing long term effects of bioremediation: soil bacterial communities 14 years after polycyclic aromatic hydrocarbon contamination and introduction of a genetically engineered microorganism. J Bioremed Biodeg 4:8
Jiménez N, Viñas M, Bayona JM et al (2007) The Prestige oil spill: bacterial community dynamics during a field biostimulation assay. Appl Microbiol Biotechnol 77:935–945
Johnsen AR, Karlson U (2005) PAH degradation of soil microbial communities-does it depend on PAH exposure? Microb Ecol 50:488–495
Johnsen AR, Styrishave B, Aamand J (2014) Quantification of small-scale variation in the size and composition of phenanthrene-degrader populations and PAH contaminants in traffic-impacted topsoil. FEMS Microbiol Ecol 88:84–93
Jones MD, Crandell DW, Singleton DR et al (2011) Stable-isotope probing of the polycyclic aromatic hydrocarbon-degrading bacterial guild in a contaminated soil. Environ Microbiol 13:2623–2632
Joye SB, Teske AP, Kostka JE (2014) Microbial dynamics following the Macondo oil well blowout across Gulf of Mexico environments. Bioscience 6:766–777

Juck D, Charles T, Whyte LG (2000) Polyphasic microbial community analysis of petroleum hydrocarbon-contaminated soils from two northern Canadian communities. FEMS Microbiol Ecol 33:241–249

Juhasz AL, Naidu R (2000) Bioremediation of high molecular weight poycyclic aromatic hydrocarbons: a review of the microbial degradation of benzo[a]pyrene. Int Biodeter Biodegrad 45:57–88

Jurelevicius D, Alvarez VM, Peixoto R et al (2012a) Bacterial polycyclic aromatic hydrocarbon ring-hydroxylating dioxygenases (PAH-RHD) encoding genes in different soils from King George Bay, Antarctic Peninsula. Appl Soil Ecol 55:1–9

Jurelevicius D, Cotta SR, Peixoto R et al (2012b) Distribution of alkane-degrading bacterial communities in soils from King George Island, Maritime Antarctic. Eur J Soil Biol 51:37–44

Jurelevicius D, Alvarez VM, Marques JM et al (2013) Bacterial community response to petroleum hydrocarbon amendments in freshwater, marine, and hypersaline water-containing microcosms. Appl Environ Microbiol 79:5927–5935

Kabelitz N, Machackova J, Imfeld G et al (2009) Enhancement of the microbial community biomass and diversity during air sparging bioremediation of a soil highly contaminated with kerosene and BTEX. Appl Microbiol Biotechnol 82:565–577

Kanaly R, Harayama S (2000) Biodegradation of high-molecular-weight polycyclic aromatic hydrocarbons by bacteria. J Bacteriol 182:2059–2067

Kanaly RA, Harayama S (2010) Advances in the field of high-molecular-weight polycyclic aromatic hydrocarbon biodegradation by bacteria. Microb Biotechnol 3:136–164

Kaplan CW, Kits CL (2004) Bacterial succession in a petroleum land treatment unit. Appl Environ Microbiol 70:1777–1786

Kappell AD, Wei Y, Newton RJ et al (2014) The polycyclic aromatic hydrocarbon degradation potential of gulf of Mexico native coastal microbial communities after the Deepwater Horizon oil spill. Front Microbiol 5:205

Karabika E, Kallimanis A, Dados A et al (2009) Taxonomic identification and use of free end entrapped cells of a new Mycobacterium sp. strain Spyr1 for degradation of polycyclic aromatic hydrocarbons (PAHs). Appl Biochem Biotechnol 159:155–167

Kasai Y, Kishira H, Sasaki T (2002) Predominant growth of *Alcanivorax* strains in oil contaminated and nutrient-supplemented sea water. Environ Microbiol 4:141–147

Kasai Y, Shindo K, Harayama S et al (2003) Molecular characterization and substrate preference of a polycyclic aromatic hydrocarbon dioxygenase from *Cycloclasticus* sp. strain A5. Appl Environ Microbiol 69:6688–6697

Kim SJ, Kweon O, Cerniglia CE (2009) Proteomic applications to elucidate bacterial aromatic hydrocarbon metabolic pathways. Curr Opin Microbiol 12:301–309

Kimes NE, Callaghan AV, Suflita JM et al (2014) Microbial transformation of the *Deepwater Horizon* oil spill—past, present, and future perspectives. Front Microbiol 5:603

King GM, Kostka JE, Hazen TC et al (2015) Microbial responses to the *Deepwater Horizon* oil spill: from coastal wetlands to the deep sea. Annu Rev Mar Sci 7:377–401

Kleindienst S, Grim S, Sgin M et al (2015) Diverse rare microbial taxa responded to the *Deepwater Horizon* deep-sea hydrocarbon plume. ISME J 10:400–415

Kostka JE, Prakash O, Overholt WA et al (2011) Hydrocarbon-degrading bacteria and the bacterial community response in Gulf of Mexico beach sands impacted by the *Deepwater Horizon* oil spill. Appl Environ Microbiol 77:7962–7974

Koukkou AI, Vandera E (2011) Hydrocarbon degrading soil bacteria. In: Koukkou AI (ed) Microbial bioremediation of non-metals. Caister Academic Press, Norfolk, UK, pp 93–117

Kristensen AH, Henriksen K, Mortensen L et al (2010) Soil physical constraints on intrinsic biodegradation of petroleum vapors in a layered subsurface. Vadose Zone J 9:137–147

Kumar M, Khanna S (2010) Diversity of 16S rRNA and dioxygenase genes detected in coal-tar-contaminated site undergoing active bioremediation. J Appl Microbiol 108:1252–1262

Labbé D, Margesin R, Schinner F et al (2007) Comparative phylogenetic analysis of microbial communities in pristine and hydrocarbon-contaminated Alpine soils. FEMS Microbiol Ecol 59:466–475

Labud V, Garcia C, Hernandez T (2007) Effect of hydrocarbon pollution on the microbial properties of a sandy and a clay soil. Chemosphere 66:1863–1871

Lacerda CMR, Reardon KF (2009) Environmental proteomics: applications of proteome profiling in environmental microbiology and biotechnology. Brief Funct Genomic Proteomic 8:75–87

Lauber CL, Hamady M, Knight R et al (2009) Pyrosequencing-based assessment of soil pH as a predictor of soil bacterial community structure at the continental scale. Appl Environ Microbiol 75:5111–5120

Le Borgne S, Paniagua D, Vazquez-Duhalt R (2008) Biodegradation of organic pollutants by halophilic bacteria and archaea. J Mol Microbiol Biotechnol 15:74–92

Leahy JG, Colwell RR (1990) Microbial degradation of hydrocarbons in the environment. Microbiol Rev 54:305–315

Leys NM, Ryngaert A, Bastiens L et al (2005) Occurrence and community composition of fast-growing *Mycobacterium* in soils contaminated with polycyclic aromatic hydrocarbons. FEMS Microbiol Ecol 51:375–388

Liang Y, Van Nostrand JD, Deng Y et al (2011) Functional gene diversity of soil microbial communities from five oil-contaminated fields in China. ISME J 5:403–413

Liang Y, Zhang X, Wang J et al (2012) Spatial variations of hydrocarbon contamination and soil properties in oil exploring fields across China. J Hazard Mater 241–242:371–378

Liang Y, Zhao H, Zhang X et al (2014) Contrasting microbial functional genes in two distinct saline-alkali and slightly acidic oil-contaminated sites. Sci Tot Environ 487:272–278

Liang Y, Zhang X, Zhou J et al (2015) Long-term oil contamination increases deterministic assembly processes in soil microbes. Ecol Appl 25:1235–1243

Liao J, Wang J, Jiang D et al (2015) Long-term oil contamination causes similar changes in microbial communities of two distinct soils. Appl Microbiol Biotechnol 99:10299–10310

Liu D, Hoynes-O'Connor A, Zhang F (2013) Bridging the gap between systems biology and synthetic biology. Front Microbiol 4:211

Lo Guidice A, Bruni V, De Domenico M et al (2010) Psychrophiles- cold-adapted hydrocarbon-degrading microorganisms. In: Timmis KN (ed) Handbook of hydrocarbon and lipid microbiology. Springer, Berlin, Heidelberg, pp 1898–1921

Lors C, Ryngaert A, Périé F et al (2010) Evolution of bacterial community during bioremediation of PAHs in a coal tar contaminated soil. Chemosphere 81:1263–1271

Löser S, Seidel H, Zehnsdorf A et al (1998) Microbial degradation of hydrocarbons in soil during aerobic/anaerobic changes and under purely aerobic conditions. Appl Microbiol Biotechnol 49:631–636

Louati H, Said OB, Got P et al (2013) Microbial community responses to bioremediation treatments for the mitigation of low-dose anthracene in marine coastal sediments of Bizerte lagoon (Tunisia). Environ Sci Pollut Res 20:300–310

Lozupone CA, Knight R (2007) Global patterns in bacterial diversity. PNAS 104:11436–11440

Lu XY, Zhang T, Fang HHP (2011) Bacteria-mediated PAH degradation in soil and sediment. Appl Microbiol Biotechnol 89:1357–1371

Ma Y, Wang L, Shao Z (2006) *Pseudomonas*, the dominant polycyclic hydrocarbons-degrading bacteria isolated from Arctic soils and the role of large plasmids in horizontal gene transfer. Environ Microbiol 8:455–465

Malik S, Beer M, Megharaj M et al (2008) The use of molecular techniques to characterize microbial communities in contaminated soil and water. Environ Int 34:265–276

Margesin R, Schinner F (2001) Biodegradation and bioremediation of hydrocarbons in extreme environments (mini-review). Appl Microbiol Biotechnol 56:650–663

Margesin R, Gander S, Zacke G et al (2003a) Hydrocarbon degradation and enzyme activities of cold-adapted bacteria and yeasts. Extremophiles 7:451–458

Margesin R, Labbé D, Schinner F et al (2003b) Characterization of hydrocarbon-degrading microbial populations in contaminated and pristine Alpine soils. Appl Environ Microbiol 69:3085–3092

Mason OU, Hazen TC, Borglin S et al (2012) Metagenome, metatranscriptome and single-cell sequencing reveal microbial response to Deepwater Horizon oil spill. ISME J 6:1715–1727

McGenity TJ (2010) Halophilic hydrocarbon degraders. In: Timmis KN (ed) Handbook of hydrocarbon and lipid microbiology. Springer, Berlin, Germany, pp 1940–1951
McGenity TJ, Folwell BD, Boyd A et al (2012) Marine crude-oil biodegradation: a central role for interspecies interactions Aquat Biosyst 8:10
McKew BA, Coulon F, Osborn AM et al (2007a) Determining the identity and roles of oil-metabolizing marine bacteria from the Thames estuary, UK. Environ Microbiol 9:165–176
McKew BA, Coulon F, Yakimov MM et al (2007b) Efficacy of intervention strategies for bioremediation of crude oil in marine systems and effects on indigenous hydrocarbonoclastic bacteria. Environ Microbiol 9:1562–1571
Militon C, Boucher D, Vachelard C et al (2010) Bacterial community changes during bioremediation of aliphatic hydrocarbon-contaminated soil. FEMS Microbiol Ecol 74:669–681
Mnif S, Chamkha M, Sayadi S (2009) Isolation and characterization of *Halomonas* sp. strain C2SS100, a hydrocarbon-degrading bacterium under hypersaline conditions. J Appl Microbiol 107:785–794
Morais D, Pylro V, Clark IM et al (2016) Responses of microbial community from tropical pristine coastal soil to crude oil contamination. Peer J 4:e1733
Moretto LM, Silvestri S, Ugo P et al (2005) Polycyclic aromatic hydrocarbons degradation by composting in a soot-contaminated alkaline soil. J Hazard Mater 126:141–148
Muckian L, Grant R, Doyle E et al (2007) Bacterial community structure in soils contaminated by polycyclic aromatic hydrocarbons. Chemosphere 68:1535–1541
Muckian LM, Grant RJ, Clipson NJW et al (2009) Bacterial community dynamics during bioremediation of phenanthrene- and fluoranthene-amended soil. Int Biodeter Biodegrad 63:52–56
Mukherjee S, Juottonen H, Siivonen P et al (2014) Spatial patterns of microbial diversity and activity in an aged creosote-contaminated site. ISME J 8:2131–2142
Newton RJ, Huse SM, Morrison HG et al (2013) Shifts in the microbial community composition of Gulf coast beaches following beach oiling. PLoS ONE 8:e74265
Ni Chadhain SM, Norman RS, Pesce KV et al (2006) Microbial dioxygenase gene population shifts during polycyclic aromatic hydrocarbon biodegradation. Appl Environ Microbiol 72:4078–4087
Niepceron M, Beguet J, Portet-Koltalo F et al (2014) Low impact of phenanthrene dissipation on the bacterial community in grassland soil. Environ Sci Pollut Res 21:2977–2987
Nikolopoulou M, Kalogerakis N (2010) Biostimulation strategies for enhanced bioremediation of marine oil spills including chronic pollution. In: Timmis KN (ed) Handbook of hydrocarbon and lipid microbiology. Springer, Berlin, Heidelberg, Germany, pp 2521–2529
Obayori OS, Salam LB (2010) Degradation of polycyclic aromatic hydrocarbons: role of plasmids. Sci Res Essays 5:4093–4106
Okoh AI (2006) Biodegradation alternative in the cleanup of petroleum hydrocarbon pollutants. Biotechnol Mol Biol Rev 1:38–50
Olajire AA, Essien JP (2014) Aerobic degradation of petroleum components by microbial consortia. J Pet Environ Biotechnol 5:195
Paisse S, Duran R, Coulon F et al (2011) Are alkane hydroxylase genes (*alkB*) relevant to assess petroleum bioremediation processes in chronically polluted coastal sediments? Appl Microbiol Biotechnol 92:835–844
Patel V, Sharma A, Lal R et al (2016) Response and resilience of soil microbial communities inhabiting in edible oil stress/contamination from industrial estates. BMC Microbiol 16:50
Peng RH, Xiong AS, Xue Y et al (2008) Microbial biodegradation of polyaromatic hydrocarbons. FEMS Microbiol Rev 32:927–955
Peng M, Zi X, Wang Q (2015) Bacterial community diversity of oil-contaminated soils assessed by high throughput sequencing of 16S rRNA genes. Int J Environ Res Public Health 12:12002–12015
Perez-de-Mora A, Engel M, Schloter M (2011) Abundance and diversity of *n*-alkane-degrading bacteria in a forest soil co-contaminated with hydrocarbons and metals: a molecular study on *alkB* homologous genes. Microb Ecol 62:959–972

Popp N, Schlömann M, Mau M (2006) Bacterial diversity in the active stage of a bioremediation system for mineral oil hydrocarbon-contaminated soils. Microbiol 152:3291–3304

Powell SM, Ferguson SH, Bowman JP et al (2006) Using real-time PCR to assess changes in hydrocarbon-degrading microbial community in Antarctic soil during bioremediation. Microb Ecol 52:523–532

Ramette A, Tiedje JM (2007) Multiscale responses of microbial life to spatial distance and environmental heterogeneity in a patchy ecosystem. PNAS 104:2761–2766

Rappé M, Giovannoni SJ (2003) The uncultured microbial majority. Annu Rev Microbiol 57:369–394

Redmond MC, Valentine DL (2012) Natural gas and temperature structured a microbial community response to the *Deepwater Horizon* oil spill. Proc Natl Acad Sci U S A 109:20292–20297

Ribeiro H, Mucha AP, Almeida MR et al (2013) Bacterial community response to petroleum contamination and nutrient addition in sediments from a temperate salt marsh. Sci Tot Environ 458–460:568–576

Rivers AR, Sharma S, Tringe SG et al (2013) Transcriptional response of bathypelagic marine bacterioplankton to the *Deepwater Horizon* oil spill. ISME J 7:2315–2329

Rodriguez-R LM, Overholt WA, Haga C et al (2015) Microbial community successional patterns in beach sands impacted by the *Deepwater Horizon* oil spill. ISME J 9:1928–1940

Rojo F (2009) Degradation of alkanes by bacteria. Environ Microbiol 11:2477–2490

Röling WFM, Milner MG, Jones DM et al (2002) Robust hydrocarbon degradation and dynamics of bacterial communities during nutrient-enhanced oil spill bioremediation. Appl Environ Microbiol 68:5537–5548

Röling WFM, Milner MG, Jones DM et al (2004) Bacterial community dynamics and hydrocarbon degradation during a field-scale evaluation of bioremediation on a mudflat beach contaminated with buried oil. Appl Environ Microbiol 70:2603–2613

Röling WFM, Ferrer M, Golyshin PN (2010) Systems approaches to microbial communities and their functioning. Curr Opin Biotechnol 21:532–538

Ron EZ, Rosenberg E (2002) Biosurfactants and oil bioremediation. Cur. Opin. Biotechnol 13:249–252

Rousk J, Baath E, Brookes PC et al (2010) Soil bacterial and fungal communities across a pH gradient in an arable soil. ISME J 4:1340–1351

Sawulski P, Clipson N, Doyle E (2014) Effects of polycyclic aromatic hydrocarbons on microbial community structure and PAH ring hydroxylating dioxygenase gene abundance in soil. Biodegrad 25:835–847

Sharpton TJ (2014) An introduction to the analysis of shotgun metagenomic data. Front Plant Sci 5:209

Shivaji S, Reddy GSN (2010) Bacterial biodiversity in Antarctica: conventional polyphasic and rRNA approaches. In: Bej AK, Aislabie J, Atlas RM (eds) Polar microbiology: the ecology, biodiversity, and bioremediation potential of microorganisms in extremely cold environments. CRC Press, Boca Raton, FL, pp 61–93

Siggins A, Gunnigle E, Abram F (2012) Exploring mixed microbial community functioning: recent advances in metaproteomics. FEMS Microbiol Ecol 80:265–280

Sihag S, Pathak H, Jaroli DP (2014) Factors affecting the rate of biodegradation of polyaromatic hydrocarbons. Int J Pure App Biosci 2:185–202

Simarro R, González N, Bautista LF et al (2013) Biodegradation of high-molecular-weight polycyclic aromatic hydrocarbons by a wood-degrading consortium at low temperatures. FEMS Microbiol Ecol 83:438–449

Simon C, Daniel R (2011) Metagenomic analyses: past and future trends. Appl Environ Microbiol 77:1153–1161

Singh OV, Nagaraj NS (2006) Transcriptomics, proteomics and interactomics: unique approaches to track the insights of bioremediation. Brief Funct Genomic Proteomic 4:355–362

Singh SN, Kumari B, Mishra S (2012) Microbial degradation of alkanes. In: Singh SN (ed) Microbial degradation of xenobiotics. Springer, Berlin, Heidelberg, pp 439–469

Singleton DR, Ramirez LG, Aitken MD (2009) Characterization of a polycyclic aromatic hydrocarbon degradation gene cluster in a phenanthrene-degrading *Acidovorax* strain. Appl Environ Microbiol 75:2613–2620

Singleton DR, Jones MD, Richardson SD et al (2013) Pyrosequence analyses of bacterial communities during simulated *in situ* bioremediation of polycyclic aromatic hydrocarbon-contaminated soil. Appl Microbiol Biotechnol 97:8381–8391

Smalla M, Wachtendorf U, Heuer H et al (1998) Analysis of BIOLOG GN substrate utilization patterns by microbial communities. Appl Environ Microbiol 64:1220–1225

Smith CB, Tolar BB, Hollibaugh JT et al (2013) Alkane hydroxylase gene (*alkB*) phylotype composition and diversity in northern Gulf of Mexico bacterioplankton. Front Microbiol 4:370

Sørensen SR, Johnsen AR, Jensen A et al (2010) Presence of psychrotolerant phenanthrene-mineralizing bacterial populations in contaminated soils from the Greenland High Arctic. FEMS Microbiol Lett 305:148–154

Stefani FOP, Bell TH, Marchand C et al (2015) Culture-dependent and—independent methods capture different microbial community fractions in hydrocarbon-contaminated soils. PLoS ONE 10(6):e0128272

Sun W, Xie S, Luo C et al (2010) Direct link between toluene degradation in contaminated-site microcosms and a *Polaromonas* strain. Appl Environ Microbiol 76:956–959

Sutton NB, Maphosa F, Morillo JA et al (2013) Impact of long-term diesel contamination on soil microbial community structure. Appl Envriron Microbiol 79:619–630

Syutsubo K, Sinthurat N, Ohashi A et al (2001) Population dynamics of an aerobic microbial consortia in thermophilic granular sludge in response to feed composition change. Water Sci Technol 43:59–66

Taketani RG, Franco NO, Rosado AS et al (2010) Microbial community response to a simulated hydrocarbon spill in mangrove sediments. J Microbiol 48:7–15

Taylor JP, Wilson B, Mills MS et al (2002) Comparison of microbial numbers and enzymatic activities in surface soils and subsoils using various techniques. Soil Biol Biochem 34:387–401

Thavamani P, Malik S, Beer M et al (2012a) Microbial activity and diversity in long-term mixed contaminated soils with respect to polyaromatic hydrocarbons and heavy metals. J Environ Management 99:10–17

Thavamani P, Megharaj M, Naidu R (2012b) Multivariate analysis of mixed contaminants (PAHs and heavy metals) at manufactured gas plant site soils. Environ Monit Assess 184:3875–3885

Tiedje JM, Cho JC, Murray A et al (2001) Soil teeming with life: new frontiers for soil science. In: Rees RM, Ball BC, Campbell CD, Watson CA (eds) Sustainable management of soil organic matter, CAB International, pp 393–425

Top EM, Springael D (2003) The role of mobile genetic elements in bacterial adaptation to xenobiotic organic compounds. Curr Opin Biotechnol 14:262–269

Törneman N, Yang X, Bååth E et al (2008) Spatial covariation of microbial community composition and polycyclic aromatic hydrocarbon concentration in a creosote-polluted soil. Environ Toxicol Chem 27:1039–1046

Torsvik V, Øvreås L (2002) Microbial diversity and function in soil: from genes to ecosystems. Curr Opin Microbiol 5:240–245

Uhlik O, Leewis MC, Strejcek M et al (2013) Stable isotope probing in the metagenomics era: a bridge towards improved bioremediation. Biotechnol Adv 31:154–165

Valentine DL, Kessler JD, Redmond MC et al (2010) Propane respiration jump-starts microbial response to a deep oil spill. Science 330:208–211

Van Hamme JD, Singh A, Ward OP (2003) Recent advances in petroleum microbiology. Microbiol Mol Biol Rev 67:503–549

Vieira PA, Vieira RB, Faria S et al (2009) Biodegradation of diesel oil and gasoline contaminated effluent employing intermittent aeration. J Hazard Mater 168:1366–1372

Viggor S, Juhanson J, Jõesaar M et al (2013) Dynamic changes in the structure of microbial communities in Baltic Sea coastal seawater microcosms modified by crude oil, shale oil or diesel fuel. Microbiol Res 168:415–427

Vila J, María Nieto J, Mertens J et al (2010) Microbial community structure of a heavy fuel oil-degrading marine consortium: linking microbial dynamics with polycyclic aromatic hydrocarbon utilization. FEMS Microbiol Ecol 73:349–362
Viñas M, Sabate J, Espuny MJ et al (2005) Bacterial community dynamics and polycyclic aromatic hydrocarbon degradation during bioremediation of heavily creosote-contaminated soil. Appl Environ Microbiol 71:7008–7018
Vitte I, Duran R, Jézéquel R et al (2011) Effect of oxic/anoxic switches on bacterial communities and PAH biodegradation in an oil-contaminated sludge. Environ Sci Pollut Res 18:1022–1032
Vitte I, Duran R, Hernandez-Raquet G et al (2013) Dynamics of metabolically active bacterial communities involved in PAH and toxicity elimination from oil-contaminated sludge during oxic/anoxic oscillations. Appl Microbiol Biotechnol 97:4199–4211
Wallisch S, Gril T, Dong X et al (2014) Effects of different compost amendments on the abundance and composition of *alkB* harboring bacterial communities in a soil under industrial use contaminated with hydrocarbons. Front Microbiol 5:96
Wang Z, Brown C (2009) Chemical fingerprinting of petroleum hydrocarbons. In: Mudge SM (ed) Methods in environmental forensics. CRC Press, Boca Raton, FL, pp 43–112
Wang W, Shao Z (2013) Enzymes and genes involved in aerobic alkane degradation. Front Microbiol 4:116
Wasmund K, Burns KA, Kurtböke I et al (2009) Novel alkane hydroxylase gene (*alkB*) diversity in sediments associated with hydrocarbon seeps in the Timor Sea, Australia. Appl Environ Microbiol 75:7391–7398
Wentzel A, Ellingsen TE, Kotlar HK et al (2007) Bacterial metabolism of long-chain *n*-alkanes. Appl Microbiol Biotechnol 76:1209–1221
Xu Y, Lu W (2010) Bioremediation of crude oil-contaminated soil: comparison of different biostimulation and bioaugmentation treatments. J Hazard Mater 183:395–401
Yakimov M, Denaro R, Genovese M et al (2005) Natural microbial diversity in superficial sediments of Milazzo Harbor (Sicily) and community successions during microcosm enrichment with various hydrocarbons. Environ Microbiol 7:1426–1441
Yang X, Beckmann D, Fiorenza S et al (2005) Field study of pulsed air sparging for remediation of petroleum hydrocarbon contaminated soil and groundwater. Environ Sci Technol 39:7279–7286
Yang S, Wen X, Zhao L et al (2014a) Crude oil treatment leads to shift of bacterial communities in soils from the deep active layer and upper permafrost along the China-Russia crude oil pipeline route. PLoS ONE 9:e96552
Yang T, Nigro LM, Gutierrez T et al (2014b) Pulsed blooms and persistent oil-degrading bacterial populations in the water column during and after the Deepwater Horizon blowout. Deep-Sea Res II. doi:10.1016/j.dsr2.2014.01.014
Yang Y, Wang J, Liao J et al (2014c) Distribution of naphthalene dioxygenase genes in crude-oil contaminated soils. Environ Microbiol 68:785–793
Yang Y, Wang J, Lian J et al (2015) Abundance and diversity of soil petroleum hydrocarbon-degrading microbial communities in oil exploring areas. Appl Microbiol Biotechnol 99:1935–1946
Yergeau E, Sanschagrin S, Beaumier D et al (2012) Metagenomic analysis of the bioremediation of diesel-contaminated Canadian high arctic soils. PLoS ONE 7:e30058
Zarraonaindia I, Smith DP, Gilbert JA (2013) Beyond the genome: community-level analysis of the microbial world Biol Philos 28:261–282
Zhang D-C, Mortelmaier C, Margesin R (2012) Characterization of the bacterial and archaeal diversity in hydrocarbon-contaminated soil. Sci Tot Environ 421–422:184–196
Zhang ZY, Zhao X, Liang YT et al (2013) Microbial functional genes reveal selection of microbial community by PAHs in polluted soils. Environ Chem Lett 11:11–17
Zhao B, Wang H, Mao X et al (2009) Biodegradation of phenanthrene by a halophilic bacterial consortium under aerobic conditions. Curr Microbiol 58:205–210
Zhang SY, Wang QF, Xie SG (2011) Microbial community changes in contaminated soils in response to phenanthrene amendment. Int J Environ Sci Technol 8(2):321-330

Chapter 10
Microbial Communities as Ecological Indicators of Ecosystem Recovery Following Chemical Pollution

Stéphane Pesce, Jean-François Ghiglione and Fabrice Martin-Laurent

Abstract 'Ecosystem recovery' is a concept that emerged from the need to preserve our environment against increasing contamination from human activity. However, ecological indicators of ecosystem recovery remain scarce, and it is still difficult to assess recovery of ecological processes at relevant spatial and temporal scales. Microbial communities hold key relevance as indicators of ecosystem recovery as they are ubiquitous among diverse ecosystems, respond rapidly to environmental changes, and support many ecosystem functions and services through taxonomic and functional biodiversity. This chapter summarizes the state-of-the-art in knowledge on the processes driving the structural and functional recovery of phototroph and heterotroph microorganisms following chemical pollution. It covers several successful case studies providing proof of principle for the relevance of using microorganisms in recovery studies in various ecosystems such as soil, freshwater and seawater. Questions remain for microbial ecotoxicologists to fully understand and predict how structural and functional recovery observed at microbial scale can reflect the recovery of an ecosystem. Moreover, new standards and norms taking into account recent advances in microbial ecotoxicology are now necessary in order to inform legislators and policymakers on the importance of considering microorganisms in environmental risk assessment, including ecological recovery monitoring.

S. Pesce (✉)
Irstea, UR MALY, Centre de Lyon-Villeurbanne, 5 rue de la Doua—CS 20244, 69625 Villeurbanne Cedex, France
e-mail: stephane.pesce@irstea.fr

J.-F. Ghiglione
Laboratoire d'Océanographie Microbienne, Observatoire Océanologique, Sorbonne Universités, CNRS, UPMC Univ Paris 06, UMR 7621, 66650 Banyuls-sur-Mer, France

F. Martin-Laurent
Agroécologie, AgroSup Dijon, INRA, Université de Bourgogne Franche-Comté, 21000 Dijon, France

C. Cravo-Laureau et al. (eds.), *Microbial Ecotoxicology*,
DOI 10.1007/978-3-319-61795-4_10

Keywords Microbial ecotoxicology · Microbial recovery · Biomonitoring · Environmental risk assessment

10.1 Relevance of Using Microbial Communities to Assess Ecosystem Recovery

The last two decades have seen a worldwide surge in environmental regulations designed to promote effective environmental management practices to reduce anthropogenic chemical impacts in ecosystems (Depledge 1998; Hering et al. 2010). Ecological restoration has thus emerged as one of the most important issues in environmental science (Montoya et al. 2012), spurring the emergence of the concept of ecosystem recovery (Fig. 10.1), which implies that a restored ecosystem evolves towards the direction of the pre-disturbance conditions to recover healthy conditions. Ecosystem recovery is built around several paradigms (Duarte et al. 2015) and driven by complex processes involving multiple biological levels over different timescales (Adams et al. 2002). It is already a challenge to define ecosystem-healthy conditions, which revolves around the concept of normal operating range (NOR) as the range in ecological metrics observed in the ecosystem's undisturbed state under natural fluctuations in environmental conditions (EFSA Scientific Committee 2016). It is also crucial to choose the appropriate ecological metrics for assessing ecosystem recovery, as they should not only inform on the structural recovery of ecosystems but also allow us to assess the recovery of ecosystem functions, including ecosystem services (Bullock et al. 2011; Montoya et al. 2012).

A few decades ago, no-one would have expected to see microbiologists play a role in the evaluation of ecosystem recovery. Today, though, the situation has reversed, as it is difficult to find a single ecosystem on earth where microorganisms have not been identified as key players in its functioning. Despite their small size, microorganisms are not only the most abundant organisms but are also recognized as major components of all biogeochemical element cycles (C, N, P, S, metals).

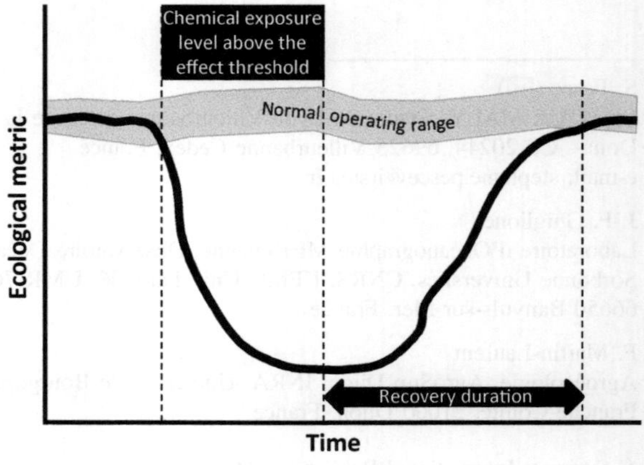

Fig. 10.1 Schematic illustration of ecosystem recovery following chemical pollution. Adapted from EFSA Scientific Committee (2016)

Important recent discoveries have advanced the genomic, biochemical, physiological and ecological bases of a variety of microbiological processes, like anaerobic methane oxidation, photosynthesis, phosphorous uptake, and many aspects of the sulfur and nitrogen cycles, from anammox reaction and dissimilatory nitrate reduction to ammonia to archaeal nitrification (Madsen 2011). Indeed, it is well acknowledged that microbial communities maintain the biosphere via the biogeochemical reactions they catalyze. Moreover, recent moves to consider microorganisms along with living animals and plants—no longer viewed as autonomous entities but rather as assemblages of different species forming ecological units called holobionts—has shaken up the life sciences (Bordenstein and Theis 2015).

Advances in microbial ecology allow us to extend the mechanistic understanding of relatively simple biological systems to complex naturally-occurring microbial communities that dwell in soils, air, sediments and waters. The emerging discipline of microbial ecotoxicology is now facing the challenge of evaluating the relevance of microbial communities for assessing ecosystem recovery after pollution (Ghiglione et al. 2016).

10.2 Structural and Functional Recovery Potential of Microbial Communities Following a Decrease in Chemical Exposure

The potential of microbial communities to recover from disturbances depends on both the internal and external recovery capacities of their constitutive populations through population growth of surviving organisms or propagules and re-colonization following passive or active dispersal, respectively (EFSA Scientific Committee 2016; Gergs et al. 2016). To gain an overview of how microbial communities can recover from chemical exposure and be able to predict recovery trajectories, it is first necessary to better understand the mechanisms underpinning internal and external recovery. Such investigations can be conducted at population and community levels using laboratory or in situ experimental studies.

10.2.1 Internal Recovery Potential of Microbial Populations: The Case of Photosynthetic Microorganisms

Among microorganisms, algae and cyanobacteria are the most intensively studied model organisms in aquatic ecotoxicology. Several studies assessing microbial recovery potential at population level have been performed using freshwater photosynthetic microbial species (Table 10.1). Vallotton et al. (2008a, b) evaluated the capacity of the Chloropyceae *Scenedesmus vacuolatus* to recover following acute

Table 10.1 Laboratory studies of the recovery potential of algal and cyanobacterial populations after exposure to various pesticides and heavy metals

Algal species	Structural metrics	Functional metrics	Contaminant(s)	Nominal concentrations	Maximal exposure/recovery duration	References
Planothidium frequentissimum *Pseudokirchneriella subcapitaa*	Teratologica l forms	Growth, viability	Metal (Cd)	20–100 μg/L	21 day/28 day	Arini et al. (2013)
Anabaena flos-aquae *Navicula pelliculosa*	Cell densities	Growth, photosynthesis	Herbicide (atrazine)	5–1000 μg/L	48 h/48 h	Brain et al. (2012)
Phaeodactylum tricornutum		Growth, phytochelatin synthesis	Metals (Cd, Pb, Zn)	112 μg/L (Cd) 207 μg/L (Pb) 65 μg/L (Zn)	8 h/24 h	Morelli and Scarano (2001)
Selenastrum capricornutum *Chlorella vulgaris*		Growth	Metal (Zn)	65 μg/L	100 day/10 day	Muyssen and Janssen (2001)
Selenastrum capricornutum	Chlorophyll a content	Growth, carbon assimilation	Metal (Cd)	30–100 μg/L	48 h/96 h	Thompson and Couture (1993)
Scenedesmus sp.		Growth, photosynthesis, respiration, uptake and assimilation of nitrate	Metals (Cu and Zn)	159–635 μg/L (Cu) 327–1635 μg/L (Zn)	48 h/96 h	Tripathi et al. (2004)
Scenedesmus sp.	Photosynthetic pigments, protein, carbohydrate and lipid content	Growth, photosynthesis, respiration, uptake and assimilation of nitrate	Metals (Cu and Zn)	158–635 μg/L (Cu) 327–1635 μg/L (Zn)	48 h/96 h	Tripathi and Gaur (2006)

(continued)

Table 10.1 (continued)

Algal species	Structural metrics	Functional metrics	Contaminant(s)	Nominal concentrations	Maximal exposure/recovery duration	References
Scenedesmus vacuolatus		Photosynthesis, growth	Herbicides (isoproturon and atrazine)	60–320 µg/L (isoproturon) 80–510 µg/L (atrazine)	25 h/48 h	Vallotton et al. (2008a)
Scenedesmus vacuolatus		Growth	Herbicide (S-metolachlor)	750 µg/L	24 h/48 h	Vallotton et al. (2008b)
Thalassiosira nordenskioeldii		Growth, phytochelatin synthesis	Metal (Cd)	0.001–10 µg/L	7 day/15 day	Wang and Wang (2011)
Microcystis aeruginosa		Growth (sensitivity tests)	Metals (Cd and Zn)	3.37 µg/L (Cd) 0.65 µg/L (Zn)	5 day/5 day	Zeng et al. (2009)

pulse exposure to various herbicides. The effective quantum yield recovered completely within 4 h after removal of atrazine and isoproturon, leading to rapid recovery of photosynthetic microorganism growth independently of the magnitude of the effects induced by these two photosystem-II inhibitors (Vallotton et al. 2008a). By testing different exposure levels to atrazine (5–1000 μg/L for 48 h), Brain et al. (2012) observed that the resulting effects on photosynthesis and growth were transient and fully reversible within 48 h in three tested photosynthetic microorganism species of chlorophyceae, cyanobacteria and diatoms, respectively. However, the recovery of *S. vacuolatus* following an acute exposure to the chloroacetanilide S-metolachlor was delayed, occurring only after 29 h, revealing that the extent and time-to-reversibility of the toxic effects may be dependent on the nature of the toxicant (Vallotton et al. 2008b).

An important parameter to consider here is the kinetics of toxicant elimination from the cells. Metals are well known to bioaccumulate in photosynthetic microorganisms. The potential of photosynthetic microbial populations to recover following metal exposure was investigated using various species belonging to the chlorophyceae (Morelli and Scarano 2001; Muyssen and Janssen 2001; Thompson and Couture 1993; Tripathi and Gaur 2006; Tripathi et al. 2004), diatoms (Arini et al. 2013; Morelli and Scarano 2001; Wang and Wang 2011) and cyanobacteria (Zeng et al. 2009). Most of these studies reported a significant decrease in intracellular concentrations of cadmium (Cd) (Arini et al. 2013; Thompson and Couture 1993; Wang and Wang 2011), copper (Cu); (Tripathi and Gaur 2006) and zinc (Zn) (Tripathi and Gaur 2006), whatever the model species. However, the extent of recovery proved variable according to the exposure conditions (duration and concentrations), parameters measured, and duration of the recovery period. This is clearly illustrated by Tripathi et al. (2004, 2006) who assessed the recovery of *Scenedesmus sp.* using a set of structural (i.e. photosynthetic pigments, protein, carbohydrate and lipid contents) and functional parameters (i.e. growth, cell viability, photosynthesis, respiration, uptake and assimilation of nitrate) following a 48 h exposure to Cu and Zn tested at two nominal concentration levels each (2.5–10 and 5–25 μM, respectively). Photosynthesis and respiration recovered quickly without any immediate change in cell density, suggesting an adaptive response for producing energy and returning to normal catabolism conditions. This functional recovery was accompanied by a slight decline in lipid contents as well as an increase in carbohydrates, which are a preferred source of energy. Nitrate reductase activity recovered much earlier than nitrate uptake, but both these processes were dependent on the recovery of photosynthesis and respiration which provide the energy required to recover microbial activities. This is consistent with the results of Tripathi et al. (2004) who observed that recovery from metal stress was slower when algae were previously exposed for 72 h to dark conditions, whereas no recovery was found in the presence of 3-(3,4-dichlorophenyl)-1,1-dimethylurea (DCMU), a transformation product of the herbicide diuron, which acts as a photosynthetic inhibitor. When photosynthesis was possible, the resulting chain of metabolic events stimulated algal growth, allowing enhanced dilution of intracellular level of metals. However, the relatively high intracellular levels of Cu or Zn maintained in algal populations exposed to the highest metal concentrations precluded complete recovery of some processes during the 96 h recovery period, which was probably too short given the concentrations tested (i.e. 10 and 25 μM, respectively). Based on a study of teratological forms, and despite complete depuration of intracellular Cd, Arini et al. (2013) also observed incomplete recovery of *Planothidium frequentissimum* diatoms, even at 23 days after removal of Cd contamination (at 20 and 100 μg/L).

Recovery assessment at population level can also be conducted by studying the adaptive processes of photosynthetic microorganisms in response to toxicant exposure. Indeed, microbial adaptation leading to the ability to tolerate toxicants is a defense strategy that generally generates an energetic cost that weakens the microorganisms' ability to cope with supplementary disturbances (Congdon et al. 2001). This means that from an ecological point of view, loss of adaptation to toxicants, at population or community level, can be perceived as an indication of microbial recovery (Pesce et al. 2013, 2016). To that effect, Morelli and Scarano (2001) and Wang and Wang (2011) studied phytochelatins, which are metal-binding thiol-containing peptides, in response to heavy metals exposure and observed a rapid decrease in the phytochelatin pool in diatoms exposed to various metals, confirming a recovery process within the populations. Another approach consists in studying the evolution of tolerance capacities of photosynthetic microorganisms towards toxicants by performing short-term toxicity tests based on functional parameters. Using this approach, and by measuring growth rates, Zeng et al. (2009) evidenced an increase in the tolerance of the cyanobacteria *Microcystis aeruginosa* towards Cd or Zn according to the pre-exposure conditions (concentration and duration) used. In the metal-free medium, an increase in sensitivity to metals was observed following 1-day recovery while a 5-day recovery led to complete loss of tolerance capacities. The same trend was observed by Muyssen and Janssen (2001) in the two chlorophyceae species *Selenastrum capricornutum* and *Chlorella vulgaris* which showed a maximal 3-fold increase in zinc tolerance (based on growth inhibition tests) after 100 days of exposure to 65 μg Zn/L followed by a significant decrease in tolerance after a 10-day recovery period in a metal-free medium. Note that the rapid decrease in the tolerance following recovery in these two studies may indicate that the tolerance involves physiological acclimatization rather than genetic adaptation, such as the production of intracellular ligands (e.g. phytochelatins or metallothioneins) which can complex or detoxify intracellular metals (Zeng et al. 2009).

10.2.2 Internal and External Recovery Potential of Microbial Communities

Even if the study of microbial recovery potential at population level may be relevant to evaluate the internal capacities of microorganisms to recover and to characterize the mechanisms involved, it is now well acknowledged that ecotoxicological studies hold stronger ecological relevance when they consider biological responses at community level, applying community ecology concepts (Clements and Rohr 2009; Geiszinger et al. 2009; Schmitt-Jansen et al. 2008). This statement also holds for recovery studies especially when the aim is to study an ecosystem's capacity to recover from disturbances (Admiraal et al. 2000; EFSA Scientific Committee 2016).

10.2.2.1 Microcosm and Mesocosm Experiments

Using microcosm or mesocosm approaches to address ecological recovery offers several advantages, including the possibilities for controlling and standardizing exposure and habitat conditions, allowing replication and statistical evaluation, and taking into consideration certain ecological interactions.

Several studies have been performed to study the potential of freshwater phototrophic microbial communities to recover following herbicide exposure (Pesce et al. 2011). Some of these works aimed specifically at evaluating short-term recovery of periphyton in herbicide-free water after an acute pulse exposure to photosystem inhibitors (i.e. *s*-triazine and substituted phenylurea herbicides), varying between 1 and 48 h (Gustavson et al. 2003; Laviale et al. 2011; Prosser et al. 2013, 2015). All of these studies showed high short-term potential recovery of photosynthesis, even after exposure to toxic concentrations significantly inhibiting this function. However, functional recovery trajectories varied according to exposure duration (Gustavson et al. 2003; Laviale et al. 2011), tested concentrations (Gustavson et al. 2003; Laviale et al. 2011; Prosser et al. 2013, 2015) and kind of toxicants, even for those having the same mode of action (Gustavson et al. 2003; Laviale et al. 2011). Gustavsson et al. (2003) also pointed out that functional recovery is sometimes not associated with structural recovery. Indeed, while the recovery of photosynthetic activity in periphyton after an acute exposure to metribuzin was almost complete after 48 h in herbicide-free water, even after exposure at the concentration of 50 μg/L where photosynthesis was inhibited by 80%, the composition of the periphyton remained impacted, even at the lowest concentration of 0.4 μg/L. This was due to the fact that chlorophytes were severely affected by exposure and failed to recover whereas diatoms and especially cyanobacteria recovered well. This report clearly illustrates that functional redundancy can contribute to the rapid recovery of some ecological functions. A delay in chlorophytes recovery following a chronic exposure to the herbicide metazachlor was also observed by Mohr et al. (2008a), confirming that different microbial populations within a complex community can exhibit different recovery trajectories following chemical exposure, due to their intrinsic properties.

However, these trajectories can also be highly influenced by the existence or not of microbial immigration processes. This was clearly demonstrated in studies by Lambert et al. (2012) and Morin et al. (2012) who observed no structural recovery of periphytic diatom communities within 6 weeks following a chronic exposure to copper when immigration process from non-exposed communities were impossible, whereas recovery was complete when these same processes were enhanced. This report was confirmed by a pollution-induced community tolerance (PICT) approach showing that the Cu phototrophic tolerance that had been induced during the exposure period was only lost when immigration was possible (Lambert et al. 2012). Moreover, photosynthesis measurements revealed that the lack of immigration precluded functional recovery of phototrophic communities (Lambert et al. 2012). Arini et al. (2012b) also suggested that the limited recovery they observed in the structure of periphytic diatom communities 8 weeks after a chronic exposure to metals may have been due, at least partially, to the difficulty of non-impacted species to invade the pre-exposed biofilms.

Nevertheless, immigration processes seem to be less important to the structural recovery dynamics of periphytic bacterial communities following a metal stress. Lambert et al. (2012) observed that, in contrast to diatoms, the structure of bacterial communities in metal-exposed samples remained quite different from controls throughout the 6-week recovery period, even when species immigration was possible. This is consistent with other reports of weak structural recovery of periphytic bacterial communities within several weeks after a decrease in metal and pesticide exposure despite the possibility of immigration of non-exposed species (Boivin et al. 2006; Dorigo et al. 2010b). Despite the lack of structural recovery, the functional recovery of heterotrophic communities (estimated from β-Glucosidase activity) was accelerated when immigration processes were possible (Lambert et al. 2012). Boivin et al. (2006) also showed that functional changes in bacterial communities (estimated from community-level physiological profiles) following Cu exposure were reversible within 28 days. All these results illustrate the crucial importance of functional redundancy acting as an ecological insurance allowing the functional recovery of microbial communities following exposure to chemicals.

Recovery in aquatic microbial communities depends not just on type of microorganisms (e.g. diatoms *vs* bacteria) and feasibility of immigration processes but also mode of life (i.e. benthic or planktonic). Mohr et al. (2008b) observed no structural recovery in periphytic phototrophic communities within 150 days following single applications of 1 and 5 μg/L of the herbicide Irgarol whereas phytoplankton recovered after just a few weeks. This suggests that Irgarol bioaccumulation in periphyton may have prolonged the exposure duration. In contrast, the recovery dynamics of phytoplankton communities generally co-occurs with toxicant dissipation in water (Brock et al. 2004; Knauert et al. 2009).

Compared to the numerous aquatic microcosm studies, soil microcosm studies assessing microbial recovery following chemical pollution are scarce. To the best of our knowledge, only a few studies have attempted to evaluate the effects of various fungicides on soil microbial communities and soil ecological processes (Bending et al. 2007; Chen and Edwards 2001; Chen et al. 2001). These studies suggest that both the magnitude of the effects and the dynamics of recovery are dependent on several factors, including kind of fungicide and soil physicochemical properties, which can be affected by management practices such as organic amendment driving soil organic matter content. For example, a significant negative effect of fungicides on dehydrogenase activity was observed only in soils exhibiting the lowest levels of organic matter and microbial biomass (Bending et al. 2007). Moreover, in these soils, chlorothalonil had a greater and more prolonged impact on the microbial community than azoxystrobin and tebuconazole. Similarly, Chen and Edwards (2001) observed only transient effects of benomyl and chlorothalonil on soil microbial activity and nitrogen dynamics while these effects were more pronounced and prolonged following captan treatment, with a significant influence of type of soil. Kostov and Van Cleemput (2001a, b) also observed that the magnitude of the inhibition of basal nitrification and N mineralization by Cu and the subsequent recovery was strongly influenced by type of soil (i.e. sandy soil *vs* sandy loam soil). Moreover, they showed that recovery of microbial activity and fertility in Cu-contaminated soils was enhanced following lime and compost amendments (Kostov and Van Cleemput 2001a, b). This may be due to the fact that compost amendment increases soil organic matter content, which improves the heavy metal binding capacity of the soil (Martinho et al. 2015). Functional recovery potential depends not just on soil physicochemical properties but also soil microbial community characteristics. For example, Griffiths et al. (2000) demonstrated that soil functional recovery can be significantly impaired by a loss of microbial diversity (estimated with a diversity index including various kinds of microorganisms, i.e. bacteria, flagellate protozoa and nematodes). This result underlines the importance of microbial diversity, which is one of the keystones of ecological insurance allowing the recovery of microbial functions following a stress.

10.2.2.2 In Situ Experiments: Translocation Studies in Lotic Ecosystems

Over the past decade, several in situ studies have set out to evaluate the potential of river periphytic communities to recover from chemical pollution using translocation approaches (Table 10.2). Translocation approaches use experimental transfers of microbial communities from a contaminated station to a reference station (i.e. pristine or less-contaminated station) to assess their trajectories of recovery. Most of these studies have focused on the capacity of phototrophic communities to recover from exposure to metals or pesticides, using structural metrics such as microbial biomass, distribution of photosynthetic microbial classes and diatom community composition (Arini et al. 2012a; Dorigo et al. 2010a, b; Fechner et al. 2012; Ivorra et al. 1999; Morin et al. 2010; Rimet et al. 2005; Rotter et al. 2011). These studies generally evidenced shifts in community structure towards the reference community following transfer from contaminated-station to reference-station, but community structure recovery times differed between studies, from a few days (Rotter et al. 2011) to a few weeks (Arini et al. 2012a; Morin et al. 2010), and also varied with type of structural metrics or indices used (Rimet et al. 2005). For example, quantitative parameters (total and photosynthetic biomasses) recovered rapidly within 4 weeks whereas biological diatom index (BDI) did not recover at all (Morin et al. 2010). Likewise, Ivorra et al. (1999) showed that diatom community compositions of biofilms transferred from metal-polluted to reference sites were still different

Table 10.2 In situ translocation studies of the recovery potential of microbial periphytic communities following a decrease in chemical exposure

Structural metrics	Functional metrics	Contaminant (s)	Exposure/recovery duration	References
Diatom community structure, teratological forms		Metals (Zn and Cd)	24 day/63 day	Arini et al. (2012b)
Microbial biomass, eukaryotic community structure	Photosynthesis (PICT approach)	Pesticides (diuron)	5 week/5 week	Dorigo et al. (2010a)
Diatom community structure, algal biomass, eukaryotic and bacterial community structure	Photosynthesis (PICT approach), respiration (PICT approach)	Pesticides (diuron), metals (Cu)	ND/9 week	Dorigo et al. (2010b)
Eukaryotic and bacterial community structure	Beta-glucosidase (PICT approach)	Metals (Cu)	23–34 day/30 day	Fechner et al. (2012)
Diatom community structure, algal biomass, microbial biomass		Metals (Zn and Cd)	7–16 day/14–18 day	Ivorra et al. (1999)
Diatom community structure, algal class composition, microbial biomass		Pesticides	4 week/8 week	Morin et al. (2010)
Diatom community structure		High organic laod	20 day/60 day	Rimet et al. (2005)
Diatom community structure, algal class composition	Photosynthesis (PICT approach)	Pesticides (prometryn)	26 day/44 day	Rotter et al. (2011)

after two weeks. Using molecular fingerprinting approaches, Dorigo et al. (2010a, b) and Fechner et al. (2012) also reported divergent results on the capacity of eukaryotic and bacterial biofilm communities to recover their reference structure within a few weeks. Indeed, while Fechner et al. (2012) observed good recovery of the genetic structure in microbial communities only 15 days after translocation, Dorigo et al. (2010a, b) observed only delayed and partial structural recovery, which was still incomplete after 9 weeks after their translocation.

These differences in time response between studies are strong evidence that in-field structural recovery trajectories of periphytic communities are influenced by a number of environmental parameters, some of which being directly related to the exposure conditions in the contaminated site, especially in terms of types of toxicants, which are more or less likely to bioaccumulate in the periphyton matrix and cells. Bioaccumulation can indeed prolong the toxicant pressure in the uncontaminated reference sites, thus delaying post-translocation microbial recovery (Dorigo et al. 2010b; Morin et al. 2010). Among toxicants, metals are well known to bioaccumulate within periphytic biofilms and several translocation studies have confirmed that depuration of metals from biofilms in reference sites can sometimes take several weeks before significant recovery becomes possible (Admiraal et al. 2000; Arini et al. 2012a; Dorigo et al. 2010b; Ivorra et al. 1999). Depuration time is influenced by several parameters, such as type of metals, microbial growth in biofilms (dilution process) and/or biofilm detachment and grazing (Arini et al. 2012a). It is also well known that following chemical exposure, the recovery of populations and communities depends on their connection to undisturbed environments conditioning migration processes (Gergs et al. 2016; Lambert et al. 2012; Morin et al. 2012). Even if lotic systems are usually well connected to undisturbed sections, allowing faster recovery than in lentic systems (Gergs et al. 2016), several authors have pointed out that recovery processes are probably facilitated in translocation studies, where exposed biofilms are directly transplanted into river sections inhabited by unexposed communities, thus facilitating migration (Arini et al. 2012a; Ivorra et al. 1999; Lambert et al. 2012). Toxicant releases and migration processes are key drivers of periphytic recovery and both are highly dependent on maturity stage of the translocated biofilms, as thicker biofilms may accumulate higher amounts of toxicants than thinner ones (Lawrence et al. 2001) while microbial immigration processes may be facilitated in early biofilm development stages (Dorigo et al. 2010b).

Some translocation studies also set out to investigate the link between structural recovery and possible changes in sensitivity towards the main pollutants identified in the contaminated sites using PICT approaches. Short-term photosynthetic bioassays applied to investigate phototrophic community recoveries after a decrease in exposure to herbicide (Dorigo et al. 2010a, b; Rotter et al. 2011) or copper (Dorigo et al. 2010b) following translocation showed a significant decrease in herbicide and copper tolerance with changes in phototrophic community composition. Likewise, PICT measurement with heterotrophic functions such as substrate-induced respiration (Dorigo et al. 2010b) and β-glucosidase activity (Fechner et al. 2012) combined with monitoring of bacterial community structure revealed that changes in community tolerance occurred concomitantly with changes in community structure. Indeed, Fechner et al. (2012) observed 15 days after translocation that the fast recovery of low tolerance levels of hetrotrophic communities towards copper was accompanied by significant modifications in bacterial community structure. Conversely, Dorigo et al. (2010b) reported limited recovery of tolerance to copper and structure in the bacterial community 9 weeks after translocation.

10.3 Case-Studies of the Use of Microbial Communities to Evaluate Ecosystem Recovery Following a Decrease in Chemical Exposure

As recently underlined by the EFSA Scientific Committee, assessing recovery in natural complex ecosystems exposed to multiple stressors and where the connection to undisturbed areas may influence recovery trajectories is far from trivial. Moreover, and in contrast to experimental studies, the lack of system replication in such approaches makes it necessary to define reference conditions for each of ecological metric measured, based on the state of the disturbed system prior to disturbance, or the state of similar but undisturbed systems, or theoretically-derived system states (Gergs et al. 2016). Nevertheless, despite these recognized weaknesses, field studies provide the most realistic assessment of 'real-life' environmental risks of chemicals. Furthermore, when conducted over a long period of time, field studies provide relevant information depicting effective ecological recovery trajectories. This section provides illustrative examples of in-field case studies designed to assess autochthonous microbial community recovery in different kinds of ecosystems.

10.3.1 Structural and Functional Recovery of Microbial Communities

Soil remediation and rehabilitation processes offer practical case-studies to assess ecosystem recovery following an improvement in chemical quality. Worldwide pollution of soils by heavy metals has prompted the development of various biotechnological strategies for remediating metal-contaminated soils, such as chemical- and bio-remediation, including phytoremediation and bioaugmentation (dos Santos et al. 2016; Kavamura and Esposito 2010). However, ultimately, the goal of soil remediation and rehabilitation is not only to eliminate the contamination but also to allow restoration of soil quality and functioning. Within this context, microbial community monitoring (e.g. Ritz et al. 2009; Schloter et al. 2003) is viewed as a way to assess the recovery of soil quality during the remediation process (Gomez-Sagasti et al. 2012). Various methods have been applied to achieve this objective, chiefly analyses of microbial biomass, basal and substrate-induced respiration, and enzymatic activities (such as urease, β-glucosidase, phosphatase, dehydrogenase, protease, invertase, etc.; Alvarenga et al. 2009; Ciarkowska et al. 2014; Epelde et al. 2008, 2009; Goupil and Nkongolo 2014; Jiang et al. 2010). These measurements of microbial abundance and activity are sometimes supplemented by the assessment of functional diversity using community-level physiological profiles (Castaldi et al. 2009; Epelde et al. 2009; Kelly and Tate 1998) and microbial community structure using phospholipid fatty acid analysis (Kelly et al. 2003) or 16S rRNA-based analyses (dos Santos et al. 2016). Taken together, these different methodologies serve to assess the recovery of soil quality supported by soil microorganisms all along the remediation and rehabilitation processes. Gomez-Sagasti et al. (2012) proposed that a better interpretation of microbial properties as indicators of soil quality could be gained by grouping microbial indicators into categories of high ecological relevance, such as soil ecosystem functions and services.

Although there is a long history of using biological indicators of anthropogenic disturbance in surface freshwater ecosystems (Kelly and Harwell 1990), this trend has really taken off over the last decade due to strong regulatory pressure exerted by the European Water Framework Directive (WFD, Directive 2000/60/EC of the European Parliament), which aims at achieving a good ecological and chemical status of surface waters. The evaluation of ecological status of water ecosystems is based on the use of several indices, including the Biological Diatom Index (Coste et al. 2009) for microbial communities. These indices, primarily based on the analysis of species characteristics such as taxonomy, abundance and identification of key species, do not reflect the ecological effects induced in response to toxicant exposure (Montuelle et al. 2010; Tlili et al. 2015). Moreover, even though the WFD was first focused on characterizing the chemical and ecological status of aquatic ecosystems, its ultimate goal is to monitor gain in ecological quality during ecological recovery following restoration measures to decrease chemical pressure (Hering et al. 2010). Surprisingly few studies have been led to assess structural and functional recovery of microbial communities in aquatic ecosystems subjected to chemical remediation (Adams et al. 2002; Arini et al. 2012c; Cherry et al. 1977). Arini et al. (2012c) assessed the ecological impact of remediation in a river subjected to an industrial contamination and did not observe significant change in periphytic diatom composition within two years due to the lack of decrease in metal accumulation (Cd and Zn) in periphyton. This study pointed out that recovery of aquatic microbial communities after industrial site remediation can sometimes be delayed. Cattaneo et al. (2004) arrived at the same conclusion after studying diatom communities along a sediment core collected in a lake with a long history of mining pollution. Indeed, by analyzing diatoms in the upper sediment layers, they detected indications of successful ecological recovery, but only 20 years after the start of remediation. However, it must be kept in mind that new diatom species can develop during the course of recovery, thus leading to the establishment of new community structures that may differ from those prevailing before disturbance (Hynynen et al. 2004). The functional consequences of these changes remain generally unknown, which highlights the limits of only assessing structural recovery of microbial communities. Adams et al. (2002) pointed out the need to combine various biological metrics to assess recovery in aquatic ecosystems. Studying recovery dynamics in a stream previously exposed to various contaminants from a nuclear weapons production facility (including heavy metals, chlorinated organics, and residual chlorine), they observed that the evolution of periphytic photosynthetic biomass (based on chlorophyll a measurement) reflected the general decrease of chlorine and mercury in the water, being more responsive than photosynthesis to recovery processes.

In marine ecosystems, there is plenty of literature on ecosystem recovery after pollution, mainly dominated by studies after oil spill. Recovery of the bacterial communities after oil pollution is closely linked to the pollution history, being much higher in ecosystems that have previously faced accidental spill or human activities compared to pristine sites (Head et al. 2006; Sauret et al. 2012). Nutrient and surfactant amendment is a widely accepted practice in oil-spill bioremediation, where resource-ratio theory (based on carbon/nitrogen/phosphorus ratios) is an important factor to determine recovery speed of the contaminated ecosystem both in terms of diversity of organisms and ecosystem functions (Delille et al. 2009; Sauret et al. 2015). Several studies used the non-specific Microtox® test based on measuring the decrease of bioluminescence of *Vibrio fisheri* to assess the toxicity stress of oil and its residues for ecosystem recovery. For example, with this test Pelletier et al. (2004) showed that intertidal sediments were still under toxicity stress one year after oil spill, whereas chemical analysis showed over 90% degradation of n-alkanes and disappearance of most light aromatics. Spectacular evidence of bacterial community resilience after pollution in marine environments comes from bacteria associated to corals. Shifts in microbiota composition often correlate with the appearance of signs of coral disease and/or bleaching, thus suggesting a causal link between microorganisms, coral health and stability of reef ecosystems (Krediet et al. 2013). For example, Garcia-Armisen et al. (2014) evidenced resilience of bacterial communities together with coral health under the influence of a sewage-polluted river. It is thus vital to evaluate both the resistance (insensitivity to disturbance) and resilience (the rate of recovery after disturbance) of microbial communities to understand the mechanisms that dictate the outcomes of host–microbial interactions and impact resilience of the host.

10.3.2 The Study of Microbial Adaptation to Toxicants for in Situ Assessment of Recovery

A major challenge in environmental risk assessment of pollutants is to establish causal relationships between chemical exposures and resulting community responses within complex ecosystems (Blanck and Dahl 1998; Tlili et al. 2015). A recent study using a large set of environmental parameters along several pollution gradients showed that this link is difficult to find, even when using multivariate statistical analysis (Sauret et al. 2016). Likewise the reliability of biological metrics for assessing recovery depends, among other things, on their causal relationships to stressors (Adams et al. 2002). Recent papers highlight the need to develop specific ecological indicators to monitor biological recovery following a decrease in toxic chemical pollution (Pesce et al. 2016; Tlili et al. 2015). As mentioned above, this need is particularly acute now that each EU member state is expected to implement the WFD, since one of the key as-yet-unresolved challenges is the evaluation of ecological recovery following water chemical quality improvement (Hering et al. 2013).

It is now well admitted that complex microbial communities are able to cope with chronic exposure to toxicants in various ecosystems through intra- or inter-specific adaptation processes. Such adaptations can lead to an increase not only in toxicant tolerance (according to the PICT concept, e.g. Pesce et al. 2010) but also in toxicant biodegradation capacities in the exposed communities in both soil and aquatic systems (Pesce et al. 2009). Given their relative specificity to various classes of toxicants (generally according to their mode of action and/or molecular structure), adaptation processes offer new insights for developing new ecological indicators to monitor microbial recovery.

Real-world case studies investigating the relevance of such approaches to evaluate community recovery from environmental contamination (i.e. in a context of long-term and progressive change in chemical quality) remain rare (Table10.2). Blanck and Dahl (1998) performed a 4-year PICT approach to assess the recovery of marine periphyton communities on the Swedish west coast after the 1989 ban on the use of tri-*n*-butyltin (TBT) in antifouling paint. The observed decrease in TBT tolerance of field-sampled periphyton communities in response to the decrease in TBT concentrations in the water confirmed that PICT approaches are suitable for assessing recovery in natural microbial communities. More recently, PICT approaches have successfully been used to assess the recovery of phototrophic microbial communities (phytoplankton and periphyton, respectively) in lake (Larras et al. 2016) and stream (Pesce et al. 2016) ecosystems in a context of chemical restoration from herbicide contamination. These studies offer evidence that PICT has potential as a powerful microbial metric to assess ecological recovery. However, prior its implementation in a regulatory framework, further work is required to standardize PICT measurement (Lambert et al. 2015; Tlili et al. 2015) and acquire baseline tolerance levels at large geographical scales (Pesce et al. 2016).

Besides PICT approaches, Pesce et al. (2013) also proposed the use microbial biodegradation potential of sediment to assess ecological recovery following a decrease in chronic exposure to organic pollutants. In a 4-year case study conducted in a small agricultural stream, the post-ban decrease in level of chronic diuron exposure in the river led to a strong decrease in sediment diuron-mineralizing capacities, revealing the recovery of the microbial community. This result brings further evidence that the study of microbial adaptation to toxicants can serve to demonstrate community recovery from environmental contamination, reflecting its relevance as an indicator in ecosystem restoration. Indeed, such approaches are generally specific to one substance, or one class of substances (according to their mode of action or their chemical structure), as shown by the results of Pesce et al. (2013, 2016) that reflected the resulting progressive decrease in diuron concentrations in the Morcille River despite the persistence of a multi-contamination context.

However, as previously stated with the PICT approach, further research is still required before the assessment of microbial biodegradation potential can be proposed as a routine protocol for evaluating ecological recovery in contaminated ecosystems. One major limitation is the use of radiorespirometry which requires specific authorization to manipulate radiolabeled contaminants. A promising alternative is the use of molecular approaches to study functional genes encoding enzymes involved in degradation pathways (Smith and Osborn 2008; Bombach et al. 2010; Monard et al. 2013), which could be potential biomarkers for the detection of organic xenobiotics (Sipilä et al. 2008). A prerequisite for applying such approaches is knowledge of the genes coding degrading enzymes, and the number of these genes known to date is still relatively limited. Rapid advances in functional genomics, such as transcriptomics and proteomics complementing traditional genetic approaches, which make it more feasible to understand gene functions, are providing methodological tools to overcome this constraint (Ortiz-Hernández et al. 2013; Karpouzas et al. 2016).

10.4 Challenges and Perspectives

As recently underlined by the EFSA Scientific Committee and touched on briefly in the first section of this chapter, the assessment of ecosystem recovery is no trivial challenge. Microbial communities are identified as major ecological engineers in the recovery of degraded ecosystems (Singh 2015) and the numerous examples cited in this chapter clearly show that microbial ecologists and ecotoxicologists have a large variety of tools and methods to study the structural and functional recovery of phototroph and heterotroph microorganisms following chemical exposure, at population and community scales and in different kinds of ecosystems. The next challenge for scientists is to translate the microbial response at ecosystem scale, or in other words to understand how structural and functional recovery observed at microbial scale can reflect wider ecosystem recovery.

Pesce et al. (2013) offers an interesting case study to illustrate the magnitude of this issue. Indeed, in their survey, although the decrease in the diuron biodegradation potential of microbial communities reflected an improvement in chemical quality of the river, it also indicated a decrease in the capacity of the microbial community to help dissipate organic toxicants. Paradoxically, this can somehow be viewed as a decrease in the efficiency of the ecosystem function supported by microbial degradation in driving natural attenuation of organic pollutants in the environment. Another point, which was raised by Gomez-Sagasti et al. (2012) and is clearly highlighted here, is that microbial properties are highly context-dependent, making each study case unique. This statement outlines the need to define ecosystem recovery targets as well as the microbial metrics needed to assess the course of recovery accordingly (Duarte et al. 2015). Such a process should be facilitated by combining (i) microbial metrics of high ecological relevance (i.e. microbial functions supporting a range of ecosystem functions and services) and (ii) microbial metrics that could serve to establish a direct link between improvement of chemical quality and microbial recovery (e.g. study of structural and functional microbial adaptation to toxicants).

Several examples cited above offer successful case studies of using microbial indicators to assess recovery following improvement in chemical quality in ecosystems ranging from soils and freshwaters to seawaters. Such case studies are particularly important to provide proof-of-principle for the relevance of considering microbial communities in recovery studies (EFSA Scientific Committee 2016). Based on this set of demonstrations, and to successfully implement a strategy for better assessing ecosystem recovery in various environments and at a larger geographical scale, there is a need to educate legislators and policymakers on the importance of considering microbial communities in environmental risk assessment, including ecological recovery monitoring.

Indeed, despite the recognized importance of microorganisms in supporting a range of ecosystemic services, they are barely protected by any regulations or legislations. For example, despite a proposal in 2006, the European Commission did not ratify the soil protection directive (Van Camp et al. 2004). Until now, only EU directive 91/414 for placing plant protection products (pesticides) on the market evaluates, at least in principle, the ecotoxicological impact of pesticides on soil microorganisms, but only using two global tests assessing their impact on the mineralization of carbon and nitrogen (EU-Regulation 1107/2009/EC). However, referring to recent work assessing the resistance and resilience of microbial communities and considering their functional redundancy, Martin-Laurent et al. (2013) suggested that carbon and nitrogen mineralization provide only a rough estimate of the possible impact of pesticides on soil microbiota. More recently, Karpouzas et al. (2016) further affirmed that these two out-of-date tests are not sensitive enough to reliably assess the impact of pesticides on the diversity and functioning of soil microbial communities and on supported ecosystemic functions. However, the tools required to monitor a range of ecosystemic functions relying on microbial communities, are still missing or remain unstandardized (e.g. Philippot et al. 2012). The absence of standardized methods means that there is no consistent dataset available that could be used to define normal operating ranges of microbial indicators, which is an important prerequisite for assessing microbial recovery in various ecosystems. This is probably due to the fact that although microbial ecologists have made huge steps forwards by developing an impressive toolbox for measuring the abundance, diversity and activity of microorganisms, they are less involved in the next-step technology knowledge transfer, mobilization and outreach to society. It thus precludes their implementation in regulatory frameworks which would better preserve environmental resources by taking into account the ecological role of microbial communities and their potential use as ecological indicators of ecosystem recovery following chemical pollution.

References

Adams SM, Hill WR, Peterson MJ, Ryon MG, Smith JG, Stewart AJ (2002) Assessing recovery in a stream ecosystem: applying multiple chemical and biological endpoints. Ecol Appl 12: 1510–1527

Admiraal W, Barranguet C, van Beusekom SAM, Bleeker EAJ, van den Ende FP, van der Geest HG, Groenendijk D, Ivorra N, Kraak MHS, Stuijfzand SC (2000) Linking ecological and ecotoxicological techniques to support river rehabilitation. Chemosphere 41:289–295

Alvarenga P, Palms P, Goncalves AP, Fernandes RM, de Varennes A, Vallini G, Duarte E, Cunha-Queda AC (2009) Organic residues as immobilizing agents in aided phytostabilization: (II) effects on soil biochemical and ecotoxicological characteristics. Chemosphere 74:1301–1308

Arini A, Feurtet-Mazel A, Maury-Brachet R, Coste M, Delmas F (2012a) Field translocation of diatom biofilms impacted by Cd and Zn to assess decontamination and community restructuring capacities. Ecol Ind 18:520–531

Arini A, Feurtet-Mazel A, Maury-Brachet R, Pokrovsky OS, Coste M, Delmas F (2012b) Recovery potential of periphytic biofilms translocated in artificial streams after industrial contamination (Cd and Zn). Ecotoxicology 21:1403–1414

Arini A, Feurtet-Mazel A, Morin S, Maury-Brachet R, Coste M, Delmas F (2012c) Remediation of a watershed contaminated by heavy metals: a 2-year field biomonitoring of periphytic biofilms. Sci Total Environ 425:242–253

Arini A, Durant F, Coste M, Delmas F, Feurtet-Mazel A (2013) Cadmium decontamination and reversal potential of teratological forms of the diatom *Planothidium frequentissimum* (Bacillariophyceae) after experimental contamination. J Phycol 49:361–370

Bending GD, Rodriguez-Cruz MS, Lincoln SD (2007) Fungicide impacts on microbial communities in soils with contrasting management histories. Chemosphere 69:82–88

Blanck H, Dahl B (1998) Recovery of marine periphyton communities around a Swedish marina after the ban of TBT use in antifouling paint. Mar Pollut Bull 36:437–442

Boivin MEY, Massieux B, Breure AM, Greve GD, Rutgers M, Admiraal W (2006) Functional recovery of biofilm bacterial communities after copper exposure. Environ Pollut 140:239–246

Bombach P, Richnow HH, Kästner M, Fischer A (2010) Current approaches for the assessment of in situ biodegradation. Appl Microbiol Biotechnol 86:839–852

Bordenstein SR, Theis KR (2015) Host biology in light of the microbiome: ten principles of holobionts and hologenomes. Plos Biol, vol 13

Brain RA, Arnie JR, Porch JR, Hosmer AJ (2012) Recovery of photosynthesis and growth rate in green, blue-green, and diatom algae after exposure to atrazine. Environ Toxicol Chem 31:2572–2581

Brock TCM, Crum SJH, Deneer JW, Heimbach F, Roijackers RMM, Sinkeldam JA (2004) Comparing aquatic risk assessment methods for the photosynthesis-inhibiting herbicides metribuzin and metamitron. Environ Pollut 130:403–426

Bullock JM, Aronson J, Newton AC, Pywell RF, Rey-Benayas JM (2011) Restoration of ecosystem services and biodiversity: conflicts and opportunities. Trends Ecol Evol 26:541–549

Castaldi P, Melis P, Silvetti M, Deiana P, Garau G (2009) Influence of pea and wheat growth on Pb, Cd, and Zn mobility and soil biological status in a polluted amended soil. Geoderma 151:241–248

Cattaneo A, Couillard Y, Wunsam S, Courcelles M (2004) Diatom taxonomic and morphological changes as indicators of metal pollution and recovery in Lac Dufault (Quebec, Canada). J Paleolimnol 32:163–175

Chen SK, Edwards CA (2001) A microcosm approach to assess the effects of fungicides on soil ecological processes and plant growth: comparisons of two soil types. Soil Biol Biochem 33:1981–1991

Chen SK, Edwards CA, Subler S (2001) Effects of the fungicides benomyl, captan and chlorothalonil on soil microbial activity and nitrogen dynamics in laboratory incubations. Soil Biol Biochem 33:1971–1980

Cherry DS, Guthrie RK, Singleton FL, Harvey RS (1977) Recovery of aquatic bacterial populations in a stream after cessation of chemical pollution. In: Water Air and Soil Pollution, vol 7, pp 95–101

Ciarkowska K, Solek-Podwika K, Wieczorek J (2014) Enzyme activity as an indicator of soil-rehabilitation processes at a zinc and lead ore mining and processing area. J Environ Manage 132:250–256

Clements WH, Rohr JR (2009) Community responses to contaminants: using basic ecological principles to predict ecotoxicological effects. Environ Toxicol Chem 28:1789–1800

Congdon JD, Dunham AE, Hopkins WA, Rowe CL, Hinton TG (2001) Resource allocation-based life histories: a conceptual basis for studies of ecological toxicology. Environ Toxicol Chem 8:1698–1703

Coste M, Boutry S, Tison-Rosebery J, Delmas F (2009) Improvements of the biological diatom index (BDI): description and efficiency of the new version (BDI-2006). Ecol Ind 9:621–650

Delille D, Pelletier E, Rodriguez-Blanco A, Ghiglione J-F (2009) Effects of nutrient and temperature on degradation of petroleum hydrocarbons in sub-Antarctic coastal seawater. Polar Biol 32:1521–1528

Depledge MH (1998) Recovery of ecosystems and their components following exposure to pollution. J Aquat Ecosyst Stress Recovery 6:199–206

Dorigo U, Berard A, Bouchez A, Rimet F, Montuelle B (2010a) Transplantation of microbenthic algal assemblages to assess structural and functional recovery after diuron exposure. Arch Environ Contam Toxicol 59:555–563

Dorigo U, Berard A, Rimet F, Bouchez A, Montuelle B (2010b) In situ assessment of periphyton recovery in a river contaminated by pesticides. Aquat Toxicol 98:396–406

dos Santos JV, Varon-Lopez M, Soares C, Leal PL, Siqueira JO, Moreira FMD (2016) Biological attributes of rehabilitated soils contaminated with heavy metals. Environ Sci Pollut Res 23:6735–6748

Duarte CM, Borja A, Carstensen J, Elliott M, Krause-Jensen D, Marba N (2015) Paradigms in the recovery of estuarine and coastal ecosystems. Estuaries Coasts 38:1202–1212

EFSA Scientific Committee (2016) Scientific opinion on recovery in environmental risk assessments at EFSA. EFSA J 14:4313

Epelde L, Becerril JM, Hernandez-Allica J, Barrutia O, Garbisu C (2008) Functional diversity as indicator of the recovery of soil health derived from *Thlaspi caerulescens* growth and metal phytoextraction. Appl Soil Ecol 39:299–310

Epelde L, Mijangos I, Becerril JM, Garbisu C (2009) Soil microbial community as bioindicator of the recovery of soil functioning derived from metal phytoextraction with sorghum. Soil Biol Biochem 41:1788–1794

Fechner LC, Versace F, Gourlay-France C, Tusseau-Vuillemin M-H (2012) Adaptation of copper community tolerance levels after biofilm transplantation in an urban river. Aquat Toxicol 106:32–41

Garcia-Armisen T, Inceoglu O, Ouattara NK, Anzil A, Verbanck MA, Brion N, Servais P (2014) Seasonal variations and resilience of bacterial communities in a sewage polluted urban river. Plos One, vol 9

Geiszinger A, Bonnineau C, Faggiano L, Guasch H, Lopez-Doval JC, Proia L, Ricart M, Ricciardi F, Romani A, Rotter S, Munoz I, Schmitt-Jansen M, Sabater S (2009) The relevance of the community approach linking chemical and biological analyses in pollution assessment. TrAC Trends Anal Chem 28:619–626

Gergs A, Classen S, Strauss T, Ottermanns R, Brock TCM, Ratte HT, Hommen U, Preuss TG (2016) Ecological recovery potential of freshwater organisms: consequences for environmental risk assessment of chemicals. In: Reviews of environmental contamination and toxicology, vol 236, v. 236, pp 259–294

Ghiglione JF, Martin-Laurent F, Pesce S (2016) Microbial ecotoxicology: an emerging discipline facing contemporary environmental threats. Environ Sci Pollut Res 23:3981–3983

Gomez-Sagasti MT, Alkorta I, Becerril JM, Epelde L, Anza M, Garbisu C (2012) Microbial monitoring of the recovery of soil quality during heavy metal phytoremediation. In: Water air and soil pollution, vol 223, pp 3249–3262

Goupil K, Nkongolo K (2014) Assessing soil respiration as an indicator of soil microbial activity in reclaimed metal contaminated lands. Am J Environ Sci, pp 403–411

Griffiths BS, Ritz K, Bardgett RD, Cook R, Christensen S, Ekelund F, Sorensen SJ, Baath E, Bloem J, de Ruiter PC, Dolfing J, Nicolardot B (2000) Ecosystem response of pasture soil communities to fumigation-induced microbial diversity reductions: an examination of the biodiversity-ecosystem function relationship. Oikos 90:279–294

Gustavson K, Mohlenberg F, Schluter L (2003) Effects of exposure duration of herbicides on natural stream periphyton communities and recovery. Arch Environ Contam Toxicol 45:48–58

Head IM, Jones DM, Roling WFM (2006) Marine microorganisms make a meal of oil. Nat Rev Microbiol 4:173–182

Hering D, Borja A, Carstensen J, Carvalho L, Elliott M, Feld CK, Heiskanen A-S, Johnson RK, Moe J, Pont D, Solheim AL, van de Bund W (2010) The European water framework directive at the age of 10: a critical review of the achievements with recommendations for the future. Sci Total Environ 408:4007–4019

Hering D, Borja A, Carvalho L, Feld CK (2013) Assessment and recovery of European water bodies: key messages from the WISER project. Hydrobiologia 704:1–9

Hynynen J, Palomaki A, Merilainen JJ, Witick A, Mantykoski K (2004) Pollution history and recovery of a boreal lake exposed to a heavy bleached pulping effluent load. J Paleolimnol 32:351–374

Ivorra N, Hettelaar J, Tubbing GMJ, Kraak MHS, Sabater S, Admiraal W (1999) Translocation of microbenthic algal assemblages used for in situ analysis of metal pollution in rivers. Arch Environ Contam Toxicol 37:19–28

Jiang J, Wu L, Li N, Luo Y, Liu L, Zhao Q, Zhang L, Christie P (2010) Effects of multiple heavy metal contamination and repeated phytoextraction by *Sedum plumbizincicola* on soil microbial properties. Eur J Soil Biol 46:18–26

Karpouzas DG, Tsiamis G, Trevisan M, Ferrari F, Malandain C, Sibourg O, Martin-Laurent F (2016) "LOVE TO HATE" pesticides: felicity or curse for the soil microbial community? An FP7 IAPP Marie Curie project aiming to establish tools for the assessment of the mechanisms controlling the interactions of pesticides with soil microorganisms. Environ Sci Pollut Res 23:18947–18951

Kavamura VN, Esposito E (2010) Biotechnological strategies applied to the decontamination of soils polluted with heavy metals. Biotechnol Adv 28:61–69

Kelly JR, Harwell MA (1990) Indicators of ecosystem recovery. Environ Manage 14:527–545

Kelly JJ, Tate RL (1998) Effects of heavy metal contamination and remediation on soil microbial communities in the vicinity of a zinc smelter. J Environ Qual 27:609–617

Kelly JJ, Haggblom MM, Tate RL (2003) Effects of heavy metal contamination and remediation on soil microbial communities in the vicinity of a zinc smelter as indicated by analysis of microbial community phospholipid fatty acid profiles. Biol Fertil Soils 38:65–71

Knauert S, Dawo U, Hollender J, Hommen U, Knauer K (2009) Effecst of photosystem II inhibitors and their mixture on freshwater phytoplankton sucession in outdoor mesocosms. Environ Toxicol Chem 28:836–845

Kostov O, Van Cleemput O (2001a) Microbial activity of Cu contaminated soils and effect of lime and compost on soil resiliency. Compost Sci Utilization 9:336–351

Kostov O, Van Cleemput O (2001b) Nitrogen transformations in copper-contaminated soils and effects of lime and compost application on soil resiliency. Biol Fertil Soils 33:10–16

Krediet CJ, Ritchie KB, Paul VJ, Teplitski M (2013) Coral-associated micro-organisms and their roles in promoting coral health and thwarting diseases. Proc R Soc B Biol Sci, vol 280

Lambert A-S, Morin S, Artigas J, Volat B, Coquery M, Neyra M, Pesce S (2012) Structural and functional recovery of microbial biofilms after a decrease in copper exposure: Influence of the presence of pristine communities. Aquat Toxicol 109:118–126

Lambert AS, Pesce S, Foulquier A, Gahou J, Coquery M, Dabrin A (2015) Improved short-term toxicity test protocol to assess metal tolerance in phototrophic periphyton: toward standardization of PICT approaches. Environ Sci Pollut Res 22:4037–4045

Larras F, Rimet F, Gregorio V, Berard A, Leboulanger C, Montuelle B, Bouchez A (2016) Pollution-induced community tolerance (PICT) as a tool for monitoring Lake Geneva long-term in situ ecotoxic restoration from herbicide contamination. Environ Sci Pollut Res 23:4301–4311

Laviale M, Morin S, Creach A (2011) Short term recovery of periphyton photosynthesis after pulse exposition to the photosystem II inhibitors atrazine and isoproturon. Chemosphere 84:731–734

Lawrence JR, Kopf G, Headley JV, Neu TR (2001) Sorption and metabolism of selected herbicides in river biofilm communities. Can J Microbiol 47(7):634–641

Madsen EL (2011) Microorganisms and their roles in fundamental biogeochemical cycles. Curr Opin Biotechnol 22:456–464

Martinho J, Campos B, Bras I, Silva E (2015) The roel of compost properties in sorption of heavy metals. Environ Prot Eng 41:57–65

Martin-Laurent F, Kandeler E, Petric I, Djuric S, Karpouzas DG (2013) ECOFUN-MICROBIODIV: an FP7 European project for developing and evaluating innovative tools for assessing the impact of pesticides on soil functional microbial diversity–towards new pesticide registration regulation? Environ Sci Pollut Res Int 20:1203–1205

Mohr S, Feibicke M, Berghahn R, Schmiediche R, Schmidt R (2008a) Response of plankton communities in freshwater pond and stream mesocosms to the herbicide metazachlor. Environ Pollut 152:530–542

Mohr S, Schroeder H, Feibicke M, Berghahn R, Arp W, Nicklisch A (2008b) Long-term effects of the antifouling booster biocide Irgarol 1051 on periphyton, plankton and ecosystem function in freshwater pond mesocosms. Aquat Toxicol 90:109–120

Monard C, Martin-Laurent F, Lima O, Devers-Lamrani M, Binet F (2013) Estimating the biodegradation of pesticide in soils by monitoring pesticide-degrading gene expression. Biodegradation 24:203–213

Montoya D, Rogers L, Memmott J (2012) Emerging perspectives in the restoration of biodiversity-based ecosystem services. Trends Ecol Evol 27:666–672

Montuelle B, Dorigo U, Berard A, Volat B, Bouchez A, Tlili A, Gouy V, Pesce S (2010) The periphyton as a multimetric bioindicator for assessing the impact of land use on rivers: an overview of the Ardières-Morcille experimental watershed (France). Hydrobiologia 657:123–141

Morelli E, Scarano G (2001) Synthesis and stability of phytochelatins induced by cadmium and lead in the marine diatom *Phaeodactylum tricornutum*. Mar Environ Res 52:383–395

Morin S, Pesce S, Tlili A, Coste M, Montuelle B (2010) Recovery potential of periphytic communities in a river impacted by a vineyard watershed. Ecol Ind 10:419–426

Morin S, Lambert A-S, Artigas J, Coquery M, Pesce S (2012) Diatom immigration drives biofilm recovery after chronic copper exposure. Freshw Biol 57:1658–1666

Muyssen BTA, Janssen CR (2001) Zinc acclimation and its effect on the zinc tolerance of *Raphidocelis subcapitata* and *Chlorella vulgaris* in laboratory experiments. Chemosphere 45:507–514

Ortiz Hernandez ML, Sánchez SE, González ED, Godínez MLC (2013) Pesticide biodegradation: mechanisms, genetics and strategies to enhance the process. In: Intech open science open mind, pp 251–287

Pelletier E, Delille D, Delille B (2004) Crude oil bioremediation in sub-Antarctic intertidal sediments: chemistry and toxicity of oiled residues. Mar Environ Res 57:311–327

Pesce S, Martin-Laurent F, Rouard N, Montuelle B (2009) Potential for microbial diuron mineralisation in a small wine-growing watershed: from treated plots to lotic receiver hydrosystem. Pest Manag Sci 65:651–657

Pesce S, Margoum C, Montuelle B (2010) In situ relationships between spatio-temporal variations in diuron concentrations and phototrophic biofilm tolerance in a contaminated river. Water Res 44:1941–1949

Pesce S, Bouchez A, Montuelle B (2011) Effects of organic herbicides on phototrophic microbial communities in freshwater ecosystems. In: Whitacre DM (ed) Reviews of environmental contamination and toxicology, vol 214, pp 87–124

Pesce S, Margoum C, Rouard N, Foulquier A, Martin-Laurent F (2013) Freshwater sediment pesticide biodegradation potential as an ecological indicator of microbial recovery following a decrease in chronic pesticide exposure: a case study with the herbicide diuron. Ecol Ind 29:18–25

Pesce S, Margoum C, Foulquier A (2016) Pollution-induced community tolerance for in situ assessment of recovery in river microbial communities following the ban of the herbicide diuron. Agric Ecosyst Environ 221:79–86
Philippot L, Ritz K, Pandard P, Hallin S, Martin-Laurent F (2012) Standardisation of methods in soil microbiology: progress and challenges. FEMS Microbiol Ecol 82:1–10
Prosser RS, Brain RA, Hosmer AJ, Solomon KR, Hanson ML (2013) Assessing sensitivity and recovery of field-collected periphyton acutely exposed to atrazine using PSII inhibition under laboratory conditions. Ecotoxicology 22:1367–1383
Prosser RS, Brain RA, Andrus JM, Hosmer AJ, Solomon KR, Hanson ML (2015) Assessing temporal and spatial variation in sensitivity of communities of periphyton sampled from agroecosystem to, and ability to recover from, atrazine exposure. Ecotoxicol Environ Saf 118:204–216
Rimet F, Cauchie HM, Hoffmann L, Ector L (2005) Response of diatom indices to simulated water quality improvements in a river. J Appl Phycol 17:119–128
Ritz K, Black HIJ, Campbell CD, Harris JA, Wood C (2009) Selecting biological indicators for monitoring soils: a framework for balancing scientific and technical opinion to assist policy development. Ecol Ind 9:1212–1221
Rotter S, Sans-Piche F, Streck G, Altenburger R, Schmitt-Jansen M (2011) Active bio-monitoring of contamination in aquatic systems-an in situ translocation experiment applying the PICT concept. Aquat Toxicol 101:228–236
Sauret C, Christaki U, Moutsaki P, Hatzianestis I, Gogou A, Ghiglione J-F (2012) Influence of pollution history on the response of coastal bacterial and nanoeukaryote communities to crude oil and biostimulation assays. Mar Environ Res 79:70–78
Sauret C, Boettjer D, Talarmin A, Guigue C, Conan P, Pujo-Pay M, Ghiglione J-F (2015) Top-down control of diesel-degrading prokaryotic communities. Microb Ecol 70:445–458
Sauret C, Tedetti M, Guigue C, Dumas C, Lami R, Pujo-Pay M, Conan P, Goutx M, Ghiglione J-F (2016) Influence of PAHs among other coastal environmental variables on total and PAH-degrading bacterial communities. Environ Sci Pollut Res 23:4242–4256
Schloter M, Dilly O, Munch JC (2003) Indicators for evaluating soil quality. Agrc Ecosyst Environ 98:255–262
Schmitt-Jansen M, Veit U, Dudel G, Altenburger R (2008) An ecological perspective in aquatic ecotoxicology: approaches and challenges. Basic Appl Ecol 9:337–345
Singh JS (2015) Microbes: the chief ecological engineers in reinstating equilibrium in degraded ecosystems. Agric Ecosyst Environ 203:80–82
Sipilä TP, Keskinen AK, Åkerman ML, Fortelius C, Haahtela K, Yrjälä K (2008) High aromatic ring-cleavage diversity in birch rhizosphere: PAH treatment-specific changes of I.E.3 group extradiol dioxygenases and 16S rRNA bacterial communities in soil. ISME J 2:968–981
Smith CJ, Osborn AM (2008) Advantages and limitations of quantitative PCR (Q-PCR)-based approaches in microbial ecology. FEMS Microbiol Ecol 67:6–20
Thompson PA, Couture P (1993) Physiology of carbon assimilation in a green alga during exposure to and recovery from cadmium. Ecotoxicol Environ Saf 26:205–215
Tlili A, Berard A, Blanck H, Bouchez A, Cássio F, Eriksson KM, Morin S, Montuelle B, Navarro E, Pascoal C, Pesce S, Schmitt-Jansen M, Behra R (2015) Pollution-induced community tolerance (PICT): towards an ecologically relevant risk assessment of chemicals in aquatic systems. Freshwater Biol
Tripathi BN, Gaur JP (2006) Physiological behavior of Scenedesmus sp during exposure to elevated levels of Cu and Zn and after withdrawal of metal stress. Protoplasma 229:1–9
Tripathi BN, Mehta SK, Gaur JP (2004) Recovery of uptake and assimilation of nitrate in Scenedesmus sp previously exposed to elevated levels of Cu^{2+} and Zn^{2+}. J Plant Physiol 161:543–549
Vallotton N, Ilda R, Eggen L, Escher BI, Krayenbuehl J, Chevre N (2008a) Effect of pulse herbicidal exposure on Scenedesmus vacuolatus: a comparison of two photosystem II inhibitors. Environ Toxicol Chem 27:1399–1407

Vallotton N, Moser D, Eggen RIL, Junghans M, Chevre N (2008b) S-metolachlor pulse exposure on the alga *Scenedesmus vacuolatus*: effects during exposure and the subsequent recovery. Chemosphere 73:395–400

Van-Camp L, Bujarrabal B, Gentile AR, Jones RJA, Montanarella L, lazabal C, Selvaradjou SK (2004) Reports of the technical working groups established under the thematic strategy for soil protection. Office for Official Publications of the European Communities, Luxembourg, EUR 21319 EN/4, 872 pages

Wang M-J, Wang W-X (2011) Cadmium sensitivity, uptake, subcellular distribution and thiol induction in a marine diatom: recovery from cadmium exposure. Aquat Toxicol 101:387–395

Zeng J, Yang L, Wang W-X (2009) Acclimation to and recovery from cadmium and zinc exposure by a freshwater cyanobacterium, *Microcystis aeruginosa*. Aquat Toxicol 93:1–10

Chapter 11
Microbial Biomarkers

Helena Guasch, Berta Bonet, Chloé Bonnineau and Laura Barral

Abstract Biomarkers, defined in ecotoxicology as functional measures of exposure to chemicals, may not be informative on the consequences of exposure at the scale of interest, which is the entire ecosystem. This drawback is because links and interactions existing between these measures and the biological system at a larger scale are not always sufficiently known. In this chapter, three different biomarkers of effect and/or exposure (i) antioxidant enzyme activities (AEA), (ii) community structure and (iii) resistance genes are discussed as potential microbial biomarkers in community ecotoxicology. First, AEA are highly sensitive to chemicals but have low specificity and their link with ecosystems' health is also unclear. The community composition changes linked to adaptation are highly informative about effects on ecosystems' health and also sensitive (responding to low but prolonged chemical exposure); however, they are not specific. Finally, resistance genes applicability is very limited since information is lacking about the genes that build resistance to the large list of chemicals of concern. In addition, their ecological implications have only been established for few chemicals, like arsenic. Similarly to investigations focused on organisms with a higher level of biological complexity, microbial biomarkers may not be sufficient to link exposure with ecosystem damage. However, these biomarkers may contribute to providing unequivocal information concerning exposure, early effects, adaptation and ecosystem damage in the framework of integrative multi-metric approaches including other chemical,

H. Guasch (✉) · B. Bonet · L. Barral
Institut d'Ecologia Aquàtica i Departament de Ciències Ambientals,
Campus Montilivi, 17071 Girona, Spain
e-mail: helena.guasch@udg.edu

B. Bonet
School of Geography, Earth and Environmental Sciences,
University of Birmingham, Birmingham, UK

C. Bonnineau
Irstea, UR MALY, Centre de Lyon-Villeurbanne, 5 rue de la Doua,
BP 32108, 69616 Villeurbanne Cedex, France

C. Cravo-Laureau et al. (eds.), *Microbial Ecotoxicology*,
DOI 10.1007/978-3-319-61795-4_11

biogeochemical and ecological metrics related to microbial metabolism such as bioaccumulation, chemical speciation, organic matter decomposition or nutrient cycling.

Keywords Antioxidant enzyme activity · Community structure · Resistance genes

11.1 Biomarkers in Microbial Ecotoxicology: From Species to Community

In environmental risk assessment, the World Health Organization defined biomarker as "almost any measurement reflecting an interaction between a biological system and a potential hazard, which may be chemical, physical, or biological. The measured response may be functional and physiological, biochemical at the cellular level, or a molecular interaction" (WHO 1993). To put emphasis on the link between a biomarker and the wider system it belongs to, we previously proposed defining biomarker as "a biological element [including biogenic chemicals] belonging to a wider system whose observation is expected to give information on this wider system based on the a priori knowledge of the links and interactions existing between the biomarker and this system" (Bonnineau et al. 2012). This definition has a wider scope than the definition given in stress ecology which is restricted to molecular, cellular, genetic, immunological and physiological measures which are expected to inform about ecological alterations (Amiard-Triquet et al. 2013).

Biomarkers are classified depending on the type of information they provide: information on the exposure to, the effects of or the susceptibility to a perturbation (Hook et al. 2014). Biomarkers of effect, the most common ones, are expected to inform on the consequences of a perturbation such as toxic exposure either because they directly measure some essential processes of the ecosystem (e.g., primary or secondary production) or because a causal relationship has been established between biomarkers' alterations and effects at higher levels of biological organization (individual, population, and ecosystem).

Biomarkers research is very common in ecotoxicology, especially in marine and freshwater environments, but still remains controversial since the links and interactions existing between biomarkers and ecosystems' health are not always sufficiently well-known (Amiard-Triquet and Berthet 2015; Hook et al. 2014; Lam 2009). It is worth mentioning that most studies are based on macro-organisms (e.g., invertebrates: Galloway et al. 2002; Prevodnik et al. 2007, fish: Colin et al. 2016; Van der Oost et al. 2003). These investigations are mainly focused on predicting the effects of pollution on populations based on biochemical responses in individuals. For instance, the inhibition of acetylcholinesterase (AChE) activity in aquatic invertebrates and fish is a biomarker of exposure to organophosphorous insecticides (Fulton and Key 2001). In *Gammarus fossarum*, inhibition of AchE was correlated with the alteration of feeding rate and locomotion (Xuereb et al. 2009), promoting the use of AchE as a biomarker of effect at population level (Chaumot et al. 2015). Fish biomarkers of effect and exposure are not only developed for their ecological relevance but also for their similarity (in terms of physiology) to humans (e.g. Wang et al. 2008).

Compared to other groups of organisms, the use microbial biomarkers in ecotoxicology is still scarce. Species-specific microbial biomarkers can be used to detect the presence of species of concern in the environment such as pathogen bacteria (e.g. Sorensen et al. 2015a, b) or toxic cyanobacteria (Zamyadi et al. 2016). For example, in toxicity tests including various metals (Ogunseitan et al. 2000), lead (Pb) was found to inhibit specifically the delta-aminolevulinate dehydratase (ALAD) activity in an environmental strain of *Pseudomonas putida*, in agreement with previous observations in multicellular eukaryotes. In addition, while ALAD activity was proportional to Pb bioavailability in *P. putida*, this activity remained stable under Pb exposure in a known metal-resistant *P. aeruginosa*. Based on these results, the authors suggested the use of ALAD activity of *P. putida* as a biomarker of exposure to Pb (Ogunseitan et al. 2000).

Though species-specific biomarkers can be useful to search for certain species or to detect specific exposure effects, microbial biomarkers measured in the environment are rather likely to refer to microbial communities including a large number of interacting species (e.g. xenobiotic degrading bacteria: Haritash and Kaushik 2009; Pesce et al. 2013).

In ecology, a community refers to an ecological unit composed of different species occupying a particular area, usually interacting with each other and their environment. Community ecology principles applied to the understanding of the effects of pollution on ecosystems, give rise to an ecotoxicology branch called community ecotoxicology (Clements and Newman 2002), which was later on applied to microbial communities (Sabater et al. 2007). We consider that this approach has a great interest in ecotoxicology since microbial biomarkers may cover a wide range of measurements from the determination of biogenic chemical concentration (e.g. phospholipid fatty acids Lupwayi et al. 2017) to the evaluation of microbial activity (e.g. photosynthetic activity; Corcoll et al. 2011) or capacity (e.g. adaptation; Pesce et al. 2016).

Moreover, the sensitivity of microbial communities can differ strongly from single species. For instance, Ricart et al. (2009) found biofilm communities to be more sensitive to the herbicide diuron than single species with an EC50 for diatom biovolume (i.e. the concentration for which diatom biovolume is reduced by 50%) of 0.09 μg diuron L^{-1}, after 8 days of exposure, while in previous single-species tests on different algal species, EC50 for growth ranged between 4 and 30 μg L^{-1} (Ma 2002; Podola and Melkonian 2005; Gatidou and Thomaidis 2007).

In this context, microbial (community) biomarkers have many advantages compared to other biomarkers used in ecotoxicology. In particular, a multiple biomarker approach at community level is likely to detect exposure (using specific biomarker of exposure) but also to estimate direct and indirect effects occurring at a population, community or ecosystem level (Bonnineau et al. 2012). Indeed, responses at a community level may also be considered as biomarkers for a set of communities or an ecosystem. It is worth mentioning that microbial communities play a key role in ecosystem functioning (e.g. biofilm community in aquatic ecosystems, Battin et al. 2016; Romaní et al. 2016) and in contaminants' dynamics. For instance, microbial detoxification reactions carried out by microbial communities mediate changes in metal speciation (Rhine et al. 2005; Hamamura et al. 2009) that may indirectly influence metal toxicity to higher trophic levels (e.g. Bouskill et al. 2006; Magellan et al. 2014). While laboratory studies allowed for identifying specific biomarkers of exposure and effects in microbial communities (e.g. Bonnineau et al. 2013; Corcoll et al. 2011; Hou et al. 2016), it is also important to note that more field studies are still needed to validate the use of such biomarkers as diagnostic tools to help identifying toxic pressures and impacts on the ecosystem (Bonet et al. 2014; Guasch et al. 2012; Nilsen et al. 2015).

In aquatic ecosystems, community ecotoxicology investigations on freshwater biofilms illustrate how a multiple biomarker approach at community level can reveal the presence and effects of pollutants at multiple scales. Biofilms (also known as periphyton) are complex and structured benthic microbial communities forming a 3D structure in a matrix composed of extracellular polymeric substances. Freshwater biofilms play a key role in river ecosystem functioning and are also likely to be affected by contaminants, consequently many studies have investigated the impact of contaminants on the structure and function of these communities (e.g. Bonet et al. 2012; Bouskill et al. 2010; Guasch et al. 2003, 2016; Lawrence et al. 2004, 2005, 2007). Microbial biomarkers specific of the different components of biofilm have been used to identify the direct but also the indirect effects of pollution (e.g. Guasch et al. 2003; Proia et al. 2011). For example, Ricart et al. (2009) highlighted the indirect effect of the herbicide diuron, inhibitor of photosynthesis, on bacteria within biofilm communities. Moreover, the use of different biomarkers may allow both the acute and chronic effects of a toxicant to be captured. Within this framework, physiological biomarkers are expected to detect acute exposure while structural biomarkers, based on community composition, may inform about the effects of chronic exposure (Guasch et al. 2010b).

Taking into account this complexity, microbial ecotoxicology investigates the effects of contaminants on target species but also the impact of contamination on (i) biotic interactions (e.g. algae/bacteria in biofilms), (ii) higher trophic levels (through the implication of microbial communities into the food chain) and (iii) important processes such as primary production, organic matter decomposition, and nutrient cycling. Thus, microbial ecotoxicology is likely to reveal indirect chemical impact not readily demonstrated by current biomonitoring tools (Bouskill et al. 2010).

Throughout this chapter, different endpoints are described as potential microbial biomarkers in the community-ecotoxicology framework. We will first present and discuss antioxidant enzymes as biomarkers of exposure and/or early effects; secondly the potential use of biomarkers of adaptation at the community level will be discussed. The chapter will also present a specific case study highlighting the use of biofilm as biomarkers of metal pollution. Finally, we will examine the perspectives in the use and application of microbial biomarkers in community ecotoxicology.

11.2 Antioxidant Enzyme Activities (AEA) as Biomarkers of Early Effects and Exposure

Antioxidant enzymes allow organisms to maintain redox homeostasis by scavenging the reactive oxygen species (ROS) resulting from normal aerobic metabolism (e.g. photosynthesis or respiration) or triggered by perturbations such as chemical contamination (Benedetti et al. 2015; Valavanidis et al. 2006). Indeed, virtually all ROS found in aquatic environments are produced by organisms through redox reactions with O_2 and they result in a common form of stress, oxidative stress. ROS can be found in several cell locations like chloroplasts in autotrophic organisms (Edreva 2005; Asada 2006), mitochondrion (Mittler 2002) or endoplasmic reticulum and microbodies like peroxisomes and glyoxysomes (Lesser 2006). Some of the most notable culprits within the ROS family are the superoxide anion radical (O_2^-), hydrogen peroxide (H_2O_2) and the hydroxyl radical (HO^-) resulting from the transfer of one, two or three electrons, respectively, to oxygen (Scandalios 1993; Mittler 2002; Edreva 2005; Wolfe-Simon et al. 2005). These ROS molecules attack lipids, proteins, nucleic acids and damage most cellular machinery, which often leads to alterations in cell structures and mutagenesis (Scandalios 1993; Mallick 2004; Wolfe-Simon et al. 2005; Lesser 2006). Thus, despite the energetic advantage gained from using O_2 as a terminal electron acceptor, cells are required to maintain an efficient defense system against O_2 by-products, like antioxidant enzymes and other non-enzymatic antioxidant mechanisms. While superoxide dismutase (SOD), catalase (CAT) or peroxidases (ascorbate peroxidase: APx or glutathione peroxidase: GPx) directly reduce the amount of ROS (Fig. 11.1), glutathione-S-transferase (GST) and glutathione reductase (GR) alleviate oxidative stress by reducing the oxidative radicals resulting of ROS reaction with cellular components (Fig. 11.1), for a complete information on antioxidant enzymes reactions see Regoli and Giuliani (2014).

Many chemicals induce oxidative stress either by directly increasing ROS production (e.g. trace metals via Fenton or Haber Weiss reactions) or by altering cellular antioxidant capacity (Lushchak 2011; Regoli and Giuliani 2014). Variations in AEA can reflect oxidative stress induced by chemicals and AEA have been used as biomarkers of contamination in ecotoxicology (Regoli and Giuliani 2014; Valavanidis et al. 2006). For instance, AEA responses to metal contamination have been widely reported in laboratory experiments and in the field, but mainly at the species level, showing their sensitivity to a large number of compounds and exposure conditions (Sauser et al. 1997; Li et al. 2006; Tripathi et al. 2006; Pereira et al. 2009). Nevertheless, both metal and organic contaminants have also been shown to alter AEA in microbial organisms at both a population and a community level in the laboratory and in the field (e.g. Bonet 2013; Bonet et al. 2014; Bonnineau et al. 2012; Geoffroy et al. 2002; Hou et al. 2016). AEA measurements at a community level are expected to reflect the tendency (activation or inhibition) observed in the majority of individuals and species within the community. In soil, many contaminants have been shown to alter CAT activity which is often used as

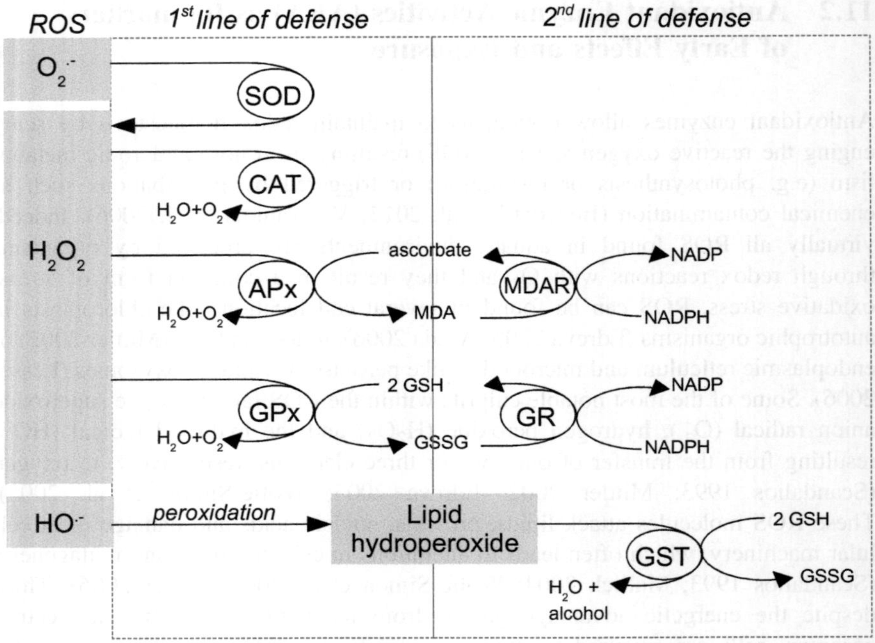

Fig. 11.1 Antioxidant enzymes contribute alleviating oxidative stress by directly scavenging ROS (*1st line of defense*), by regenerating antioxidant compounds or by scavenging oxidative radicals (*2nd line of defence*). *ROS* reactive oxygen species; *SOD* superoxide dismutase; *CAT* catalase; *APx* ascorbate peroxidase; *GPx* glutathione peroxidase; *MDA* monodehydroascrobate; *MDAR* MDA reductase; *GSH* reduced glutathione; *GSSG* oxidized glutathione; *GR* glutathione reductase; *GST* glutathione-S-transferase

an indicator of microbial activity (Rao et al. 2014a, b). For instance, the application of the pesticide chlorpyrifos (as Lorsban®4E) at the highest recommended dose provoked a strong decrease in soil catalase activity (about 22% lower than control) in a mesocosm study (Sanchez-Hernandez et al. 2017). AEA modifications induced by organic contaminants, metals and nanoparticules were found different for each antioxidant enzyme tested but also depended on the exposure intensity (i.e. contaminant concentration and exposure duration). For example, in an indoor microcosm experiment, 24 h of exposure to Zn (400 μg Zn L^{-1}) was sufficient to induce APx activity in periphyton whereas in the same experiment CAT activity was induced after 21 days of exposure (Bonet et al. 2012). Hou et al. (2016) also highlighted differences in CAT and SOD responses to the exposure to zinc nanoparticules (ZnO–NPs) in a microcosm study. In particular, CAT activity was induced by low ZnO–NPs concentrations (from 1 to 100 mg L^{-1}) during the first 3 days of exposure while SOD activity was stimulated mainly by higher ZnO–NPs concentrations (30 and 100 mg L^{-1}) and remained stimulated during longer exposure (up to 21 days of exposure). Another example pointed out the concomitant increase of metallothionein, a low molecular-weight cysteine-rich protein acting to efficiently bind labile metals, and the antioxidant Mn-cofactor superoxide dismutase gene (*sodA*) in response to trace metal toxicity in microorganisms (*Pseudomonas aeruginosa*) isolated from polluted marine sediments (Bouskill et al. 2007).

These examples illustrate how the oxidative stress induced by chemicals is commonly assessed. However, AEA responses are not specific, on the contrary they respond to a large number of compounds (organic and inorganic) and exposure conditions, in agreement with observations in animals.

In addition, it is also important to highlight that AEA responses are faster (i.e. can be detected before) and more sensitive (i.e. detectable at lower concentrations of chemicals) than traditional biomarkers (e.g. Geoffroy et al. 2004; Dewez et al. 2005; Guasch et al. 2010b). For instance, in the microalgae *Scenedesmus obliquus,* the herbicide flumioxazin induced CAT activity after 6 h of exposure to 3 $\mu g\ L^{-1}$ whereas 24 h of exposure were necessary to observe inhibition of the photosynthetic activity (Geoffroy et al. 2004). Bonnineau et al. (2013) also highlighted the early response (after 6 h of exposure) of CAT activity to the pesticide oxyfluorfen in periphytic biofilm (exposed to 15 $\mu g\ L^{-1}$). Consequently AEA may be considered as early indicators of stress of microbial communities providing an early warning of the occurrence of a stressful event (Bonet et al. 2012; Bonnineau et al. 2012, 2013).

To better understand the impact of changes in AEA at a higher scale of biological organisation and to gain in ecological relevance, AEA responses to seasonal variations (Bonet et al. 2013; Closa and Goicoechea 2010) and to various perturbations (including chemical contamination) have also been investigated in complex microbial communities forming biofilms using both field and microcosm approaches (Bonet et al. 2012, 2014; Bonnineau et al. 2010, 2011; Guasch et al. 2010a). Some field studies showed seasonal variations of CAT activity from microbial communities of an unmanaged forested soil and of CAT, APx, GR and GST activities from periphytic biofilms in a non-polluted river (Bonet et al. 2013; Closa and Goicoechea 2010). This natural variability is likely to impair the establishment of direct links between the absolute value of one AEA and contaminants exposure or effects. To overcome this limitation, AEA measurements should be included in a wider biomarker set in an integrated biomarkers approach. Beliaeff and Burgeot (2002) suggested representing normalized response of biomarkers in star plots to integrate the response of various biomarkers in a unique visualisation. Furthermore, the star plot area can be used to summarise the changes of the whole set of biomarkers tested in an integrated biomarker response (IBR) index (Beliaeff and Burgeot 2002; Devin et al. 2014; Sanchez et al. 2013). For example, Hou et al. (2016) calculated the IBR index integrating the response of CAT, SOD, GR and GPx from periphytic biofilms exposed to ZnO–Nps, in a microcosm study. IBR indices increased with exposure concentrations; in addition IBR indices increased strongly in the first 3 days of exposure and remained stable afterwards.

Finally, the impact of AEA variations on microbial community structure and function is poorly understood and only a few studies have investigated the link between AEA and oxidative stress damage leading to contradictory conclusions. For instance, high levels of AEA (SOD, CAT) in native phytoplankton from a lake contaminated by metals and PAHs (Lago Menor, Italy) were associated with low levels of oxidative damage (measured as proteins radical) whereas laboratory exposure to these contaminants provoked oxidative damage in similar phytoplankton communities (Vega-López et al. 2013). Due to the high reactivity of ROS and radicals, measuring oxidative damage in environmental samples is challenging, however recent advances (Grisham 2013; Aksmann et al. 2016) provide new and useful tools to better understand the impact of antioxidant enzymes modulation on oxidative damage in microbial communities. Further studies should also focus on the links between microbial AEA and communities capacities to compensate oxidative damage and to maintain their structure and functions.

11.3 Microbial Biomarkers Based on Community Adaptation

It is well known that toxic compounds pose a real risk to natural communities, even at low concentrations, since long-term exposure will erode the pool of species leading to the selection of the most tolerant ones (Liess et al. 2013). The selected species will perform better under the prevailing polluted conditions, which is commonly called community adaptation. This has been demonstrated in aquatic invertebrates (e.g. Van den Brink et al. 2009) but also in microbenthic algal communities (e.g. Guasch et al. 2003; Sabater et al. 2007). This is the conceptual framework behind the use of pollution indices for biomonitoring, which are based on the presence or absence of tolerant species (e.g. Kelly et al. 1998), or the PICT (Pollution Induced Community Tolerance) which is based on the display of an overall increase in tolerance of a community to the toxicant compared to a reference community as a result of the selection of the most tolerant organisms as well as to the establishment of mechanisms for detoxification (Tlili et al. 2016).

If microbial communities are found almost everywhere, this is due to their high degree of metabolic flexibility, physiological tolerance and diversity (Allison and Martiny 2008). This adaptation may appear at a genetic level (e.g. Morgante et al. 2015), through an elevated expression of the genes building this tolerance and also at a community level by a shift in the composition of the community selecting the most tolerant species (e.g. Wright et al. 2006).

Therefore, we propose that any metrics informing about community adaptation to toxicants be considered as potential biomarkers of chronic exposure. The question is how to discern among the large set of endpoints related with the adaptation of microbial communities to chemical pollution those that better fit as biomarkers of adaptation. Here we discuss the pros and cons of three different types: (i) the genes of resistance (ii) the species composition (iii) the acquisition of tolerance to oxidative stress.

11.3.1 Genes of Resistance

The study of genes of pollutants resistance in microbiology is rather common. However, most research focuses on pollutants which can be sequestered, transformed or degraded by microbial communities such as metals or aromatic hydrocarbons (Lloyd and Lovley 2001; Choi et al. 2013; Das et al. 2016) rather than on their use as biomarkers. In addition, most knowledge on resistance genes is based on cultured microorganisms (Pal et al. 2014), thus limited to the small percentage that can be cultivated. However, recent studies confirm the potential of the metagenomics approach to increase this knowledge, since it allows the assessment of the diversity of pollutant resistance genes in the entire gene pool of environmental samples (e.g. Golebiewski et al. 2010; Li et al. 2015; Morgante et al. 2015). The question here is whether the genes of resistance could be used as reliable biomarkers of pollution or not. To answer this challenging question we will focus on metal and antibiotic resistance genes research findings. More specifically on their diversity and specificity, searching for studies showing direct links between exposure (i.e. environmental occurrence and bioavailability) and the prevalence of the expected resistance genes in these contaminated environments. It is assumed that when concentrations inside the cells are too elevated, microorganisms will react by the expression of specific resistance systems. Therefore, microorganisms displaying such resistance systems will be selected in contaminated environments (Roosa et al. 2004).

11.3.1.1 Metal Resistance

Metal resistance systems are present in nearly all microorganisms. They arose because microorganisms exist in an environment that has always contained metals (Bruins et al. 2000). Research carried out since the early 1970s identified several microorganisms as being resistant to certain toxicants. These investigations led to the general idea that: (i) many microorganisms had successfully adapted to the presence of several metals by the use of chromosomal plasmid or transposon-encoded metal-resistance mechanisms; ii) that the most common mechanism was the presence of ion-selective ATPase (adenosine tri-phosphate) pumps and iii) that these mechanisms were highly specific (Buirns et al. 2000). Moreover, many genes involved in microbial resistance to most essential (Cu, Cu, Ni and Zn) and non-essential metals (Ag, Al, Cd, Pb and Hg) were identified (Buirns et al. 2000). However, recent studies indicate that metal-defense mechanisms are less selective than it was previously thought. For instance, Radnieki et al. (2008) demonstrated that relatively low concentrations of zinc chloride caused the up-regulation of the gene *merA* (mercury reductase) in *Nitrosomonas europaea*, concomitant with the demonstration of tolerance to Zn. This gene was previously identified as a specific gene of resistance to mercury, at relatively low concentration questioning the specificity of this response and its use as biomarker of effect and/or exposure. Another argument supporting this point is the presence, in many bacterial species, of mechanisms of resistance which are not selective. This is the case of a heavy-metal efflux pump of the cation diffusion facilitator (CDF) family. This effluent pump, responsible for removal of structurally unrelated toxic metals like cadmium, cobalt, silver, zinc and copper from the cytoplasm, was found in the genome of some Fe-bacteria (Emerson et al. 2013).

In order to illustrate this point, we will describe with more detail the genes that convey As resistance, which have been well characterized in the literature. Arsenate-Resistant Microorganisms (ARMs) have been isolated from As-contaminated soils and mine tailings. ARMs reduce As^{V}–As^{III} as a mean of resistance allowing them to cope with high arsenic levels in their environment (Oremland and Stolz 2005). As^{V} detoxification is mostly carried out by the cytoplasmic arsenate reductase (ArsC) system composed of a cytoplasmic arsenate reductase enzyme and As^{III} efflux pump (ArsB), encoded by three genes (*arsR*, *arsB* and *arsC*) in an *ars* operon, which encode a cytoplasmic arsenate reductase enzyme and As^{III} efflux pump (ArsB) (Amend et al. 2014; Patel et al. 2007; Wang et al. 2015), or may stay sequestered inside the cell, bound to cysteine residues in enzymes (Páez-Espino et al. 2009). In addition to the common three genes *arsRBC* found on *ars* operons, two additional genes, *arsD* and *arsA*, can be found on this operon, such as the *arsRDABC* operon in *Escherichia coli* plasmid R773 (Morgante et al. 2015). Cells expressing the *arsRDABC* operon are more resistant to As^{V} and As^{III} than those expressing *arsRBC* operons. ArsA forms a complex with ArsB that catalyses ATP-driven As^{III}/Sb^{III} efflux, while ArsD increases the affinity of ArsA for As^{III}, increasing its efflux and resistance at environmental concentrations of arsenic (Li et al. 2006). A similar process mediated also by glutathione is carried out by microalgae at intracellular level (Rahman and Hassler 2014). This process allows microalgae surviving in aquatic habitats with high As^{V} levels, due to their ability to rapidly excrete arsenic from the cells (Wang et al. 2015). In cyanobacteria (specifically, *Synechocystis* sp.), an operon containing three genes similar to the bacterial genes mediating As resistance (*acr3*, *arsH* and *arsC*) was identified. These three genes encode a cytoplasmic arsenate reductase (ArsC) enzyme and an As^{III} efflux pump, named Acr3 (López-Maury et al. 2003). ArsH is a flavin mononucleotide (FMN)-containing protein of unknown function and a member of the family of NADPH-dependent FMN reductases. Despite the nature of its final electron acceptor and the potential role of ArsH in the response to oxidative stress induced by arsenite (Hervás et al. 2012), the role of ArsH in the resistance to arsenic remained to be clarified.

Ford et al. (2005) research focused on the presence of specific genes that convey resistance to toxic metals using As-resistance genes as a first step towards the use of resistance or catabolic genes as microbial biomarkers. The genes *arsA*, *arsB* and *arsC* were found in isolates obtained from polluted sediments. However, it was not possible to establish a direct link between the presence of *ars* genes and As tolerance, bringing into question its application as biomarkers of adaptation (Ford et al. 2005). It is worth mentioning that this relationship was previously shown in the acidophilic iron oxidizing archaeon *Ferroplasma acidarmanus*

Gihring et al. (2003) and confirmed in the cultured bacteria *Pseudomonas putida* (Fernández et al. 2014).

Recent advances in metagenomics techniques allow a more comprehensive assessment of the microbial biodiversity and functioning and contribute to improving our understanding of metal-resistance. Ecotoxicogenomics, the application of "omics" technology in ecotoxicology (defined in the late 1990s as the study of gene and protein expression integrating transcriptomics, proteomics, and metabolomics into ecotoxicology, Neumann and Galvez 2002), is further promoting this development (Martyniuk and Simmons 2016).

Several studies have used a metagenomics approach to identify the ubiquitous distribution of metal resistance genes (e.g. Roosa et al. 2004; Li et al. 2015; Bengtsson-Palme et al. 2016). As an example, the metagenome of microbial communities from the Tinto River (South of Spain), a natural acid mine drainage (AMD) environment, was explored to search for novel genes involved in As resistance (Morgante et al. 2015). In this case, a functional metagenomics approach by direct cloning of AMD metagenomic DNA and its subsequent heterologous expression in *E. coli* cells allowed the identification of 13 functionally active genes. Most genes (11) encoded proteins not previously related to heavy-metal resistance, and only two were related to heavy metal resistance. Moreover, some of the genes identified were found to confer resistance to other stress and lethal conditions. This study (Morgante et al. 2015) supports the idea that metal-resistance is common but less specific and more complex than it was previously thought. This previous dogma is now seriously disputed thanks to the advances in new high throughput sequencing technologies (Martyniuk and Simmons 2016).

11.3.1.2 Influence of Metals on Antibiotic Resistance

It is nowadays accepted that human activity is responsible for a general environmental increase of antibiotic resistance genes from all classes of antibiotics. One of the greatest concerns about the presence of antibiotics in the environment is the emergence and dissemination of antibiotic resistance genes (ARGs). However, new technological techniques, especially metagenomics, have also contributed to question the linkage between antibiotics occurrence and resistance (e.g. Bengtsson-Palme et al. 2016). Indeed, many of these genes can be mobilized and transferred to other bacteria (see the review about antibiotics resistance by Galan et al. (2013) for more details).

The question of whether non-antibiotic compounds might increase antibiotic resistance was already posed by Alonso et al. (2001) in terrestrial environments. This author suggested an association between metal and antibiotic resistances since resistance to both could be transferred among organisms through mobile DNA elements. This was clearly demonstrated in terrestrial bacteria by Berg et al. (2005). This investigation showed that Cu-amendment of soils significantly increased the amount of Cu-resistant bacteria. Furthermore, Cu-resistant isolates from those Cu-contaminated soils had significantly higher incidence of resistance to several antibiotics. Therefore, metal pollution (Cu) not only selected for Cu-resistance but also indirectly for antibiotic-resistance in Cu-resistant bacteria. This observation is supported by other investigations showing that the *tcrB* gene, which confers Cu resistance, was physically linked, on a single transferable plasmid, to the *vanA* gene cluster and *ermB* gene, which are responsible for macrolide resistance (see the review paper by Baker-Austin et al. 2006). Interestingly, Berg et al. (2005) also highlighted effects of Cu on bacterial community structure.

Co-selection for antibiotics resistance resulting from metal-exposure was also shown by Wright et al. (2006) in stream communities. In this case, metal and antibiotic tolerance increased in non-cultured bacteria sampled along a gradient of metal contamination demonstrating a direct link between real exposure and community tolerance.

In summary, studies reported in the literature support that bacteria may have intrinsic tolerance and that co-resistance, cross-resistance and co-regulation of resistance to antibiotics are very common in microorganisms (Baker-Austin et al. 2006). For instance, microbial exposure to the sulphonamide sulfamethoxazole, in bioreactors, led to significant changes in microbial community composition however the genes of resistance (*sulI*) were stable during all the experiment (Collado et al. 2013). This stability was attributed to the high mobility of the plasmids carrying these resistance genes. While Huerta el al. (2013) found difficult to link antibiotics occurrence with the relative abundance of antibiotic resistance genes, Marti et al. (2013) could observe an increase in the relative abundance of antibiotic resistance genes in biofilm samples collected downstream of a waste water treatment plant. In a comprehensive investigation of resistance genes of environmental samples (metagenomes) against antibiotics, biocides and metals and their co-selection potential in sewage treatment plants (STP), Bengtsson-Palme et al. (2016) could not demonstrate any direct selection for resistance genes to any particular antibiotic classes in the STP. Furthermore, they found limited support for co-selection between antibiotics and/or biocides and metals on the genetic level. However, some specific resistance genes enriched during sludge treatment and the abundance of mobile elements were not significantly reduced through water treatment.

All together indicate that an overall increase in the expression of microbial genes of resistance is expected in polluted environments, and this overall increase can be indicative of effects of and exposure to chemicals. However, the specific cause and implications of these alterations are very uncertain. Ecotoxicogenomic techniques, especially metagenomics approaches as those reported above (Bengtsson-Palme et al. 2016), may contribute to overcoming some of these limitations.

11.3.2 Community Structure

Community ecology principles state that the set of species found in a community will always respond to and be selected by the prevailing environmental conditions (e.g. Miltelbach 2012). Focusing on microbial communities' ecotoxicology, a biomarker of pollution could stem from community composition changes resulting from the selection pressure exerted by a pollutant. At the beginning of exposure, this selection pressure is not expected to affect the pool of species but their relative abundances, since it will favour the most tolerant ones and inhibit the growth of the most sensitive ones without influencing the overall species richness which may even increase. Species richness may finally decrease if the intensity and/or duration of exposure cause the extinction of the sensitive species, only the more tolerant ones remaining. While these expectations are theoretically correct, we wonder whether it is possible to unequivocally relate a specific type of pollution to the set of species found in the community. This potential link has been widely discussed in the literature. For instance, in the review paper by Sabater et al. (2007), a large number of investigations addressing the effects of chemicals on biofilm communities were already reported. At that time, most examples were focused on metal toxicity and few on other toxic compounds (i.e. pesticides). It is worth mentioning that the majority of investigations (89%) included species composition analyses. Remarkably, a direct link between pollution and the community composition was seen in most of the cases (14 up to 17). In a later update (Guasch et al. 2012), a greater number of investigations were reported including studies dealing with metals, pesticides and emerging pollutants. This larger dataset revealed that the co-occurrence of different types of pollution and stressors may make it difficult to elucidate the causes of the observed species composition changes in many cases. Indeed, a clear link between pollution and the species composition was mainly found in highly metal-polluted rivers. Investigations performed by Morin et al. (2012), which aimed at describing the global patterns of diatom community response to metal contamination in fluvial systems, exemplify this difficulty. In these studies, the concomitant loss of sensitive species with the development of more resistant ones allowed for a clear discrimination of three different metal-pollution categories: low or no-metal pollution; moderate metal pollution and high metal pollution. However, they could not differentiate situations of low metal pollution from the non-polluted ones.

As mentioned above, the co-occurrence of different types of stressors may prevent elucidating the causes of the observed species composition found in the field, particularly if pollution is low. However, this does not mean that a specific effect did not occur. For instance, Barral-Fraga et al. (2016), found a clear effect of a low arsenate concentration (130 $\mu g\ L^{-1}$) on diatom species composition after 13 days of exposure in indoor channels. Arsenate exposure caused a selection for metal-tolerant diatom species, indicating that this exposure level may alter the biofilm community structure in the field. In another investigation, Ponsatí et al. (2016) explored the relative importance of environmental factors and organic micro-pollutants (herbicides, insecticides, industrial organic compounds, personal care products, antibiotics and pharmaceuticals) on determining the structure and function of fluvial biofilms using a large dataset. By applying multivariate analyses, the authors were able to identify a gradient of pollution (sites with high concentrations of industrial organic compounds, herbicides and pharmaceuticals, as well as organic and inorganic nutrients) and this gradient matched with the diatom community structure, i.e. higher levels of pollution co-occurred with pollution-tolerant diatom taxa.

Not only diatom community composition but also bacterial community composition can be affected by contaminants. With this frame of reference, Corcoll et al. (2015) showed that a mixture of pharmaceuticals (a mixture including antibiotics, β-blockers, an anti-inflammatory, a psychiatric, a lipid regulator and a diuretic drug) can affect different components of biofilm communities in indoor channels. As for the autotrophic component, the biomass and taxa richness decreased and the algal community composition was affected. Moreover the structure of the bacterial community (based on denaturing gradient gel electrophoresis of amplified 16S rRNA genes) changed and showed a reduction of the operational taxonomic units (OTUs) richness. Nevertheless, this cause-and-effect relationship was less clear in the field (Corcoll et al. 2014). The authors demonstrated that algal and microbial communities of biofilms growing downstream a WWTP acquired tolerance to pharmaceuticals (ibuprofen and diclofenac) and that these changes were caused by the replacement of sensitive species by others tolerant to the new conditions. However, pharmaceuticals were not the exclusive factor shaping the response of biofilm communities to the WWTP effluent, since stream water was not only enriched with pharmaceuticals but also with organic and inorganic nutrients. Similarly, Huerta et al. (2013) found a link between antibiotic occurrence and the overall bacterial community patterns in water samples taken from water supply reservoirs. However variation partitioning analysis results indicated that not only antibiotics concentrations (specifically enrB and blaTEM) contributed to explain the bacterial community but also the geographical location of the reservoirs and some of the measured environmental variables (water temperature, pH and the light extinction coefficient). In contrast with these studies, a clearer link between antibiotics and bacterial community structure was described by Collado et al. (2013) in bioreactors. Cell numbers of bacteria tolerant to the antibiotic increased, whereas other bacterial classes were found to be more vulnerable to the antibiotic load and disappeared. It is important to mention that the dose applied was in this case very high.

The examples detailed above, confirm that while possible, the unequivocal link between pollution and the set of species found in microbial communities is rather difficult. This difficulty stems from our lack of information about the bio-indicative value of each species for most types of pollution. Again, ecotoxicogenomics, especially comprehensive metagenomic approaches analysing changes in genes (of resistance and defense) in relation to taxonomic changes, may contribute to overcoming these limitations.

11.3.3 Acquisition of Tolerance to Oxidative Stress

As detailed in the Sect. 11.2, AEA measurements at a community level are expected to reflect the tendency (activation or inhibition) observed in the majority of individuals and species within the community. Therefore, AEA measurement can also be used to reveal the acquisition of tolerance resulting from chronic exposure of biofilm to oxidative stress (Bonet et al. 2013, 2014; Bonnineau et al. 2013). For instance, in a microcosm study on periphytic biofilms, the pattern of CAT activity in response to acute exposure to a range of oxyfluorfen, a pesticide inducing oxidative stress, allowed discriminating chronically exposed biofilms from non-exposed ones (Bonnineau et al. 2013). By contrast, CAT activity was similar between control biofilms and chronically exposed biofilms in absence of acute challenge. Indeed, exposure to oxidative stress is likely to provoke community adaptation which can be revealed by the community capacity to cope with a larger range of oxidative stress levels during an acute exposure (Fig. 11.2). This methodology has already been successfully applied in microbial ecotoxicology to identify contaminants exposure (Pesce et al. 2016; Tlili et al. 2016).

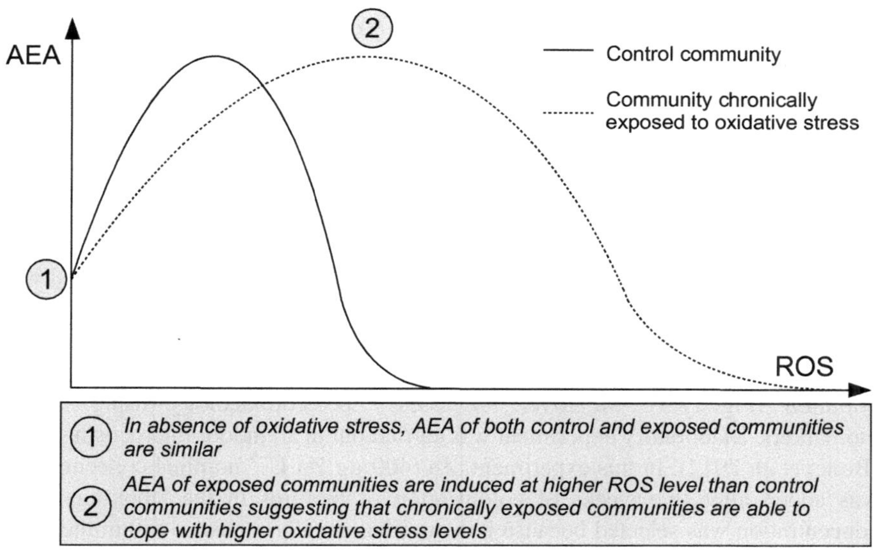

Fig. 11.2 Theoretical framework illustrating the acquisition of tolerance to oxidative stress by microbial communities. AEA from microbial communities chronically exposed to oxidative stress (*in dotted line*) are expected to be stimulated on a higher range of ROS levels than non-exposed communities (*control in plain line*). *AEA* antioxidant enzyme activities; *ROS* reactive oxygen species

11.4 Case Study on the Use of Biofilm Biomarkers to Assess Zn Ecotoxicology

The use of biofilm biomarkers in field experiments, despite their high complexity (because several parameters such as temperature, light or metal concentrations cannot be controlled but monitored), are crucial to explore the potential of AEA and other biomarkers as ecosystems "health" biomarkers. Moreover, microcosm experiments (more controlled even though less ecologically relevant) are also important to better understand biomarkers responses to single-stress factors (i.e. a specific type of pollution). This case study is based on field experiments conducted in two mine areas which were chosen in purpose to investigate biofilm responses to metal pollution. In both cases, Zn was the most toxic metal that exceeded concentration thresholds marked by the European Water Framework Directive (WFD 2000). The first study-site, the Riou Mort, is a small tributary of the River Lot located in the metal industry basin of Decazeville (SW, Bordeaux, France). Three sites with different types and levels of metal pollution were selected: Decazeville (D) upstream of a Zn factory, located in the urban zone and with a background level of Zn pollution, Le Grange (G) with the highest Zn values because of its location downstream of the Zn factory and the confluence of a non-Zn polluted tributary (Riou Viou) and, further downstream, Joanis (J), with moderate levels of Zn pollution (Fig. 11.3). The second study-site, the Riera d'Osor is a small tributary of the River Ter located in the north-east of Catalonia (NE Spain). This stream, in comparison with the Riou Mort, is relatively well preserved and with lower concentrations of Zn (Fig. 11.3). It has well developed riparian vegetation and low urban pressures although hydrology has been altered due to the diversion of part of the stream discharge for electric power production. Four sites with different Zn concentrations were selected: a reference site located upstream, before the Zn mine source (Up); further downstream but still upstream of the Zn main source and with low Zn pollution, a site designated as mining 1 (M1); the third site was located just after the mine source (referred to as MS) with continuous inputs of Zn provoking higher Zn pollution (mining 2 (M2)); and, finally, mining 3 (M3) with lower Zn pollution (Fig. 11.3). Moreover, to assess Zn ecotoxicology using biofilm biomarkers, a laboratory experiment was carried out in an indoor microcosm system (Bonet et al. 2012). In this experiment, Zn (400 μg Zn L^{-1} nominal concentration) was added after five weeks of colonization of biofilms in the laboratory. This concentration was selected because it is environmentally realistic, within the range found in the Riou Mort and the Riera d'Osor (Fig. 11.3).

In all approaches, field and microcosm experiments, we investigated several biofilm biotic responses to acute (up to 24 h) and chronic (from 1 week on) exposure to assess metal toxicity, mainly Zn toxicity.

Short-term Zn exposure led to a very fast transfer of metals from the media to biofilms (Fig. 11.3). Within a short period of time, accumulation of Zn reached 150–225 μg g DW^{-1} in microcosm, up to 200–500 μg g DW^{-1} in Riera d'Osor and one order of magnitude higher, around 5000–6000 μg g DW^{-1} in Riou Mort.

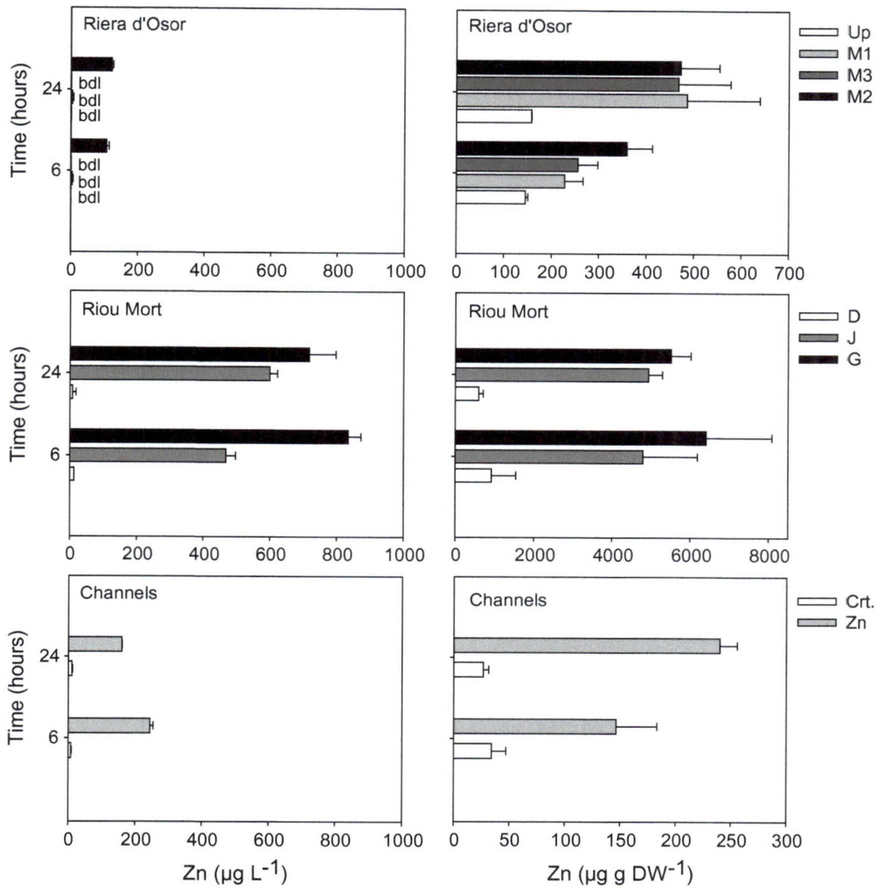

Fig. 11.3 Summary of average and standard deviations of total metal concentration dissolved in water (μg L^{-1}) in the *left side*, and accumulated in biofilm (μg gDW^{-1}) in the *right side*, after 6 and 24 h of exposure in Riera d'Osor, Riou Mort and a microcosm experiment. *bdl* below detection limit. *Gray intensity* indicate the increase of Zn pollution at each site: upstream (*Up*), mining 1 (*M1*), mining 3 (*M3*), mining 2 (*M2*), Decazeville (*D*), Joanis (*J*), Le Grange (*G*), control channels (*Crt.*) and Zn channels (*Zn*). *Source* Bonet's Thesis (2013)

Most AEA with the exception of CAT followed predictable, but different patterns of variation (Fig. 11.4). APx increased linearly reaching the highest values at the maximum Zn accumulation values measured. SOD followed a reverse pattern, decreasing with Zn accumulation. GR pattern followed a sigmoidal model, whereas GST reached the maximum values at intermediate Zn accumulation values (Figs. 11.3 and 11.4). Overall, this set of AEA, although having differences in sensitivity depending on the magnitude of exposure, were proposed as effect-based monitoring tools for the detection of the impairment caused by metal contamination episodes of short duration like an accidental metal spill (Fig. 11.4).

To better understand the impact of changes in AEA at a higher scale of biological organisation and to gain in ecological relevance, AEA responses to seasonal variations were assessed upstream, in the reference site (Bonet et al. 2013, 2014). In this site, CAT, APx and GR activities showed a seasonal pattern: increasing under warm and high light conditions when green algae were abundant in the biofilm

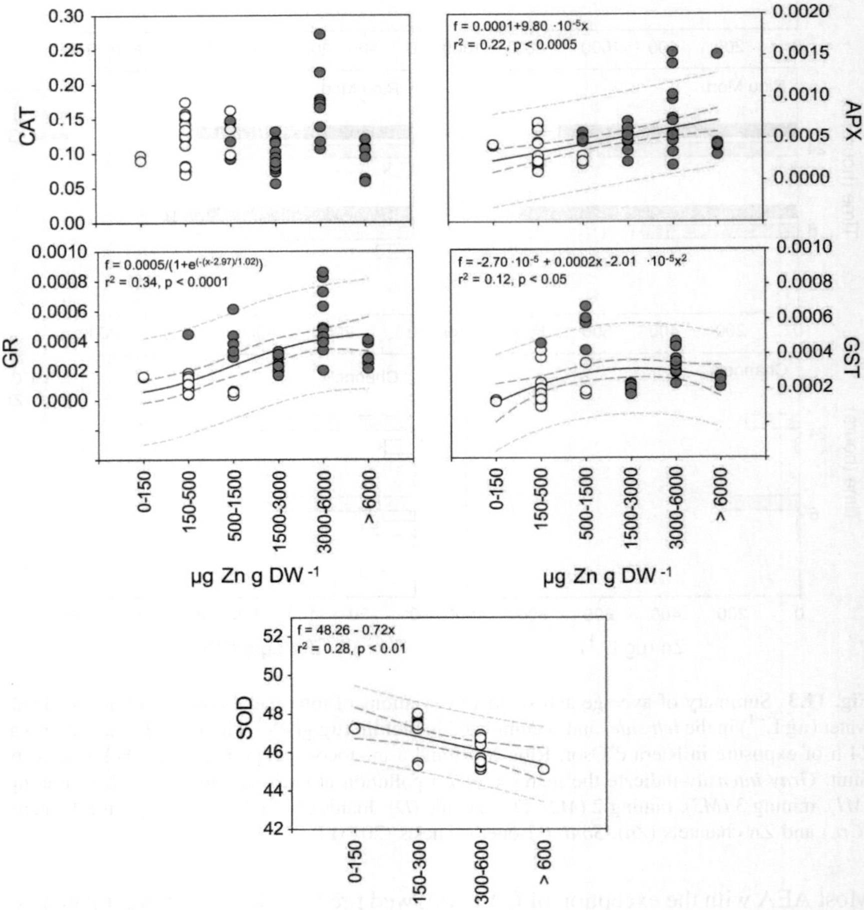

Fig. 11.4 Acute AEA responses in function of Zn accumulation (µg Zn g DW^{-1}) in Riera d'Osor (*in white*) and Riou Mort (*in gray*). Catalase (*CAT*) did not fit to any model. Ascorbate peroxidase (*APx*) and superoxide dismutase (*SOD*) followed a linear model, GR a sigmoidal and GST a quadratic one. CAT is expressed in µmol H_2O_2 µg $protein^{-1}$ min^{-1}, APX in µmol Ascorbate µg $protein^{-1}$ min^{-1}, GR in µmol NADPH $µg^{-1}$ protein min^{-1}, GST µmol CDNB conjugate $µg^{-1}$ protein min^{-1} and SOD in U µg $protein^{-1}$. *Medium-dash dark gray lines* indicate the 95% of confidence while *short-dash light gray lines* indicate the 95% of prediction. *Source* Bonet Thesis (2013)

community (Table 11.1). On the contrary, it is worth highlighting that CAT and APx remained almost constant in microcosm experiments, where environmental variability is controlled and not expected to cause stress under the experimental conditions established (Bonet et al. 2012).

As for chronic metal exposure, AEA responses were of lower magnitude than those seen in short-term exposure (Bonet et al. 2013, 2014). However, the analysis of data allowed discriminating metal exposure effects from environmental variability. Additionally, GST was clearly inhibited by metal pollution which affected the temporal patterns, masking the expected seasonality (i.e. AEA response to temperature and light changes in summer vs. winter). AEA contributed to coping with the oxidative stress caused by the environment (Table 11.1) and this important functionality was lost, or at least masked by the effects caused by metal exposure (Fig. 11.5). Since environmental factors like water temperature and light intensity or community composition may cause similar effects as metals on CAT, APx and GR, Bonet et al. (2013) highlighted that these parameters should also be measured and taken into account for the comparison between reference and impacted biofilm communities. In addition, environmental parameters can not only act as direct stressors on AEA but can also influence AEA response to toxicity. Light stress has been shown to influence directly AEA in biofilms but also to modify biofilm sensitivity to organic (glyphosate) and inorganic (Cu) pollutants, constraining its capacity to cope with further stress factors (Bonnineau et al. 2011).

In contrast to biofilm AEA, the photon yield, a parameter related to the photosynthetic performance of the community, did not show differences after chronic exposure, and this was attributed to community adaptation to metal pollution (Corcoll et al. 2012). Tlili et al. (2011) and Corcoll et al. (2012) reported a clear decrease in diatom abundance in the most metal-polluted site of the Riera d'Osor in favour of cyanobacteria and green algae. Furthermore, not only was a reduction in diatom diversity and biovolume, and an increase in malformed diatoms described under these metal conditions (Corcoll et al. 2012; Morin et al. 2012), but also an increase in biofilm community tolerance to metal pollution based on PICT tests

Table 11.1 Summary of environmental and biological parameters that influenced the natural variability of AEA in non-metal polluted sites

	Light	Temperature	Fo (Gr)
CAT	✓	✓	✓
APx	✓	✓	✓
GR	✓	✓	✓
GST	Ø	Ø	Ø
SOD	Ø	✓	Ø

The results were obtained from an annual cycle (Bonet et al. 2013) and a translocation experiment (Bonet et al. 2014) performed in Riera d'Osor. Fo (Gr) refers to the fluorescence linked to green algae

Source Bonet's Thesis (2013)

An ✓ indicates a relationship (either positive or negative) and the intensity (in bold) the magnitude. An Ø indicates the lack of relationship

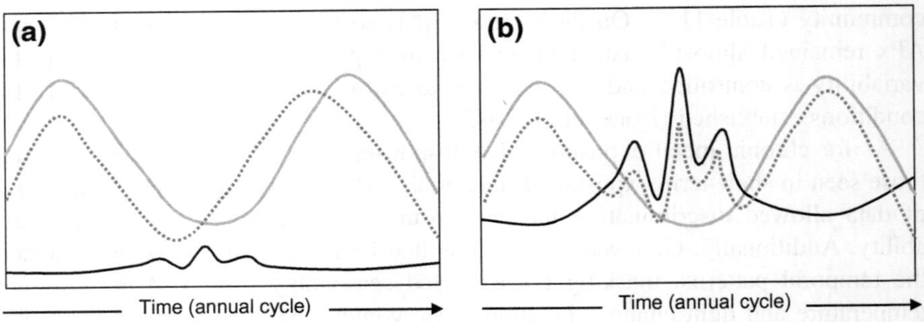

Fig. 11.5 Hypothetical responses of antioxidant enzyme activities (*AEA*) over a year in a fluvial system. *The grey line* represents environmental conditions, the *black line* represents metal concentration and the *grey dotted line* represents antioxidant enzyme activity response. The only difference between A and B is the increase in metal concentration in the stream from non-metal polluted or low metal concentrations to moderate concentrations. *Panel A* shows a pattern of AEA affected by environmental conditions under non-metal or low metal polluted conditions. *Panel B* shows the effect of moderate metal polluted conditions over the AEA pattern (similar to those reported in the Osor stream) (Bonet et al. 2013)

(Tlili et al. 2011). Remarkably, diatoms were strongly affected by metal pollution. Biofilm communities from site M2 (the most Zn polluted site) were largely different to those from the other sites (Table 11.2).

Among the 132 diatom taxa identified, the ones in M2 site were characterised by an association of *Achnanthidium minutissimum*, *A. pyrenaicum*, *Cocconeis placentula varieties* (placentula, lineata and euglypta, mainly), *Eolimna minima* and *Planothidium frequentissimum*. Additionally diatoms were characterized by small-sized species (low biovolume) with deformities, either pioneer, substrate-adherent species which are more metal-tolerant or species that are frequently reported in metal-contaminated environments (Morin et al. 2012). Moreover, the "*Indice de Polluosensibilité Spécifique*" (a diatom-based biotic index) values recorded at this site were on the border of poor water quality according to the 5 IPS categories recognised by the Water Framework Directive: very good (>16), good (16–13) moderate (12–9), poor (8–5) and bad (<5). And it was the same for the Riou Mort where mean values of the IPS index ranged from 7.2 to 10.7 indicating poor to moderate water quality (Morin et al. 2012). Furthermore, combining both data on cell size and abundances of deformities allowed to discriminate different levels of metal stress, low metal stress (only deformities) and moderate metal stress (where deformities are associated to very low biovolumes).

To sum up the lessons learnt from this case study, we would like to remark that: (i) biofilm metal contents are good indicators of metal pollution since biofilms are able to accumulate metals even when water metal concentrations are below detection limits; (ii) AEA show a sensitive and predictable but non-linear response after acute exposure and that each AEA has a different pattern of variation iii) AEA patterns of variation are masked by the effects caused by environmental variability,

Table 11.2 Diatom metrics from biofilm translocated to the reference site with background metal concentration, to very-low, low and moderate metal-pollution after 3 and 5 weeks of exposure in the Riera d'Osor

Metric	Site code (metal pollution category)				ANOVA	
	Up (background)	M1 (very-low)	M3 (low)	M2 (moderate)	F (3, 8)	*p*
Number of species	38 ± 1^{a}	35 ± 2^{ab}	39 ± 2^{a}	29 ± 1^{b}	7.173	0.002
Teratoforms (%)	0.54 ± 0.02^{a}	2.39 ± 0.56^{b}	2.52 ± 0.25^{b}	3.35 ± 0.65^{b}	10.33	<0.001
Biovolume (μm^3)	1358 ± 49^{a}	1830 ± 119^{a}	1574 ± 84^{b}	225 ± 26^{b}	178.8	<0.001
IPS	15.2 ± 0.2^{a}	14.2 ± 0.3^{ab}	13.5 ± 0.2^{b}	8.6 ± 0.2^{c}	159.4	<0.001

Each value corresponds to the average and standard error (n = 6). For each parameter, different letters indicate significant differences ($p < 0.05$) between sites after ANOVA and Tukey's HSD test. *Source* Modified from Corcoll's Thesis (2011)

whereas biomarkers of adaptation discriminate different levels of metal stress. Thus, and according to European guidelines (e.g. E.C., Guidance document n. 25, 2010), we strongly recommend for environmental risk assessment to use the triad approach (Chapman 1990) which is a combination of the three assessment methods (i) chemicals like metal concentration in water and biofilms, (ii) biomarkers like AEA and (iii) species composition analyses. Finally, despite the progress made by the scientific community in developing and adapting new biomarkers (e.g. molecular and biochemical approaches), further work is still needed to better understand biomarkers responses and build up a bridge between pollution and ecological status, with known pollutants and emerging ones which appear in our freshwater systems every day due to human activities.

11.5 Perspectives in the Development and Use of Microbial Biomarkers in Ecotoxicology

In ecotoxicology, a good microbial biomarker should a priori accomplish three main premises, first be specific of a type of pollution, second be sensitive and third provide information on its final effects on the ecosystem. Other attributes such as its proportionality between dose and response, reproducibility or technical difficulty and cost of the analysis are also important but less essential than the previous ones. Along the chapter, we have focused on AEA, community structure and the resistance genes in microbial communities as biomarkers of exposure, effects and adaptation of chemicals in the environment. Based on our review, it is clear, up to now, that none of them may fulfil the three main premises reported above.

The examples detailed in previous sections highlight that several AEA might be highly sensitive to environmental exposure to chemicals (metals), but have low specificity since their response to other environmental stress factors (e.g. light or water temperature) is also important, thereby overlapping with their response to metal toxicity. The link between AEA responses and ecosystem health is also unclear. Indeed, rapid AEA changes in response to a perturbation can be used as early warning of exposure but cannot inform on persistent ecosystem damage since the mechanisms of defense (i.e. AEA) may be sufficient to cope with chemical stress allowing the affected organisms to recover after exposure. It is noteworthy to add that resistance of microbial communities to a range of oxidative stress levels may also be used to reveal community adaptation and therefore exposure. Nevertheless, since most AEA have a bell-shape response towards a gradient of oxidative stress, more investigation and statistical development are required to determine the optimal design (e.g. number of concentrations tested, number of replicates, etc.) of the acute test to ensure a correct estimation of the dose-response curve of AEA. More field studies investigating AEA response in contrasting environments would be particularly helpful to optimize their use in environmental risk assessment.

Metrics informing about the adaptive response of microbial communities in terms of their species composition changes (i.e. community structure) are very sensitive to low but chronic pollution. Algal taxonomy has been largely used to study toxicant-induced selection in biofilm communities, due to its tradition but also to its high sensitivity. Specific morphological endpoints, e.g. teratologies or cell sizes, have also proved to detect metal pollution successfully. However, our lack of information about the bio-indicative value of most species for most types of pollution prevents, in many cases, the unequivocal link between pollution and community structure.

As for resistance genes, they were considered as specific biomarkers directly linked with pollution; however recent studies denied this paradigm. Moreover, there is a lack of information about the genes which confer resistance to the large set of chemicals of environmental concern.

As mentioned above, ecotoxicogenomics, especially comprehensive metagenomic approaches analysing changes in genes (of resistance and defense) in relation to taxonomic changes, may contribute to overcoming these limitations. For instance, metagenomics have been described as effective tools to assess the prevalence of specific gene sequences in tolerant communities and their taxonomic affinities in natural biofilms thus bridging the gap between exposure to a specific type of pollution (shown with the prevalence of specific genes) and community structure (Eriksson et al. 2009).

In addition, the link between changes in microbial biomarkers caused by chemicals and their direct and indirect effects on the ecosystem is often missing. For example, Tuulaikhuu et al. (2016) showed that exposure of periphyton and fish to arsenic (28–120 μg L^{-1}) for 60 days in mesocosms (indoor channels) not only affected the microbial community (by inhibiting biofilms potential ability to use organic phosphorus and to oxygenate the aquatic environment) but also fish health (detected as an increase in CAT activity), due to both direct exposure of fish to arsenic and the damage that arsenic caused to microbial functions affecting the fish environment.

This review envisages a set of metrics of interest as microbial biomarkers in ecotoxicology. However, these biomarkers may not be sufficient to link exposure with ecosystem damage. It is therefore necessary to use an integrated approach including, for each specific case, the most appropriate set of variables which provide unequivocal information concerning exposure, early effects, adaptation and ecosystem damage (Fig. 11.6). This was shown in our case study (as mentioned in Sect. 11.4) where all the different biofilm metrics came together to provide the toolbox needed to detect the impact caused by metal pollution at different scales of time and different exposure levels. It must not be overlooked that toxicants are expected to affect biogeochemical processes associated with bacterial metabolism, such as organic matter decomposition, nutrient cycling or chemical speciation in a non-specific manner. With this regard, these microbial processes commonly addressed in ecology and biogeochemistry should also be considered as sensitive tools for assessing ecotoxicological effects of pollutants on natural communities and ecosystems (Guasch et al. 2016). Beliaeff and Burgeot (2002) suggested representing normalized response of biomarkers in star plots to integrate the response of various biomarkers in a unique visualisation. Such approaches are not limited to AEA and can integrate a large variety of biomarkers. Integrated biomarker (IBR) response indices were generally correlated with in situ perturbations such as groundwater contamination (e.g. Si et al. 2016) or agricultural practices (e.g. Parelho et al. 2016). In this context, we suggest following an integrated approach not only including microbial biomarkers but other functional and ecological variables to provide what could be called integrated biomarker and ecological (IBER) response indices.

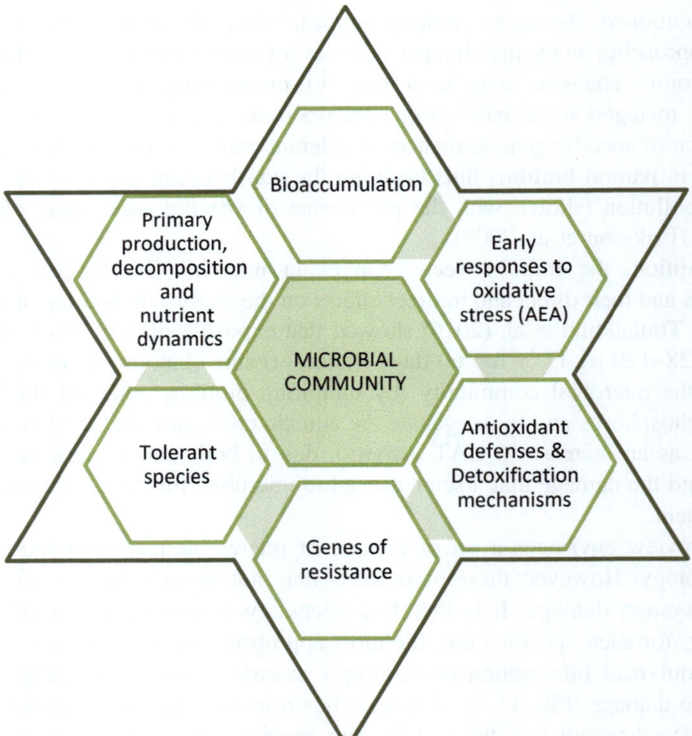

Fig. 11.6 Star diagram illustrating a set of metrics that could build an integrated microbial toolbox. The unequivocal link between exposure and ecosystem health is based on the results of a large set of microbial metrics related with exposure, early-effects or responses, adaptation and ecosystem functioning

References

Aksmann A, Pokora W, Baścik-Remisiewicz A, Dettlaff-Pokora A, Tukaj Z (2016) High hydrogen peroxide production and antioxidative enzymes expression in the *Chlamydomonas reinhardtii* cia3 mutant with an increased tolerance to cadmium and anthracene. Phycol Res 64:300–311. doi:10.1111/pre.12147

Alonso A, Sánchez P, Martínez JL (2001) Environmental selection of antibiotic resistance genes. Environ Microbiol 3(1):1–9

Allison SD, Martiny JBH (2008) Resistance, resilience, and redundancy in microbial communities. PNAS 105(1):11512–11519

Amend JP, Saltikov C, Lu GS, Hernandez J (2014) Microbial arsenic metabolism and reaction energetics. Rev Mineral Geochem 79(1):391–433

Amiard-Triquet C, Amiard J-C, Rainbow PS (2013) Ecological biomarkers: Indicators of Ecotoxicological Effects. CRC Press, Taylor & Francis Group, p 450

Amiard-Triquet C, Berthet B (2015) Individual biomarkers. In: Aquatic ecotoxicology: advancing tools for dealing with emerging risks, pp 153–182. doi:10.1016/B978-0-12-800949-9.00007-3

Asada K (2006) Production and scavenging of reactive oxygen species in chloroplasts and their functions. Plant Physiol 141:391–396

Baker-Austin C, Wright MS, Stepanauskas R, McArthur JV (2006) Co-selection of antibiotic and metal resistance. Review Trends Microbiol 14(4):176–182

Barral-Fraga L, Morin S, Rovira MDM, Urrea G, Magellan K, Guasch H (2016) Short-term arsenic exposure reduces diatom cell size in biofilm communities. Environ Sci Pollut Res 23:4257–4270. doi:10.1007/s11356-015-4894-8

Battin TJ, Besemer K, Bengtsson MM, Romaní AM, Packmann AI (2016) The ecology and biogeochemistry of stream biofilms. Nat Rev Microbiol 14:251–263

Benedetti M, Giuliani ME, Regoli F (2015) Oxidative metabolism of chemical pollutants in marine organisms: molecular and biochemical biomarkers in environmental toxicology. Ann N Y Acad Sci 1340:8–19. doi:10.1111/nyas.12698

Beliaeff B, Burgeot T (2002) Integrated biomarker response: a useful tool for ecological risk assessment. Environ Toxicol Chem 21:1316–1322

Bengtsson-Palme J, Hammarén R, Pal Ch, Östmanc M, Björlenius B, Flach CF, Fick J, Kristiansson E et al (2016) Elucidating selection processes for antibiotic resistance in sewage treatment plants using metagenomics. Sci Total Environ 572:697–712

Berg J, Tom-Petersen A, Nybroe O (2005) Copper amendment of agricultural soil selects for bacterial antibiotic resistance in the field. Lett Appl Microbiol 40:146–151

Bonet B, Corcoll N, Guasch H (2012) Antioxidant enzyme activities as biomarkers of Zn pollution in fluvial biofilms. Ecotox Environ Safe 80:172–178

Bonet B, Corcoll N, Acuña V, Sigg L, Behra R, Guasch H (2013) Seasonal changes in antioxidant enzyme activities of freshwater biofilms in a metal polluted Mediterranean stream. Sci Total Environ 444:60–72

Bonet B (2013) Antioxidant enzyme activities in fluvial biofilms as biomarkers of metal pollution. Doctoral Thesis, University of Girona, Catalonia (Spain)

Bonet B, Corcoll N, Tlili A, Morin S, Guasch H (2014) Antioxidant enzyme activities in biofilms as biomarker of Zn pollution in a natural system: an active bio-monitoring study. Ecotox Environ Safe 103:82–90

Bonnineau C, Guasch H, Proia L, Ricart M, Geiszinger A, Romaní AM, Sabater S (2010) Fluvial biofilms: a pertinent tool to assess beta-blockers toxicity. Aquat Toxicol 96:225–233

Bonnineau C, Bonet B, Corcoll N, Guasch H (2011) Catalase in fluvial biofilms: a comparison between different extraction methods and example of application in a metal-polluted river. Ecotoxicol 20:293–303

Bonnineau C, Moeller A, Barata C, Bonet B, Proia L, Sans-Piché F, Schmitt-Jansen M, Guasch H, Segner H (2012) Advances in the multibiomarker approach for risk assessment in aquatic ecosystems. In: Guasch H, Ginebreda A, Geiszinger A (eds) Emerging and priority pollutants in rivers. The handbook of environmental chemistry. Springer, Berlin, pp 147–179

Bonnineau C, Tlili A, Faggiano L, Montuelle B, Guasch H (2013) The use of antioxidant enzymes in freshwater biofilms: temporal variability vs. toxicological responses. Aquat Toxicol 136:60–71

Bouskill NJ, Handy RD, Ford TE, Galloway TS (2006) Differentiating copper and arsenic toxicity using biochemical biomarkers in *Asellus aquaticus* and *Driessena polymorpha*. Ecotox Environ Saf 65:342–349

Bouskill NJ, Barnhart EP, Galloway TS, Handy RD, Ford TE (2007) Quantification of changing *Pseudomonas aeruginosa* soda, htpX and mt gene abundance in response to trace metal toxicity: a potential in situ biomarker of environmental health. FEMS Microbiol Ecol 60:276–286

Bouskill NJ, Barker-Finkel J, Galloway TS, Handy RD, Ford TE (2010) Temporal bacterial diversity associated with metal-contaminated river sediments. Ecotoxicology 19:317–328

Bruins MK, Kapil S, Oehme FW (2000) Microbial resistance to metals in the environment. Rev Ecotox Environ Saf 45:198–207

Chapman PM (1990) The sediment quality triad approach to determining pollution-induced degradation. Sci Total Environ 97(98):815–825

Chaumot A, Geffard O, Armengaud J, Maltby L (2015) Gammarids as reference species for freshwater monitoring. In: Aquatic ecotoxicology: advancing tools for dealing with emerging risks, pp 253–280. doi:10.1016/B978-0-12-800949-9.00011-5

Choi EJ, Jin HM, Lee SH, Math KR, Madsen EL, Jeoana CH (2013) Comparative genomic analysis and Benzene, Toluene, Ethylbenzene, and o-, m-, and p-Xylene (BTEX) degradation pathways of *Pseudoxanthomonas spadix* BD-a59. Appl Environ Microbiol 79(2):663–671

Clements WH, Newman MC (2002) Community ecotoxicology. Wiley, West Sussex, p 336

Closa I, Goicoechea N (2010) Seasonal dynamics of the physicochemical and biological properties of soils in naturally regenerating, unmanaged and clear-cut beech stands in northern Spain. Eur J Soil Biol 46:190–199. doi:10.1016/j.ejsobi.2010.04.003

Collado N, Buttiglieri G, Marti E, Ferrando-Climent L, Rodriguez-Mozaz S, Barceló D, Comas J, Rodriguez-Roda I (2013) Effects on activated sludge bacterial community exposed to Sulfamethoxazole. Chemosphere 93:99–106

Colin N, Porte C, Fernandes D, Barata C, Padrós F, Carrassón M et al (2016) Ecological relevance of biomarkers in monitoring studies of macro-invertebrates and fish in Mediterranean rivers. Sci Total Environ 540:307–323. doi:10.1016/j.scitotenv.2015.06.099

Corcoll N (2011) The use of pulse amplitude modulated fluorescence techniques for metal toxicity assessment in fluvial biofilms. Doctoral Thesis. University of Girona, p 222

Corcoll N, Bonet B, Leira M, Guasch H (2011) Chl-a fluorescence parameters as biomarkers of metal toxicity in fluvial biofilms: an experimental study. Hydrobiologia 673:119–136. doi:10.1007/s10750-011-0763-8

Corcoll N, Bonet B, Morin S, Tlili A, Leira M, Guasch H (2012) The effect of metals on photosynthesis processes and diatom metrics of biofilm from a metal-contaminated river: a translocation experiment. Ecol Indic 18:620–631

Corcoll N, Acuña V, Barceló D, Casellas M, Guasch H, Huerta B et al (2014) Pollution induced community tolerance to non-steroidal anti-inflammatory drugs (NSAIDs) in fluvial biofilm communities affected by WWTP effluents. Chemosphere 112:185–193

Corcoll N, Casellas M, Huerta B, Guasch H, Acuña V, Rodríguez-Mozaz S et al (2015) Effects of flow intermittency and pharmaceutical exposure on the structure and metabolism of stream biofilms Sci Total Environ 503–504:159–170

Das S, Dash HR, Chakraborty J (2016) Genetic basis and importance of metal resistant genes in bacteria for bioremediation of contaminated environments with toxic metal pollutants. Appl Microbiol Biotechnol 100:2967–2984

Devin S, Burgeot T, Giambérini L, Minguez L, Pain-Devin S (2014) The integrated biomarker response revisited: Optimization to avoid misuse. Environ Sci Pollut Res 21:2448–2454. doi:10.1007/s11356-013-2169-9

Dewez D, Geoffroy L, Vernet G, Popovic R (2005) Determination of photosynthetic and enzymatic biomarkers sensitivity used to evaluate toxic effects of copper and fludioxonil in alga *Scenedesmus obliquus*. Aquat Toxicol 74:150–159

Edreva A (2005) Generation and scavenging of reactive oxygen species in chloroplasts: a submolecular approach. Agr Ecosyst Environ 106:119–133

Emerson D, Field EK, Chertkov O, Davenport KW, Goodwin L, Munk C, Nolan M, Woyke T (2013) Comparative genomics of freshwater Fe-oxidizing bacteria: implications for physiology, ecology, and systematics. Front Microbiol 4(254):1–17

Eriksson KM, Antonelli A, Nilsson RH, Clarke AK, Blanck H (2009) A phylogenetic approach to detect selection on the target site of the antifouling compound irgarol intolerant periphyton communities. Environ Microbiol 11:2065–2077

Fernández M, Udaondo Z, Niqui J-L, Duque E, Ramos J-L (2014) Synergic role of the two ars operons in arsenic tolerance in *Pseudomonas putida* KT2440. Environ microbiol reports 6 (5):483–489

Ford TE, Jay J, Patel A, Kile M, Prommasith P, Galloway T, Sanger R, Smith K, Depledge M (2005) Use of ecotoxicological tools to evaluate the health of New Bedford Harbor sediments: a microbial biomarker approach. Environ Health Perspect 113:186–191

Fulton MH, Key PB (2001) Acetylcholinesterase inhibition in estuarine fish and invertebrates as an indicator of organophosphorus insecticide exposure and effects. Environ Toxicol Chem 20:37–45

Galan J-K, González-Candelas F, Rolain J-M, Cantón R (2013) Antibiotics as selectors and accelerators of diversity in the mechanisms of resistance: from the resistome to genetic plasticity in the β-lactamases world. Review Article. Front Microbiol 4(9):1–17

Galloway TS, Millward N, Browne MA, Depledge MH (2002) Rapid assessment of organophosphorous/carbamate exposure in the bivalve mollusc *Mytilus edulis* using combined esterase activities as biomarkers. Aquat Toxicol 61(3):169–180

Gatidou G, Thomaidis NS (2007) Evaluation of single and joint toxic effects of two antifouling biocides, their main metabolites and copper using phytoplankton bioassays. Aquat Toxicol 85:184–191

Geoffroy L, Teisseire H, Couderchet M, Vernet G (2002) Effect of oxyfluorfen and diuron alone and in mixture on antioxidative enzymes of *Scenedesmus obliquus*. Pestic Biochem Physiol 72:178–185

Geoffroy L, Frankart C, Eullaffroy P (2004) Comparison of different physiological parameter responses in Lemna minor and *Scenedesmus obliquus* exposed to herbicide flumioxazin. Environ Pollut 131:233–241

Gihring TM, Bond PL, Peters SC et al (2003) Arsenic resistance in the archaeon "*Ferroplasma acidarmanus*": new insights into the structure and evolution of the ars genes. Extremophiles 7:123–130. doi:10.1007/s00792-002-0303-6

Golebiewski M, Deja-Sikora E, Tretyn A (2010) Direct cloning of heavy metal resistance genes from metagenomics DNA. J Biotechnol Special Abstracts 150S:S1–S576

Grisham MB (2013) Methods to detect hydrogen peroxide in living cells: Possibilities and pitfalls. Comp Biochem Physiol A Mol Integr Physiol, First Int Conf Oxidative Stress Aquat Ecosyst 165:429–438. doi:10.1016/j.cbpa.2013.02.003

Guasch H, Admiraal W, Sabater S (2003) Contrasting effects of organic and inorganic toxicants on freshwater periphyton. Aquat Toxicol 64:165–175

Guasch H, Serra A, Corcoll N, Bonet B, Leira M (2010a) Metal ecotoxicology in fluvial biofilms: potential influence of water scarcity. In: Sabater S, Barceló D (eds) Water scarcity in the mediterranean: Perspectives under global change. Springer, Berlin, Germany, pp 41–54

Guasch H, Atli G, Bonet B, Corcoll N, Leira M, Serra A (2010b) Discharge and the response of biofilms to metal exposure in Mediterranean rivers. Hydrobiologia 657:143–157

Guasch H, Bonet B, Bonnineau C, Corcoll N, López-Doval J, Muñoz I, Ricart M, Serra A, Clements W (2012) How to link field observations with causality? In: Guasch H, Ginebreda A, Geiszinger A (eds) Field and experimental approaches linking chemical pollution with ecological alterations. Emerging and priority pollutants in rivers. Springer, Berlin, vol 19, 181–218

Guasch H, Artigas J, Bonet B, Bonnineau C, Canals O et al (2016) The use of biofilms to assess the effects of chemicals on freshwater ecosystems in aquatic biofilms. In: Romaní AM, Guasch H, Balaguer M (eds) Ecology, water quality and wastewater treatment. Horizon Scientific Press-Caister Academic Press. ISBN 978-1-910190-17-3

Hamamura N, Macur RE, Korf S, Ackerman G, Taylor WP, Kozubal M, Reysenbach AL, Inskeep WP (2009) Linking microbial oxidation of arsenic with detection and phylogenetic analysis of arsenite oxidase genes in diverse geothermal environments. Environ Sci Technol 11:421–431

Haritash AK, Kaushik CP (2009) Biodegradation aspects of polycyclic aromatic hydrocarbons (PAHs): a review. J Hazard Mater 169:1–15. doi:10.1016/j.jhazmat.2009.03.137

Hervás M, López-Maury L, León P, Sánchez-Riego AM, Florencio FJ, Navarro JA (2012) ArsH from the cyanobacterium *Synechocystis* sp. PCC 6803 is an efficient NADPH-dependent quinone reductase. Biochemistry 51:1178–1187

Hook SE, Gallagher EP, Batley GE (2014) The role of biomarkers in the assessment of aquatic ecosystem health. Integr Environ Assess Manag 10:327–341. doi:10.1002/ieam.1530

Hou J, You G, Xu Y, Wang C, Wang P et al (2016) Antioxidant enzyme activities as biomarkers of fluvial biofilm to ZnO NPs ecotoxicity and the integrated biomarker responses (IBR) assessment. Ecotoxicol Environ Saf 133:10–17. doi:10.1016/j.ecoenv.2016.06.014

Huerta B, Marti E, Gros M, López P, Pompêo M, Armengol J et al (2013) Exploring the links between antibiotic occurrence, antibiotic resistance, and bacterial communities in water supply reservoirs. Sci Total Environ 456–457:161–170

Kelly MG, Cazaubon A, Coring E, Dell'uomo A et al (1998) Recommendations for the routine sampling of diatoms for water quality assessments in Europe. J Appl Phycol 10:215–224

Lam PKS (2009) Use of biomarkers in environmental monitoring. Ocean Coast Manag 52:348–354. doi:10.1016/j.ocecoaman.2009.04.010

Lawrence JR, Chénier M, Roy R, Beaumier D, Fortin N, Swerhone GDW, Neu TR, Greer CW (2004) Microscale and molecular assessment of the impacts of nickel, nutrients, and oxygen level on river biofilm communities. Appl Environ Microbiol 70:4326–4339

Lawrence JR, Swerhone GDW, Wassenaar LI, Neu TR (2005) Effects of selected pharmaceuticals on riverine biofilm communities. Can J Microbiol 51:655–669

Lawrence JR, Swerhone GDW, Topp E, Korber DR, Neu TR, Wassenaar LI (2007) Structural and functional responses of river biofilm communities to the nonsteroidal anti-inflammatory diclofenac. Environ Toxicol Chem 26:573–582

Lesser MP (2006) Oxidative stress in marine environments: biochemistry and physiological ecology. Annu Rev Physiol 68:253–278

Li M, Hu C, Zhu Q, Chen L, Kong Z, Liu Z (2006) Copper and zinc induction of lipid peroxidation and effects on antioxidant enzyme activities in the microalga *Pavlova viridis* (Prymnesiophyceae). Chemosphere 62:565–572

Li A-D, Li L-G, Zhang T (2015) Exploring antibiotic resistance genes and metal resistance genes in plasmid metagenomes from wastewater treatment plants. Front in Microbiol 6(1025):1–11

Liess M, Foit K, Becker A, Hassold E, Dolciotti I et al (2013) Culmination of low-dose pesticide effects. Environ Sci Technol 47:8862–8868

López-Maury L, Florencio FJ, Reyes JC (2003) Arsenic sensing and resistance system in the cyanobacterium *Synechocystis* sp. Strain PCC 6803. J Bacteriol 5363–5371

Lloyd JR, Lovely DR (2001) Microbial detoxification of metals and radionuclides. Environ biotechnol 12:248–253

Lupwayi NZ, Larney FJ, Blackshaw RE, Kanashiro DA, Pearson DC (2017) Phospholipid fatty acid biomarkers show positive soil microbial community responses to conservation soil management of irrigated crop rotations. Soil Tillage Res 168:1–10. doi:10.1016/j.still.2016.12.003

Lushchak VI (2011) Environmentally induced oxidative stress in aquatic animals. Aquat Toxicol 101:13–30. doi:10.1016/j.aquatox.2010.10.006

Ma J (2002) Differential sensitivity to 30 herbicides among populations of two green algae *Scenedesmus obliquus* and *Chlorella pyrenoidosa*. B Environ Contam Tox 68:275–281

Magellan K, Barral-Fraga L, Rovira M, Srean P, Urrea G, García-Berthou E, Guasch H (2014) Behavioural and physical effects of arsenic exposure in fish are aggravated by aquatic algae. Aquat Toxicol 156C:116–124. doi:10.1016/j.aquatox.2014.08.006

Mallick N (2004) Cooper-induced oxidative stress in the chlorophycean microalgae *Chloeralla vulgaris*: response of the antioxidant system. J Plant Physiol 161:591–597

Miltelbach GG (2012) Community ecology. Sinauer Associates, p 400

Marti E, Jofre J, Balzacar JL (2013) Prevalence of antibiotic resistance genes and bacterial community composition in a river influenced by a wastewater treatment plant. PLoS ONE 8: e78906

Martyniuk CJ, Simmons DB (2016) Spotlight on environmental omics and toxicology: a long way in a short time. Comp Biochem Physiol Part D Genomics Proteomics 19:97–101

Mittler R (2002) Oxidative stress, antioxidant and stress tolerance. Plant Sci 7:405–410

Morgante V, Mirete S, de Figueras CG, Postigo Cacho M, González-Pastor JE (2015) Exploring the diversity of arsenic resistance genes from acid mine drainage microorganisms. Environ Microbiol 17(6):1910–1925

Morin S, Cordonier A, Lavoie I, Arini A, Blanco S, Duong TT, Tornés E et al (2012) Consistency in diatom response to metal-contaminated environments. In: Guasch H, Ginebreda A, Geiszinger A (eds) Emerging and priority pollutants in rivers. Springer, Berlin, pp 117–146 Hdb. Env. Chem

Neumann NF, Galvez F (2002) DNA microarrays and toxicogenomics: applications for ecotoxicology. Biotechnol Adv 20(5–6):391–419

Nilsen EB, Rosenbauer RJ, Fuller CC, Jaffe BJ (2015) Sedimentary organic biomarkers suggest detrimental effects of PAHs on estuarine microbial biomass during the 20th century in San Francisco Bay, CA, USA. Chemosphere 119:961–970. doi:10.1016/j.chemosphere.2014.08.053

Ogunseitan O, Yang S, Ericson J (2000) Microbial δ-aminolevulinate dehydratase as a biosensor of lead bioavailability in contaminated environments. Soil Biol Biochem 32(13):1899–1906

Oremland S, Stolz JF (2005) Arsenic, microbes and contaminated aquifers. Trend Microbiol 13 (2):45–49

Páez-Espino D, Tamames J, de Lorenzo V, Cánovas D (2009) Microbial responses to environmental arsenic. Biometals 22(1):117–130

Pal Ch, Bengtsson-Palme J, Rensing C, Kristiansson E, Larsson DGJ (2014) BacMet: antibacterial biocide and metal resistance genes database. Nucleic Acids Res 42:737–743

Parelho C, Rodrigues AS, Barreto MC, Ferreira NGC, Garcia P (2016) Assessing microbial activities in metal contaminated agricultural volcanic soils—an integrative approach. Ecotoxicol Environ Saf 129:242–249. doi:10.1016/j.ecoenv.2016.03.019

Patel PC, Goulhen F, Boothman C, Gault AG, Charnock JM, Kalia K, Lloyd JR (2007) Arsenate detoxification in a Pseudomonad hypertolerant to arsenic. Arch Microbiol 187(3):171–183

Pereira P, De Pablo H, Rosa-Santos F, Pacheco M, Vale C (2009) Metal accumulation and oxidative stress in Ulva sp. substantiated by response integration into a general stress index. Aquat Toxicol 91:336–345

Pesce S, Margoum C, Rouard N, Foulquier A, Martin-Laurent F (2013) Freshwater sediment pesticide biodegradation potential as an ecological indicator of microbial recovery following a decrease in chronic pesticide exposure: a case study with the herbicide diuron. Ecol Indic 29:18–25. doi:10.1016/j.ecolind.2012.12.014

Pesce S, Margoum C, Foulquier A (2016) Pollution-induced community tolerance for in situ assessment of recovery in river microbial communities following the ban of the herbicide diuron. Agric Ecosyst Environ 221:79–86. doi:10.1016/j.agee.2016.01.009

Podola B, Melkonian M (2005) Selective real-time herbicide monitoring by an array chip biosensor employing diverse microalgae. J Appl Phycol 17:261–271

Ponsatí L, Corcoll N, Petrović M, Picó Y, Ginebreda A, Tornés E, Guasch H, Barceló D, Sabater S (2016) Multiple-stressor effects on river biofilms under different hydrological conditions. Freshw Biol 61:2102–2115

Prevodnik A, Lilja K, Bollner T (2007) Benzo[a]pyrene up-regulates the expression of the proliferating cell nuclear antigen (PCNA) and multixenobiotic resistance polyglycoprotein

(P–gp) in Baltic Sea blue mussels (*Mytilus edulis* L.) Comp Biochem Physiol Part C 145(2): 265-274

Proia L, Morin S, Peipoch M, Romani AM, Sabater S (2011) Resistance and recovery of river biofilms receiving short pulses of Triclosan and Diuron. Sci Total Environ 409:3129–3137

Radnieki TS, Semprini L, Dolan ME (2008) Expression of merA, amoA and hao in continuously cultured *Nitrosomonas europaea* cells exposed to zinc chloride additions. Biotechnol Bioeng 102(2):546–553

Rahman MA, Hassler C (2014) Is arsenic biotransformation a detoxification mechanism for microorganisms? Aquat Toxicol 146:212–219

Rao MA, Scelza R, Acevedo F, Diez MC, Gianfreda L (2014a) Enzymes as useful tools for environmental purposes. Chemosphere 107:145–162. doi:10.1016/j.chemosphere.2013.12.059

Rao MA, Scelza R, Acevedo F, Diez MC, Gianfreda L (2014b) Enzymes as useful tools for environmental purposes. Chemosphere 107:145–162. doi:10.1016/j.chemosphere.2013.12.059

Regoli F, Giuliani ME (2014) Oxidative pathways of chemical toxicity and oxidative stress biomarkers in marine organisms. Mar Environ Res 93:106–117. doi:10.1016/j.marenvres.2013.07.006

Rhine ED, Garcia-Dominguez E, Phelps CD, Young LY (2005) Environmental microbes can speciate and cycle arsenic. Environ Sci Technol 39:9569–9573

Ricart M, Barcelo D, Geiszinger A, Guasch H, Alda ML, Romaní AM, Vidal G, Villagrasa M, Sabater S (2009) Effects of low concentrations of the phenylurea herbicide diuron on biofilm algae and bacteria. Chemosphere 76:1392–1401

Romaní AM, Guasch H, Balaguer M (eds) (2016) Aquatic biofilms: ecology, water quality and wastewater treatment. Horizon Scientific Press- Caister Academic Press. ISBN: 978-1-910190-17-3

Roosa S, Wattiez R, Prygiel E, Ludovic Lesven L, Billon G, Gillan DC (2004) Bacterial metal resistance genes and metal bioavailability in contaminated sediments. Environ Poll 189:143–151

Sabater S, Guasch H, Ricart M, Romaní A, Vidal G, Klünder C, Schmitt-Jansen M (2007) Monitoring the effect of chemicals on biological communities. The biofilm as an interface. Anal Bioanal Chem 387:1425–1434

Sanchez W, Burgeot T, Porcher J-M (2013) A novel "integrated biomarker response" calculation based on reference deviation concept. Environ Sci Pollut Res 20:2721–2725. doi:10.1007/s11356-012-1359-1

Sanchez-Hernandez JC, Sandoval M, Pierart A (2017) Short-term response of soil enzyme activities in a chlorpyrifos-treated mesocosm: use of enzyme-based indexes. Ecol Indic 73:525–535. doi:10.1016/j.ecolind.2016.10.022

Sauser KR, Liu JK, Wong TY (1997) Identification of a copper-sensitive ascorbate peroxidase in the unicellular green algae *Selenastrum capricornutum*. Biometals 10:163–168

Scandalios JG (1993) Oxygen stress and superoxide dismutases. Plant Physiol 101:7–12

Si W, He X, Li A, Liu L, Li J, Gong D, Liu J, Liu J, Shen W, Zhang X (2016) Application of an integrated biomarker response index to assess ground water contamination in the vicinity of a rare earth mine tailings site. Environ Sci Pollut Res 23:17345–17356. doi:10.1007/s11356-016-6728-8

Sorensen JPR, Lapworth DJ, Read DS, Nkhuwa DCW, Bell RA et al (2015a) Tracing enteric pathogen contamination in sub-Saharan African groundwater. Sci Total Environ 538:888–895. doi:10.1016/j.scitotenv.2015.08.119

Sorensen JPR, Lapworth DJ, Read DS, Nkhuwa DCW, Bell RA et al (2015b) Tracing enteric pathogen contamination in sub-Saharan African groundwater. Sci Total Environ 538:888–895. doi:10.1016/j.scitotenv.2015.08.119

Tlili A, Corcoll N, Bonet B, Morin S, Montuelle B, Bérard A, Guasch H (2011) In situ spatio-temporal changes in pollution-induced community tolerance to zinc in autotrophic and heterotrophic biofilm communities. Ecotoxicology 20:1823–1839

Tlili A, Berard A, Blanck H, Bouchez A, Cássio F, Eriksson KM et al (2016) Pollution-induced community tolerance (PICT): towards an ecologically relevant risk assessment of chemicals in aquatic systems. Freshw Biol 61:2141–2151. doi:10.1111/fwb.12558

Tripathi BN, Mehta SK, Amar A, Gaur JP (2006) Oxidative stress in *Scenedesmus* sp. during short-and long-term exposure to Cu^{2+} and Zn^{2+}. Chemosphere 62:538–544

Tuulaikhuu B-A, Bonet B, Guasch H (2016) Effects of low arsenic concentration exposure on freshwater fish in presence of fluvial biofilms. Sci Total Environ 544:467–475

Valavanidis A, Vlahogianni T, Dassenakis M, Scoullos M (2006) Molecular biomarkers of oxidative stress in aquatic organisms in relation to toxic environmental pollutants. Ecotox Environ Saf 64:178–189

Van den Brink PJ, Crum SJH, Gylstra R, Bransen F, Cuppen JGM, Brock TCM (2009) Effects of a herbicide–insecticide mixture in freshwater microcosms: risk assessment and ecological effect chain. Environ Polluti 157:237–249

Van der Oost R, Beyer J, Vermeulen NPE (2003) Fish bioaccumulation and biomarkers in environmental risk assessment: a review. Environ Toxicol Pharmacol 13:57–149. doi:10.1016/S1382-6689(02)00126-6

Vega-López A, Ayala-López G, Posadas-Espadas BP, Olivares-Rubio HF, Dzul-Caamal R (2013) Relations of oxidative stress in freshwater phytoplankton with heavy metals and polycyclic aromatic hydrocarbons. Comp Biochem Physiol A Mol Integr Physiol, First Int Conf Oxidative Stress Aquat Ecosyst 165:498–507. doi:10.1016/j.cbpa.2013.01.026

Wang RL, Bencic D, Biales A, Lattier D, Kostich M, Villeneuve D, Ankley GT, Lazorchk J, Toth G (2008) DNA Microarray-based ecotoxicological biomarker discovery in a small fish model species. Environ Toxicol and Chem 27(3):664–675

Wang Y, Wang S, Xu P, Liu C, Liu M, Wang Y, Ge Y (2015) Review of arsenic speciation, toxicity and metabolism in microalgae. Rev Environ SciBiotechnol 14(3):427–451

WFD 2000/60/EC (Water Framework Directive) (2000) Available at

Wolfe-Simon F, Grzebyk D, Schofield O, Falkowski PG (2005) The role and evolution of superoxide dismutases in algae. J Phycol 41:453–465

World Health Organization (1993) Biomarkers and risk assessment: concepts and principles. Environ Health Crit 155

Wright MS, Loeffler Peltier G, Stepanauskas R, McArthur V (2006) Bacterial tolerances to metals and antibiotics in metal-contaminated and reference streams. FEMS Microbiol Ecol 58:293–302

Xuereb B, Lefèvre E, Garric J, Geffard O (2009) Acetylcholinesterase activity in *Gammarus fossarum* (*Crustacea Amphipoda*): linking AChE inhibition and behavioural alteration. Aquat Toxicol 94:114–122. doi:10.1016/j.aquatox.2009.06.010

Zamyadi A, Choo F, Newcombe G, Stuetz R, Henderson RK (2016) A review of monitoring technologies for real-time management of cyanobacteria: recent advances and future direction. Trac-Trends Anal Chem 85:83–96. doi:10.1016/j.trac.2016.06.023

Chapter 12
Bacterial Bioreporter Applications in Ecotoxicology: Concepts and Practical Approach

Sulivan Jouanneau, Marie-José Durand, Ali Assaf, Marine Bittel and Gérald Thouand

Abstract Bioreporters are widespread in the ecotoxicological field, and they are used in concert with a range of physico-chemical methods for environmental characterization. These biological methods allow the assessment of other parameters that are not otherwise accessible. Among the broad diversity of available bioreporters, bacterial approaches are particularly interesting for their simple implementation, low cost and timeliness of their response (because of their metabolic kinetics and their growth rate). In this chapter, we are interested in the primary proposed strategies from one initial assumption: the use of one bioreporter to test one parameter. This strategy rapidly reached its limits (for lack of specificity or representativity), thus opening the way for other biological approaches that were more reliable but also more complex (implemented technology and data treatment).

Keywords Bacterial bioreporter · Overall parameters · Specific detection · Environmental monitoring · Ecotoxicology · Approach limits

12.1 Introduction

The physico-chemical approaches, used to monitor the environmental quality, are particularly specific and accurate when the parameters in question are known (for example, the quantification of a specific chemical in river water). In the environment, the diversity and concentration of chemicals are highly variable from one sample to another. Consequently, physico-chemical methods are quickly limited to characterizing the relevant complexity of a sample (Fig. 12.1). Methods based on bioreporters open the vision field to new metrological possibilities in this context.

S. Jouanneau (✉) · M.-J. Durand · A. Assaf · M. Bittel · G. Thouand
University of Nantes, UMR CNRS GEPEA 6144, 18 Bd Gaston Defferre, 85000 La Roche sur Yon, France
e-mail: Sulivan.jouanneau@univ-nantes.fr

M. Bittel
TRONICO Vigicell, 18 Bd Gaston Defferre, 85000 La Roche sur Yon, France

C. Cravo-Laureau et al. (eds.), *Microbial Ecotoxicology*,
DOI 10.1007/978-3-319-61795-4_12

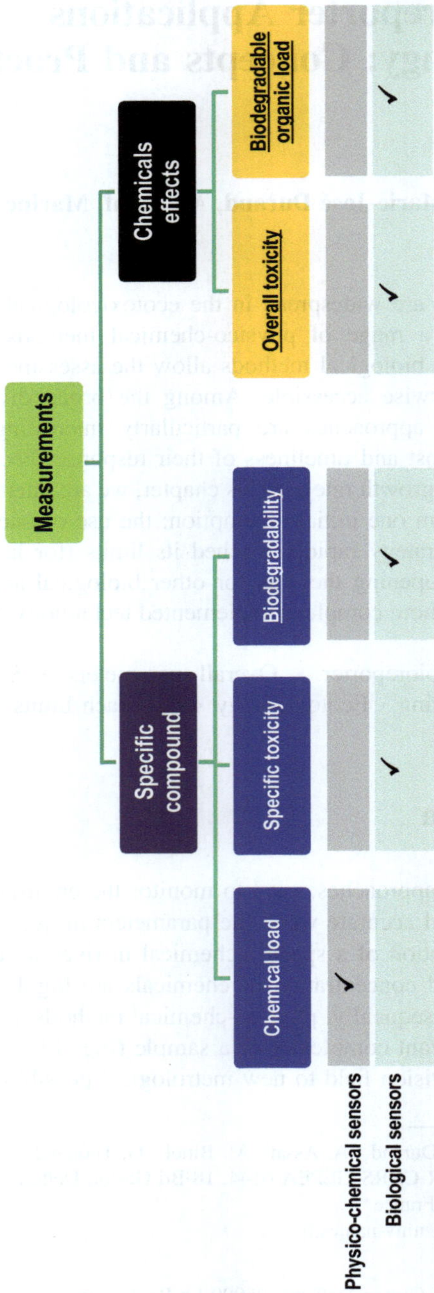

Fig. 12.1 Metrological complementarity of the physico-chemical and biological methods

They allow the improved characterization of the toxic potential of samples via several biological parameters such as the bioavailable concentration and the environmental persistence of specific chemicals or the overall toxicity of the study sample.

Nevertheless, according to Bartell (Bartell 2006; Burger and Gochfeld 2001), a biological indicator must meet certain attributes such as biological relevance (low natural variability, measurable biological signal, and having a response that is attributable to specific stress conditions), methodological relevance (measurement reliability, low cost, and straightforward data interpretation), and societal relevance (social demand).

Among the wide diversity of available natural bioreporters, bacteria fulfill the required specifications to be relevant biological indicators. These microbial reporters have been the subject of a large number of publications over the last decade (Su et al. 2011; Sun et al. 2015; Ponomareva et al. 2011; Dai and Choi 2013). However, the list of potential targets has remained limited (primarily in terms of overall parameters). In addition, to enlarge the detection range of these natural bioreporters, scientists can employ genetic engineering to produce new bacterial strains to detect some specific parameters (Park et al. 2013; Lei et al. 2006; Durand et al. 2003; Ivask et al. 2009; Hakkila et al. 2004; Charrier et al. 2011; Gu and Gil 2001).

In this chapter, the primary strategies proposed in the context of environmental monitoring and their respective limits are detailed point-by-point.

12.2 The Dream of Using "One Cell to Measure One Parameter": The Monoparametric Approach

As shown above, the primary objective was to develop methodological alternatives based on bacterial cells that could be complementary to physico-chemical monitoring methods. The initial ambition was to propose relevant bioreporters in this context (natural ones or those obtained by genetic engineering). For that purpose, the first strategy under consideration (as detailed in the following paragraphs) was to implement bioreporters that could provide a specific response. In other words, one bacterial strain was identified or developed to measure one analyte or one parameter (for measurements of the overall toxicity from the environmental persistence or the bioavailable concentration).

12.2.1 Detection of an Overall Parameter

One of the best-known microbial bioreporters is the natural bioluminescent bacterial strain *Aliivibrio fischeri* (formerly known as *Vibrio fischeri*). This strain is implemented in a bioassay to allow the assessment of water toxicity (for overall acute toxicity measurement). This strategy was incorporated into an international standard (ISO 11348) and commercialized into kits under the trade denominations Microtox® (Modern Water, UK), ToxAlert® (Merck, Germany), BioTox™ (Aboatox, Finland) or LUMIStox™ (Hach Lange, US) (Jennings et al. 2001; Turner et al. 2010). The underlying principle consists of the exposure, over a short time period, of bioluminescent cells to a test sample. According to the inhibition level of the biological signal, it is possible to deduce the acute toxicity of the sample. The simplicity and robustness of this method have contributed greatly to its democratization. An "online version" of this bioassay for field applications has also been proposed by Microlan (iTox-Toxcontrol).

This assessment strategy (monitoring the inhibition of a biological signal) was enlarged to the other natural bioluminescent strains (Peinado et al. 2002) as well as some constitutive bioreporters (Peinado et al. 2002; Chang et al. 2004) (i.e., the biological signal is not controlled by induction or repression). In this latter case, the biological signal is not always bioluminescent but can be fluorescent or colorimetric.

To assess chronic overall toxicity, the principle is similar to that of an acute bioassay, but the exposure duration is significantly increased. The primary parameter that is monitored in this case is relative to the cellular growth rate; several indicators can be used, such as the cellular density (Radix et al. 2000; Gellert et al. 1999), as described in the ISO 10712:1995 standard (for a bioassay based on the growth inhibition of *Pseudomonas putida*), or the bioluminescence (Gellert 2000; Menz et al. 2013).

Within this global framework of environmental monitoring, another overall parameter was studied, namely the biodegradable organic load. The primary bioassay that was developed to assess this parameter is detailed in ISO 5815 standard (ISO 2003a, b) (the measurement of the biochemical oxygen demand-BOD). The principle underlying this rudimentary method (Great Britain 1908) consists of exposing a water sample (carbon sources) to an unknown environmental inoculum over a 5-day period under controlled conditions (in terms of temperature and a dark room) and with a limited concentration of dissolved oxygen (saturation concentration in water). The monitoring of the oxygen consumption (as a marker of metabolic activity) during this period allows the quantification of biodegradation activity by microorganisms, and consequently, to assess the biodegradable organic load in the sample under analysis. The primary limit of this approach concerns the environmental diversity of the inoculum, which inevitably induces significant variability between assays (Jouanneau et al. 2014).

Many studies have proposed alternative methods to allow this problematic limit, and for that reason, the analytical strategies are based on the implementation of only one biological indicator in most cases (Jouanneau et al. 2014; Yoshida et al. 2001; Chee et al. 2005; Raud et al. 2012). Other markers of metabolic activity were proposed to replace oxygen consumption (the limiting factor of the reference method), such as mediator redox (Yoshida et al. 2001; Pasco et al. 2004; Dudal et al. 2006).

The ease-of-use of these methods, both at the implementation and data analysis levels, has greatly contributed to the deployment of these analytical approaches

Fig. 12.2 Lack of inherent representativity of a unique bioreporter in a monitoring context of an overall parameter

within the scientific community in addition to physico-chemical methods. Over the two last decades, many publications have been dedicated to this topic (Su et al. 2011; Lei et al. 2006; Ivask et al. 2009; D'Souza 2001).

In both cases (assessment of overall toxicity and environmental persistence), mono-parametric (only one bioreporter) approaches have led to the same metrological limitations, namely, a lack of representativity of the provided biological information. The field of view is relative to the selected biological indicator (biological model) and is consequently too narrow to be easily extrapolated to another organism. As a result, the operational domain is restricted by the intrinsic properties of bioreporters such as the metabolic abilities required to assimilate some organic compounds (environmental persistence) or their natural robustness in resisting environmental stress (Fig. 12.2).

12.2.2 Measurement of Specific Compounds

This strategy was also deployed for the detection and/or quantification of specific targets (e.g., specific toxicity, chemicals). In this case, the metrological strategy was based on the implementation of genetically engineered bioreporters, which were modified to acquire some specific detection properties (Durand et al. 2003; Ivask et al. 2007, 2009; Hakkila et al. 2004; Charrier et al. 2011; Belkin 2003, 2006; Gueune et al. 2009; Roda et al. 2011; King et al. 1990).

Two methodologies could be used to obtain specific bacterial bioreporters, as depicted in Fig. 12.3. First, the bacterial genes involved in the degradation pathway of an organic substance or in the mechanism of resistance against metal compounds are described; in this case, reporter genes (*lux CDABE*, the *luc* gene from fireflies or *gfp* genes) are inserted downstream from a promoter of these known genes. In the second case, the reporter genes are inserted randomly within the chromosome of the

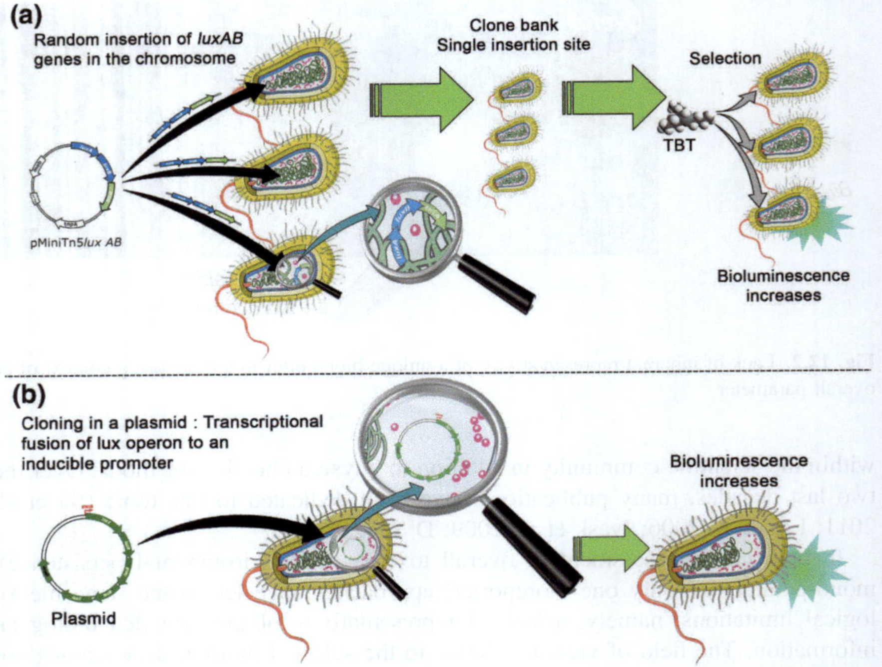

Fig. 12.3 Strategies for designing specific recombinant bacteria. **a** The luciferase genes (*luxAB*) are randomly inserted into the chromosome of *Escherichia coli*. Each clone from the bank is tested against a particular compound (TBT, or tributyltin) and selected when bioluminescence is induced. Decanal (aldehyde) must be added to the medium when *luxAB* genes are used instead of *luxCDABE*. **b** The reporter genes (*lux, luc,* and *gfp*) are inserted downstream of the promoter of a known gene (in *red*). When the bacteria are exposed to the analyte (M), the bioluminescence increases

bacteria (*Escherichia coli*, in most cases) to obtain a clone bank. Each clone is tested against a substance and selected when bioluminescence increases.

12.2.2.1 Example of Specific Strains for Organometallic or Organic Compound Analysis

A recombinant *E. coli* for sensing organometallic compounds has been obtained by using a strategy that was developed by Guzzo and Dubow (Guzzo et al. 1991) for identifying uncharacterized *E. coli* genes. A unique insertion of the *luxAB* genes was made in the *E. coli* chromosome to obtain a library, and the library was screened for bioluminescence in the presence and absence of organotin (Fig. 12.4). One clone called TBT3 was selected, and it displayed augmented luminescence in a dose-dependent manner upon exposure to tributyltin (TBT). Its sensitivity to TBT was 0.08 μM (26 μg/L) and 0.1 nM (0.03 μg/L) for dibutyltin (DBT, one of the

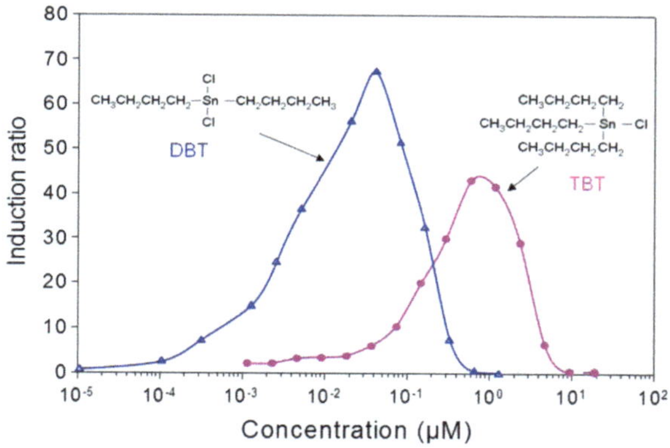

Fig. 12.4 Detection of tributyltin and dibutyltin in sea water with the TBT3 strain (Durand et al. 2003). Induction radio: The bioluminescence of TBT3 was recorded after 1 h of incubation with the compound/background bioluminescence of TBT3

degradation products of TBT), and this strain was not induced by other organotin or other compounds (Durand et al. 2003; Gueune et al. 2008, 2009). A simple bioassay for detecting TBT in paint and wastewater from shipyards has been developed (Gueune et al. 2009).

Organic compounds could be detected with recombinant bacteria by cloning the gene involved in the degradation pathway upstream of the reporter genes. Benzene and atrazine detection will be used as examples.

The TOL plasmid of *Pseudomonas putida* contains the genes of enzymes that are involved in degrading benzene and its derivatives. These genes have been used to build a plasmid with the luciferase gene in fireflies (pTSN316). An *E. coli* H10 that was transformed with this plasmid produces luminescence in the presence of aromatic compounds (Nakazawa et al. 1980). Nevertheless, applications of this bacterium are limited because of the accumulation of toxic derivatives in the cell for benzene concentrations higher than 0.5 ppm (Berno et al. 2004).

The development of recombinant strains for organic pollutant detection requires a good knowledge of the genes that are involved in the degradation pathway. Few of these constructions have been dedicated to field applications for the monitoring of BTEX (Benzene, Toluene, Ethylbenzene, and Xylenes), toluene or naphthalene (Xu et al. 2013; Fernández-Piñas et al. 2014).

The s-triazine family of herbicides has been used intensively in Europe and is still applied in some countries. These herbicides have various toxic effects ranging from aquatic organisms to humans as well as a low degradation rate leading to the accumulation of cyanuric acid compounds, which are common byproducts of s-triazine. Triazine monitoring in the environment is performed by chemical analysis or immunoassay. Nevertheless, Hua et al. (2015) proposed a bacterial bioreporter to monitor atrazine and cyanuric acid. They used the plasmid p-ADP-1, which is borne by *Pseudomonas* sp. and is involved in the atrazine degradation pathway for the genetic construction of recombinant *E. coli*. Atrazine degradation is encoded by *atz* genes, which are divided into two sets of (i) the three constitutive genes *atzA*, *atzB* and *atzC*, which are involved in the degradation of atrazine in cyanuric acid compounds, and (ii) the mineralization of cyanuric acid, which

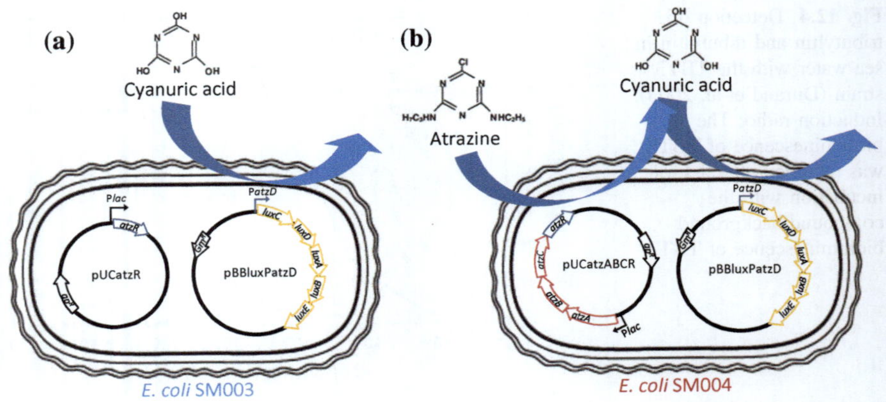

Fig. 12.5 Bioreporter designed to specifically detect cyanuric acid and atrazine (Hua et al. 2015)

requires the inducible *atzDEF* operon, the expression of which is dependent on its regulator *atzR*. Because the first set (i) of genes is constitutively expressed, the promoter cannot be used as a bioreporter. The strategy was to create two complementary bioluminescent strains, with one induced only by cyanuric acid (Fig. 12.5a) and the other for atrazine detection after its degradation in cyanuric acid (Hua et al. 2015) (Fig. 12.5b). The detection range reported for atrazine was from 0.22 to 15 μM and from 7.83 to 2.89 mM for cyanuric acid. Nevertheless, field applications are still in development.

12.2.2.2 Bacteria for the Detection of Inorganic Compounds (Metal Trace Elements, or MTE)

The construction of a metal bioreporter requires knowledge of MTE regulation by the cells. The resistance mechanisms of non-essential MTE efflux pumps are effective for exporting MTE to the outside of the cell, and these transporters could be encoded chromosomally or by mobile genetic elements (Ma et al. 2009). The genes involved in these mechanisms are used to construct recombinant strains for MTE detection.

Approximately thirty bioluminescent bacteria have been described in the literature, and generally, the promoters of genes involved in MTE resistance are cloned upstream of reporter gene(s), in a low copy plasmid inside a host strain (*E. coli*, *Pseudomonas* sp., *Bacillus subtilis*, *Staphylococcus aureus*, and others). Many studies have been performed with standard metal solutions, and the induction of strains was dose-responsive after an incubation period of 60–180 min. Some of these recombinant strains are very sensitive, for example, to construction with the *merR* promoter for which the detection limits reported for Hg^{2+} ranged from 0.03 to 0.003 μg/L (Durand et al. 2015).

The experimental procedure for monitoring MTE detection by recombinant bacteria are usually conducted with batch culture cells or freeze-dried cells, or they are included in a biosensor. Applications of environmental samples with bioluminescent bacteria (bioassay or biosensors) have been intensively reviewed in recent years, and most of them concern wastewater or leachate from polluted soil (Xu et al. 2013; Durand et al. 2015; Eltzov and Marks 2011; Checa et al. 2012). For example, one application of bacterial bioreporters concerns the monitoring of arsenic in water samples. Arsenic is an abundant element in Earth's crust, and its concentration could be enhanced by anthropogenic activities or by hydrogeological conditions, leading to the serious contamination of groundwater in some parts of the world. Because of its high toxicity, the WHO (World Health Organization) has recommended a 10 µg/L limit for inorganic arsenic in drinking water. Nevertheless, despite the use of water treatment systems and the available chemical analysis, the contamination of drinking water in developing countries (Bangladesh, Laos, Cambodia and Vietnam) has remained a problem because of the high cost of these techniques. It could be useful to monitor arsenic alternative methods based on bacterial bioreporters (Kaur et al. 2015; Merulla et al. 2013). More than 20 bacteria that were engineered for arsenic detection with different reporter genes have been created over the last 15 years because their resistance system is well-characterized (Diesel et al. 2009). Resistance against arsenic is provided by the *ars* operon, which is found on the plasmid or in the chromosome (the *arsRDABC* operon for *E. coli*), and the expression of this operon is controlled by ArsR protein. In the absence of arsenic, ArsR is bound to a specific DNA sequence (operator) and represses the transcription of defense genes (*arsD, arsC*, *arsA* and *arsD*). In the presence of arsenic, ArsR loses its affinity for the operator, and *arsDCAB* genes are transcribed. The bioreporter for arsenic detection bears a plasmid with a second copy of the operator-promotor sequence of ArsR that is transcriptionally fused with genes for a reporter protein (luciferase, GFP, β-galactosidase, or others). When arsenic enters bioreporter cells and binds to ArsR, the *arsR* promotor is derepressed, leading to the transcription of reporter genes.

Siegfried et al. (2012). have developed a bioassay for field applications, and validations of this assay have been performed in Bangladesh under realistic conditions; more than 2000 bioreporter analyses have been conducted. A cross analysis with 24 groundwater samples by ICP-MS and the bioassay kit are consistent for arsenic concentrations below 150 µg/l and more than 10 µg/L. Because of its low cost, this bioassay is a useful tool for developing countries to monitor arsenic.

12.2.2.3 Limits of These Strategies for Monitoring a Specific Parameter

One of the primary limits of using bioreporters to monitor one specific compound is their lack of specificity. None of these reporters are specific to one substance, even for organic substances, because the genes involved in the degradation pathway or in MTE resistance could be induced by more than one substance (Fig. 12.6).

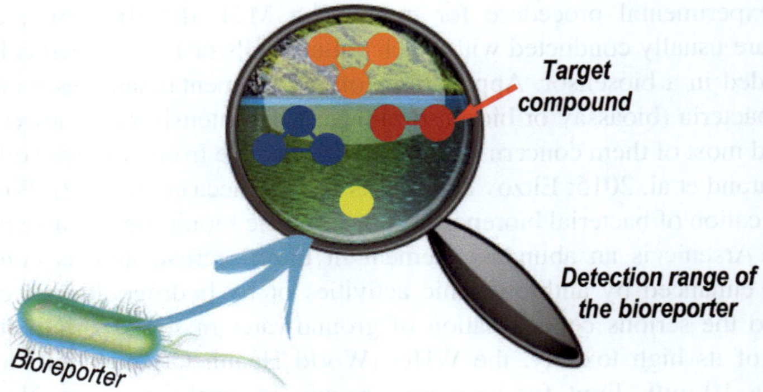

Fig. 12.6 Primary limit (lack of specificity) of approaches that are based on an unique biological indicator

Several tactics were proposed to increase the performance of the MTE molecular manipulation detection (Yagur-Kroll and Belkin 2014). Hynninen et al. (2010) reported an improvement of the detection threshold consecutive to an altering of the metal efflux system of *Pseudomonas putida*-based Cd/Zn/Pb-bioreporters (genetic modification of four metal efflux transporter genes). These metabolic changes cause intracellular accumulation of MTE enhancing, at the same time, cellular sensitivity. Another strategy to enhance sensitivity as developed by Merulla and Van der Meer (2016) was to decrease the background level of the strains. These investigators used the ArsR-operator system after adding a second copy of the operator for ArsR. A decrease in the reporter gene expression is observed in the absence of arsenite, but inducible control in the presence of effectors is preserved.

Improving the specificity of the strain remained difficult, so the identification of MTE in the environmental sample could be performed by cross-detecting a panel of the strain. This strategy has been applied by Jouanneau et al. (2011) (see Sect. 12.3.2). The concept was extended both for the assessment of overall parameters and for the measurement of specific chemicals.

12.3 The Multicellular Approach to Measuring One Parameter

Despite a clear benefit relative to the implementation of these mono-parametric approaches (ease of use and simplified interpretation of biological data), these latter approaches have important bottlenecks (limited representativity and lack of specificity) that significantly restrict their potential development and/or commercialization (except in the case of the bioassay based on bioluminescent bacteria, as in ISO standard 11348).

To circumvent these limitations, a strategic evolution was proposed that consisted in a multiplication of the number of bioreporters used by the parameters. The primary aim was to improve the quality of the selected biological information. This concept was extended for both the assessment of the overall parameters and for the measurement of specific chemicals.

12.3.1 Assessment of Overall Parameters

To make up for the absence of representativity in the mono-parametric strategy as induced by the limited natural properties of the implemented bioreporters, a second-generation strategy was proposed. This latter strategy is based on the implementation of several biological reporters and permits improvements in the metrological mesh (Fig. 12.7).

12.3.1.1 Microbial Assay for Overall Toxicity Assessment

In accordance with this concept, two multi-parametric assays were listed in the literature for applications to the field of toxicity assessment. These bioassays are commercialized by the NCIMB Company (UK) under the trade name MARA and LumiMARA. The MARA assay is based on the growth inhibition (chronic test; 18-h duration) of 11 microbial strains (*Microbacterium* sp., *Brevundimonas diminuta, Citrobacter freundii, Comamonas testosteroni, Enterococcus casseliflavus, Delftia acidovorans, Kurthia gibsonii, Staphylococcus warnerii, Pseudomonas aurantiaca, Serratia rubidaea,* and *Pichia anomala* yeast) (Fai and Grant 2010; Wadhia 2008; Wadhia et al. 2007) (Fig. 12.8).

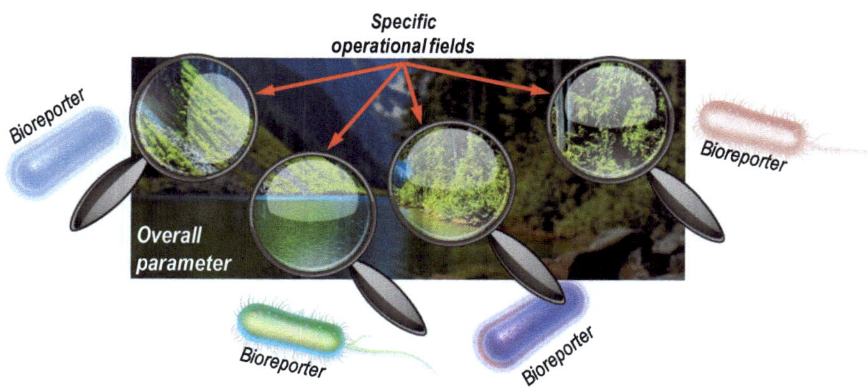

Fig. 12.7 Methodological improvements were implemented to increase the representativity of bioreporter information

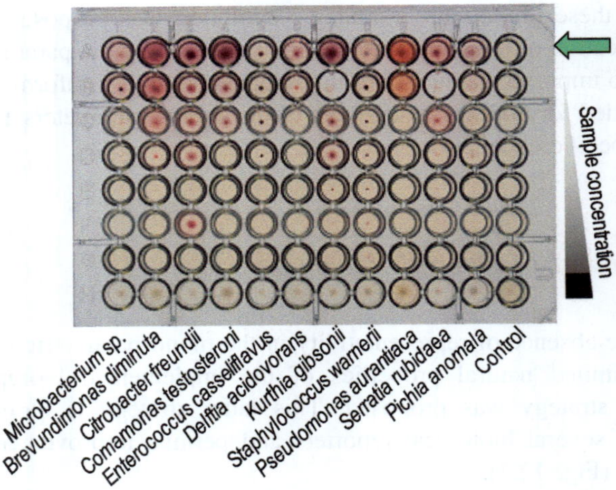

Fig. 12.8 Implementation of a MARA assay using a chemical

Table 12.1 Bioluminescent strains implemented in the LumiMARA assay

LumiMARA strains	Reference
Aliivibrio fischeri	NCIMB 30268
Aliivibrio fischeri	NCIMB 30274
Photobacterium leiognathi	NCIMB 30266
Photobacterium leiognathi	NCIMB 30269
Photobacterium phosphoreum	NCIMB 30267
Photobacterium phosphoreum	NCIMB 30270
Photobacterium phosphoreum	NCIMB 30271
Photorhabdus asymbiotica	NCIMB 30276
Photorhabdus luminescens	NCIMB 30275
Vibrio harveyi	NCIMB 30272
Vibrio harveyi	NCIMB 30273

The toxic effect is highlighted by assessing of the bioluminescence inhibition emitted by cells

The inhibition of the growth of the micro-organisms is measured well-by-well by assessing of the intensity the redox dye (2,3,5-Triphenyl-tetrazolium chloride—red dye—marker of metabolic activity).

The LumiMARA assay resembles the bioluminescence tests described in Sect. 12.2.1, but it is based on a set of 11 bacterial bioreporters (Table 12.1). It is dedicated to the assessment of acute toxicity (for exposure durations of 15–30 min) (Jung et al. 2015).

Conceptually, these strategies allow the enlargement of metrological points of view by comparing them with mono-parametric approaches (for a better representativity of the overall toxicity). Nevertheless, a technological obstacle persists in relation to interpretations of the biological information. In fact, each bioreporter provides individual data without overall integration, which should result in an easy reading. In other words, these assays look more like a compilation of several biological measurements than an overall approach to toxicity.

The strategy described below proposes to remove this limit by implementing a data analysis algorithm.

12.3.1.2 Biodegradable Organic Matter Detection by Using a Bacterial Set

As explained in Sect. 12.2.1, the most common method (ISO 5815) for assessing biodegradable organic matter is based on the implementation of environmental inocula, which can be highly variable, and at its origin, it can have a significant lack of methodological robustness. The mono-parametric approaches have reached their limits, and thus it was pertinent to consider other methodological paths based on several bioreporters. In addition, to address the lack of representativity in the mono-parametric approaches (cf. Sect. 12.2.1), or, conversely, the strong variability generated by the wide diversity of inocula, the use of several biological reporters was considered. The primary difficulty is to determine the size of the bacterial set.

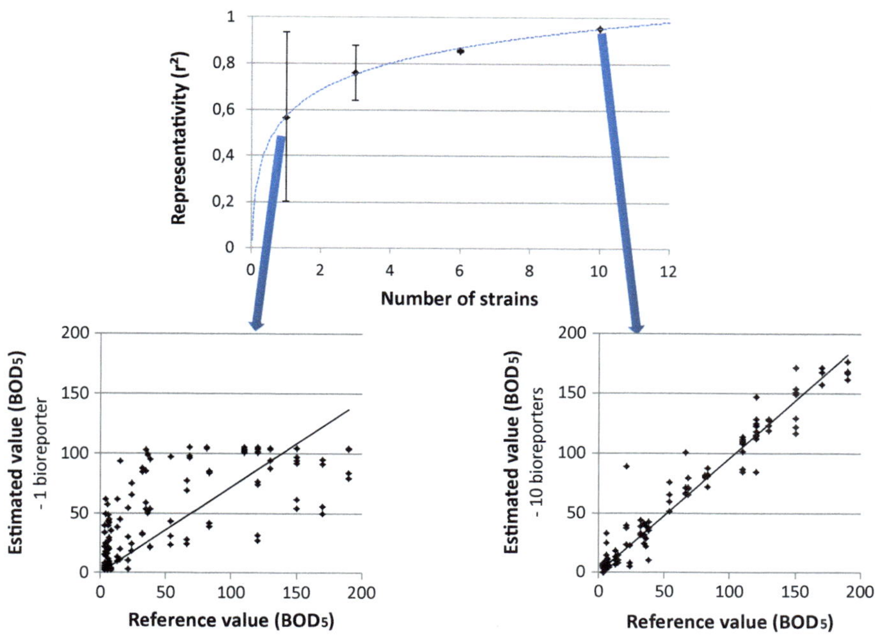

Fig. 12.9 Representativity of biological information (according to the number of implemented strains) in comparison with the reference method (from experimental data from work in progress)

Indeed, an insufficient number of strains risks generating similar limitations as those found with the mono-parametric approach (the lack of representativity), but too many strains can increase the assay costs without a significant gain of the representativity (Fig. 12.9).

Two pathways are described in the literature. First, the different biological indicators are mixed to form an artificial inoculum (Catterall et al. 2003). That approach allows the representativity of the given measurement to increase while ensuring an easy reading of the biological information (the bacterial inoculum provides only one piece of global information). This strategy, which is simple to implement, still poses some limits, such as the control of the cellular proportions (the fraction of each strain in the inoculum) during the assays, which can induce a significant level of variability.

The second strategy consists in using several strains separately (Raud and Kikas 2013) (Jouanneau et al. work in progress). The individual control of each bacterial strain over the entire analysis duration leads to a clear improvement in the reproducibility of results. However, each strain provides specific information; in others words, each biological datum is a component of the global parameter. Consequently, a supplementary step is required to analyze the biological data. To that end, complex statistical tools for data mining are needed (decision trees, neural networks, etc.). These empirical models are designed from an existing database consisting of biological information that is obtained from known samples (and characterized according to the reference method).

12.3.2 Specific Detection of Metals

As widely shown in the past (cf. Sect. 12.2.2.3), the primary limitation of approaches based on only one bioreporter is the lack of specificity. To illustrate this limit, Table 12.2 shows MTE detection by different bioluminescent strains that were transformed with a plasmid containing the bioluminescence genes *luxCDABE* from *Aliivibrio fischeri* under the control of heavy metal-inducible promoters. With the exception of some strains (Durand et al. 2003; Hua et al. 2015), it is not rare when one "specific" strain really detects several chemical targets (Ivask et al. 2009; Jouanneau et al. 2011).

To overcome this problem, a multi-parametric strategy was proposed on the basis of an analysis of crossed data from several bioreporters (Jouanneau et al. 2011; Elad et al. 2008). Contrary to the overall approach, the aim is not to improve the representativity but to increase the specificity of analytical strategies, as shown in Fig. 12.10.

The example that was proposed to illustrate this strategy is based on the works of Jouanneau et al. (2011), which were dedicated to the detection and quantification of four potentially mixed metals in environmental samples. The heart of the strategy is based on five bioluminescent bioreporters (genetic engineering), each of which had a specific plasmid (four inducible plasmids and one constitutive one). The lack of

Table 12.2 Detection of MTE with various strains of recombinant *E. coli*

Metals	Detection limits (μM)					
Ag	*E. coli* K12MG1655 pBarslux[a]	*E. coli* K12MG1655 pBcoplux[a]	*E. coli* K12MG1655 pBmerlux[a]	*E. coli* K12MG1655 pBzntlux[a]	*E. coli* K12MG1655 pBpbrlux[b]	*E. coli* K12MG1655 pBgollux[b]
Ag		2.75				4.27
$As^{(III)}$	0.256		15.6	28.52		
$As^{(V)}$	0.3		12.65	9.32		
Cd	5.9		0.011	0.0045	0.16	
$Cr^{(III)}$						
$Cr^{(VI)}$				597.2		
Co				0.22		
Cu		90.5		16.92		250
Fe			16.1	4.34		
Mn						
Hg			$1.70 \cdot 10^{-7}$	0.01		
Ni				4.4		
Au		10				50
Pb	4.16			2.2	3.8	
Zn				1.7	3.59	
Sn				12.95		

[a]Charrier et al. (2011), Jouanneau et al. (2011); [b]Hua A, unpublished results

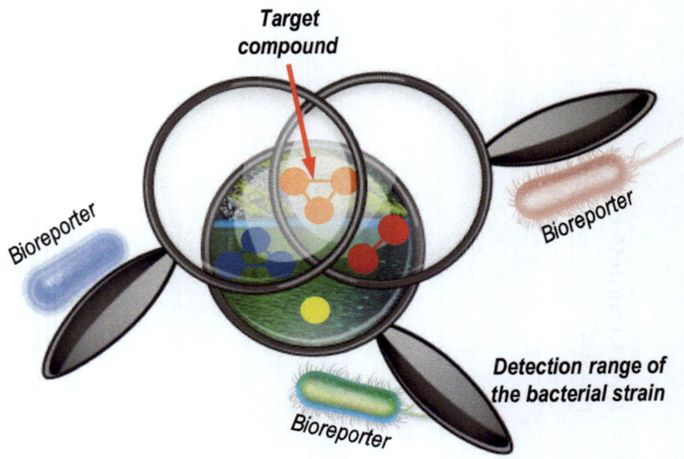

Fig. 12.10 Methodological strategy that was implemented to increase the specificity of approaches based on bacterial reporters

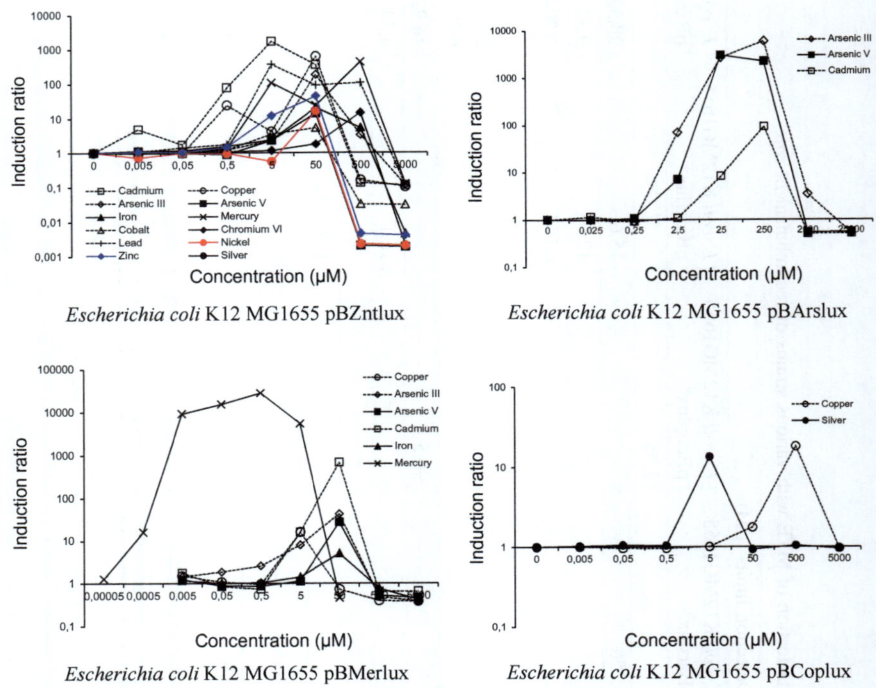

Fig. 12.11 Detection ranges obtained with the inducible strains (Jouanneau et al. 2011)

individual specificity did not allow the reliable detection of the metals in the samples (several metals were detected by the same strain), as shown in Fig. 12.11.

To improve the selectivity of the strains, the authors (Jouanneau et al. 2011) proposed simultaneously mining the different bioreporter data to cross the biological information. For that purpose, data mining algorithms are needed to analyze these complex multi-parametric data and to build biological "fingerprints" of pollution. The selected model was based on decision trees (with a step-by-step classification algorithm) and was designed from an existing database of known metal mixtures. Each bioreporter is involved, if required, in the classification model at different levels according to the investigated chemical (Fig. 12.12) to assess its concentration in study samples. In this example (Fig. 12.12) the first branch separates the data into two groups according to the segmentation value of the variable 'Zntlux' (1.425—bioluminescence value produced by the bioreporter *E. coli* K12 MG1655 pBZntlux). If the bioluminescence value 'Zntlux' is higher than this first threshold, the cadmium concentration is estimated between 250 nM and 50 μM. Otherwise a second classification level is performed until the decision tree allows a determination of the concentration range.

Thanks to this crossing strategy, the reliable identification and quantification of chemicals (in this case, 4 metals) from "specific" bioreporters has been made possible at a reliability superior to 94%.

Although the results are promising, this strategy nevertheless has some drawbacks. As explained below, the design of interpretation algorithms requires an existing database. The size of this database is directly dependent on the number of chemical targets. For example, the addition of one supplementary metal (at three concentration levels) requires a database that will be three times larger than the database used in this study (at 192 tested conditions). Moreover, the development of new models requires the implementation of other "specific" bioreporters. That implementation implies two primary bottlenecks. The first one concerns the multiplication of the strain number, inducing an extra cost of implementation (growth, preparation, storage, etc.). Second, the catalog of available strains is limited, and the development of a new bioreporter is therefore required, which is generally time-consuming, complex and costly.

12.4 Detection of One Analyte by Using Several Parameters in One Cell: The Multi-parametric Approach

As seen before, the first biological assessment methods were intended to follow a single bioreporter (monoparametric). The collected information provides a first analytical dimension, that is, the partial evaluation of the searched parameter (overall or specific parameter). To improve their performances, a second dimension came with the concept of multicellular analysis. The aim was to increase the

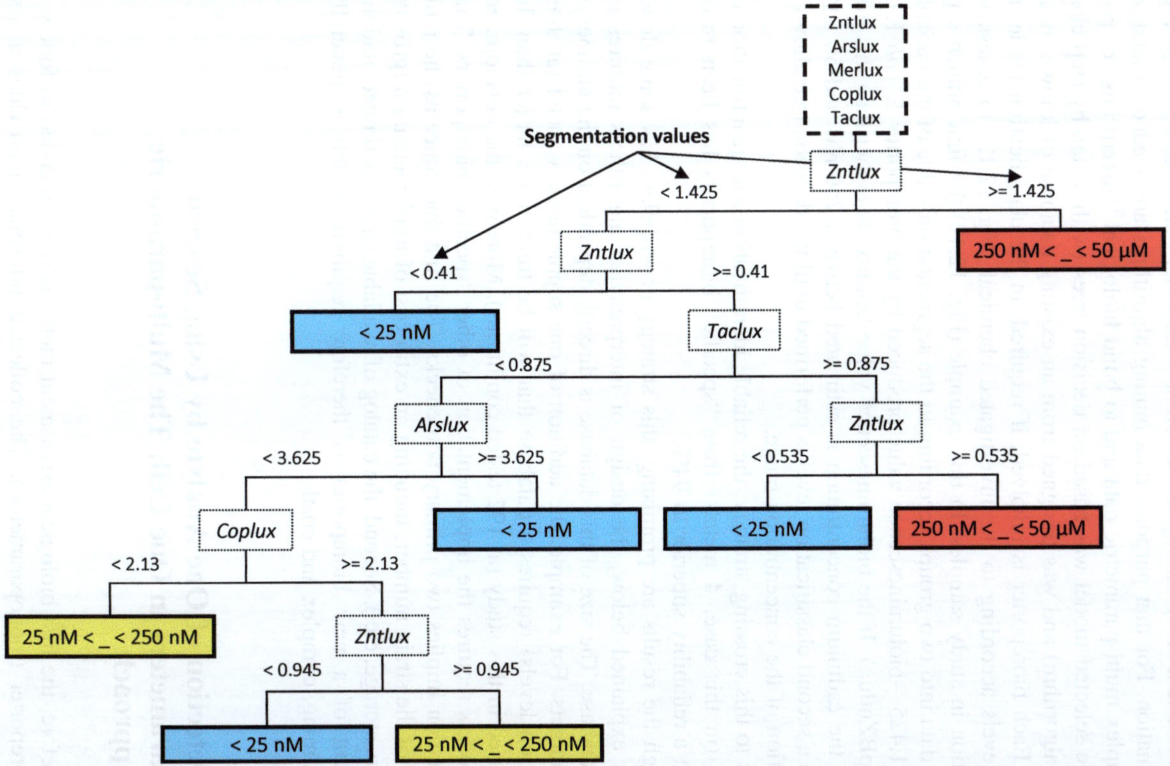

Fig. 12.12 Determination algorithm (decision tree) designed to assess the cadmium concentration (He et al. 2013). *Zntlux, Arslux, Merlux, Coplux* and *Taclux*: variable names relative to the biological data provided by the 5 bioluminescent bioreporters. The segmentation values correspond to the thresholds allowing a classification of biological information

representativity or specificity through the simultaneous observation of several microbial bioreporter responses in the studied sample. However, although the results were promising, these techniques can involve complex genetic manipulations (in the case of specific detection), and they are time-consuming and generally costly.

To circumvent these difficulties, a third generation of approaches was considered to limit the number of biological indicators in use, and, in the same time, to ensure the reliability of the biological information. This last generation was primarily implemented within the framework of specific detection.

12.4.1 Strains Containing Several Genetic Constructions

The first strategy under consideration consists in the construction of bioreporters that can emit different signals depending on the analytes that are in the sample. This strategy was proposed by Roda et al. (2011). To limit the number of necessary bio-indicators, the authors designed a bacterial strain harboring a DNA sequence that could highlight copper (*Photinus pyralis* wild-type luciferase—λ_{max} = 557 nm, inducible promoter: *copA*) and another sequence to assess the overall toxicity (a red-emitting mutant *P. pyralis* with thermostable luciferase—λ_{max} = 618 nm, constitutive regulation) of the analyzed sample (internal viability control) (Fig. 12.13). Thanks to this double genetic construction, it is easy to discriminate between the quantification/detection area and the toxic area.

The constitutive signal allows to monitor the overall toxicity level in the analysed sample. At the same time, the analyte-induced signal allows to detect specifically copper.

This approach nevertheless has similar limits to the strategies based on several bioreporters. Indeed, the strain catalog is extremely restricted, and the complexity of genetic construction makes the design of new strains even more arduous.

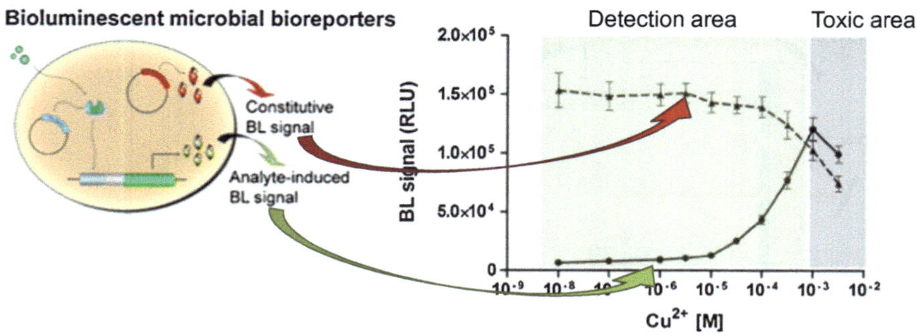

Fig. 12.13 Schematic representation of the designed microbial bioreporter to allow the simultaneous assessment of the copper concentration and overall toxicity (Roda et al. 2011) (with permission of Springer)

12.4.2 Detection by Raman Spectroscopy

Biological indicators interact with an exposed sample as a whole and set up a metabolic response that is adapted to different environmental stimuli. Despite this adaptation, most bioreporters provide only one or two biological data, which are converted into a specific or overall parameter. This strategy, which has been used up until the present, allows researchers to obtain a limited view of the global interactions between the bioreporter and the analyzed sample.

Within this framework, Raman spectroscopy offers a multi-parametric approach that provides an overview of these physiological changes as caused by the toxic agent. This technique is an optical method based on the interaction of light with matter. The difference between laser energy and inelastic scattered photons corresponds to the vibrational energy levels of chemical bonds present in the sample (Pence and Mahadevan-Jansen 2016). Raman spectra can be considered as the molecular fingerprint of the analyzed sample (Fig. 12.14).

The importance of this technique is its status as a non-invasive action and a rapid way to access a large amount of information (it is a multiparametric approach). The resulting Raman spectrum offers an overall view that is very useful for understanding the analyzed sample. One measurement of a few seconds is able to provide the molecular composition of a bacterial cell, allowing, for example, the determination of its physiological state (DNA/RNA bands) and/or its biological molecule content (carbohydrates, proteins, and lipids). However, the richness of the Raman spectroscopy signal is both an advantage and a drawback of the method. The differences in the spectra are difficult to distinguish by simple visual inspection, and the sought-after effects are often hidden among all the information. Consequently,

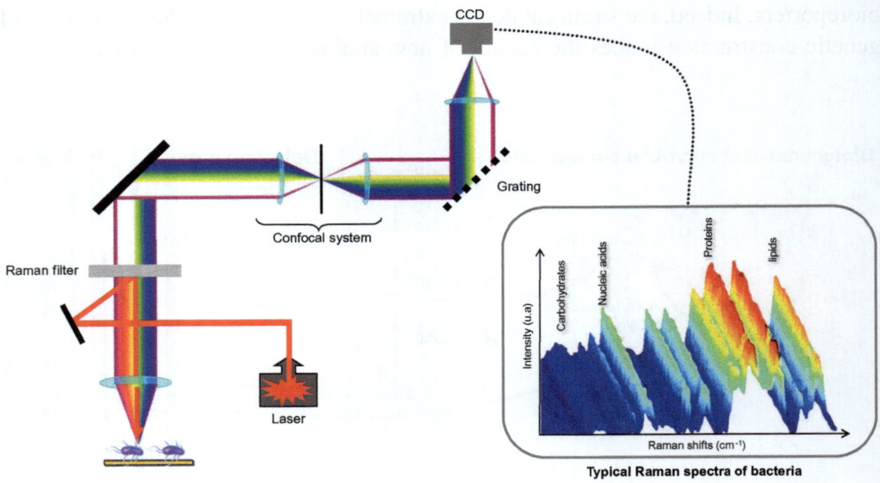

Fig. 12.14 Metrological strategy of Raman spectroscopy, which provides a molecular fingerprint of the bioreporter/sample interaction

statistical tools such as multivariate algorithms are needed (Rae et al. 2014) to highlight the differences between spectra.

Raman applications are numerous and varied, and they extend to all domains of the sciences, from physics and chemistry to the analysis of biological species at the nanoscale (Marshall and Marshall 1922; Fan et al. 2010; Marshall et al. 2010; Vankeirsbilck et al. 2002). The latest progress in Raman spectroscopy (several sources of irradiation and very sensitive detectors) opens new research perspectives. Alongside its use as an alternative optical method for microbial detection, it has also been shown that physiological variations can be observed through Raman spectra. The effects of toxic molecules such as antibiotics can be highlighted, and the observed differences can be matched with physiological events such as the fragmentation of DNA in dying cells or decreasing protein synthesis (Choo-Smith et al. 2001; Huang et al. 2010; Neugebauer et al. 2006; Schuster et al. 2000). By allowing, in one shot, a multiparametric overview of pollutant toxicity targets, this method could allow the rapid identification of toxicity mechanisms towards the better monitoring of pollutant impacts.

This third analytical dimension permits the identification and understanding of toxicity mechanisms, which is necessary to understand their potential effects on the environment and human health. However, this information is currently out of reach without complex and expensive metabolomics methods (Fig. 12.15).

12.4.2.1 Example of Arsenic Toxicity in *E. Coli*

Over the past decade, some researchers have started to investigate these approaches in relation to environmental applications and toxicity assessments. Most of the time, the results have shown significant variations in the observed microorganism spectra following pollutant exposure. However, the primary challenges come from the difficulty in linking significant changes in Raman bands with actual toxicity impacts on cells (Singer et al. 2005; Tian et al. 2012). In fact, the richness of the Raman spectroscopy signal is both an advantage and a drawback to the method, and researched effects are often hidden among all the information. Thus, these data require important analytical work and the implementation of statistical tools (Rae et al. 2014; Jarvis and Goodacre 2005; Lasch 2012). In this context, recent work on the use of Raman spectroscopy for the monitoring of arsenic toxicity shows interesting results (Bittel et al. 2015) (Fig. 12.16).

Spectra for *E. coli* bacteria that were exposed to four arsenic concentrations were collected and analyzed. The first results show the differences between the spectral variations in the molecular footprint of the sample. However, it was difficult, without analytical tools, to evaluate the significance of those differences and to establish clear links with the toxicity. To highlight the specific effects of the metal, special attention has been paid to the statistical data analysis. A spectra "quality-test" based on Independent Component Analysis (Rutledge and Jouan-Rimbaud Bouveresse 2013) has been designed and has allowed a significant

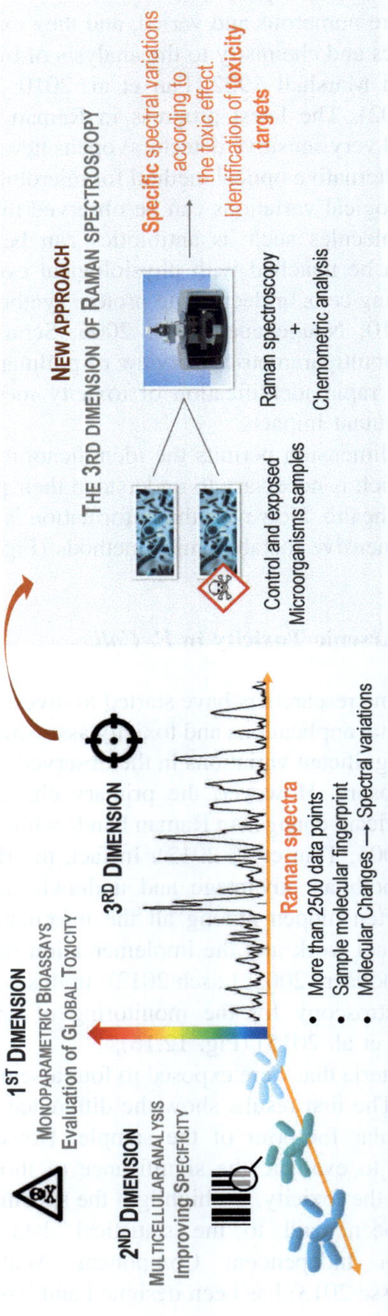

Fig. 12.15 Toxicity assessment at 3 complimentary dimensions

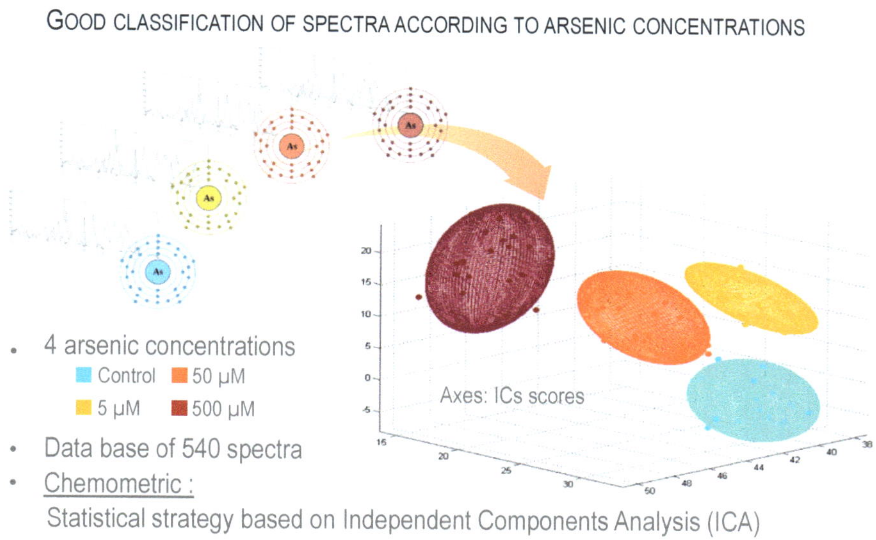

Fig. 12.16 Observation of an arsenic "dose-response" effect on *E. coli* according to Raman spectra[80]

improvement in the correct classification of spectra according to the tested arsenic concentrations. The results were finally consistent with a dose-response effect.

However, the primary advantage of using Raman spectroscopy as an observation method for bioassays and biosensors is its analysis of a multiparametric signal rather than the monoparametric measurements that are traditionally obtained with other techniques. Thus, a chemical interpretation of the results has been performed to link the observed variations to the physiological event that was induced by arsenic exposure. For this purpose, the most significant Raman bands have been identified. They were determined during the classification process, and they illustrate that the toxicity impact is significant on related biomolecules.

For example, some band's variations could be linked to the production of specific membrane transporters for the evacuation of the toxic compounds or the accumulation of arsenic in the bacterial cell wall (465–525 cm^{-1}). Some others can be associated with a denaturation phenomenon or the decrease of protein production accompanying a cell growth slowdown (1200–1400 cm^{-1}). As a last example, variations in a band that correspond to the phosphodioxy groups PO_2^- that are characteristic of nucleic acids (1100 cm^{-1}) were particularly interesting. Indeed, arsenic is known to be a structural analog of phosphate and to damage DNA by breaking phosphate bonds. Thus, the highlighted variations were consistent with known toxicity mechanisms, thus allowing the definition of "spectral signatures" of the metal's effect on the bacteria (Fig. 12.17).

Thus, the multiparametric aspect of Raman spectroscopy permits an overview of toxicity impacts on microorganisms. It helps to identify pollutant toxicity targets

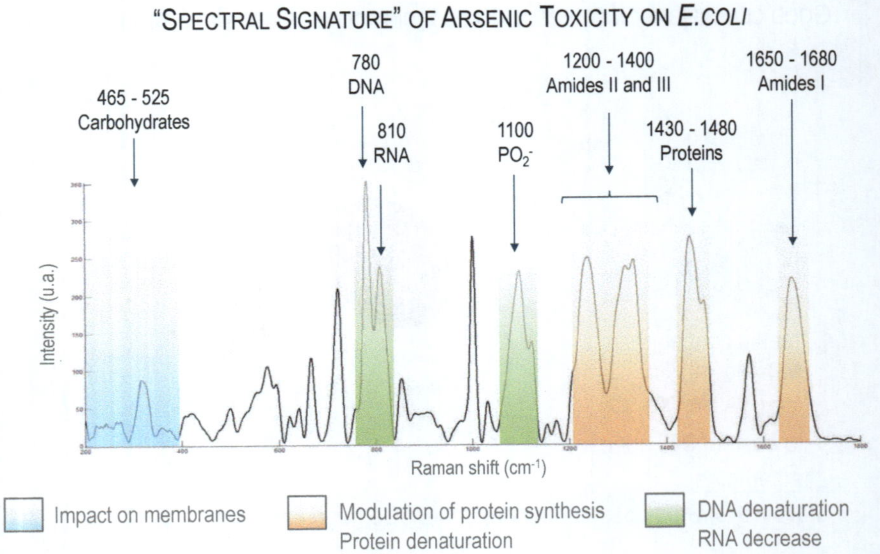

Fig. 12.17 Model spectra from *E. coli*. The highlighted Raman bands are particularly impacted by arsenic toxicity

without the need for complex and expensive metabolomic methods. When coupled with classical monoparametric bioassays involving different types of microorganisms, it could provide a fast and reliable biological alternative method to improve toxicity assessments.

12.5 Conclusions

Driven by both an omnipresent regulatory pressure and a will for environmental preservation, biological approaches were proposed in addition to physico-chemical methods. The primary objective was to focus on a biological characterization of analyzed samples. Some parameters, such as the overall or specific toxicity, the environmental persistence or the bioavailable fraction of chemicals are not measurable via physico-chemical approaches.

Many analytical methods were proposed by using living organisms (bioreporters) from different trophic levels, but among the wide spectra of implemented biological candidates, the bacterial cells were particularly interesting. The notable benefits of these biological indicators (easy to produce, high growth rate, limited biological demands) have widely contributed to their democratization in environmental monitoring and more broadly, to the field of metrological methods.

Initially, these bacterial strategies were based on an ideal vision, namely to use one bioreporter per parameter. However, these first strategies quickly reached their limits for overall or specific parameters. Although they are very easy to put into practice and they are inexpensive, these methods lacked significant robustness in the desired analytical context. From this observation, the proposed strategies have continued to evolve to increase their robustness, becoming more and more complex. The proposed strategies are based on the multiplication of collected biological information by increasing the number of implemented bioreporters or biomarkers. In these two cases, a supplementary data-mining step was required to interpret the multi-parametric information. This latest is generally ensured via statistical approaches such as PCA, neural networks, decision trees, and others.

Thus, through these overall approaches (biological recognition and statistical data mining), the biological methods based on bioreporter(s) appear now as pertinent methodologies in the stringent framework of environmental analysis.

References

Bartell SM (2006) Biomarkers, bioindicators, and ecological risk assessment—a brief review and evaluation. Environ Bioindic 1(1):60–73

Belkin S (2003) Microbial whole-cell sensing systems of environmental pollutants. Curr Opin Microbiol 6(3):206–212

Belkin S (2006) Genetically engineered microorganisms for pollution monitoring. In: Twardowska I, Allen HE, Häggblom MM, Stefaniak S (eds) Soil and water pollution monitoring, protection and remediation, vol 69. NATO Science Series: IV: Earth and Environmental Sciences. Springer, Netherlands, pp 147–160

Berno E, Pereira Marcondes DF, Ricci Gamalero S, Eandi M (2004) Recombinant *Escherichia coli* for the biomonitoring of benzene and its derivatives in the air. Ecotoxicol Environ Saf 57 (2):118–122

Bittel M, Cordella CBY, Assaf A, Jouanneau S, Durand MJ, Thouand G (2015) Potential of raman spectroscopy to monitor arsenic toxicity on bacteria: insights toward multiparametric bioassays. Environ Sci Technol 49(20):12324–12332

Burger J, Gochfeld M (2001) On developing bioindicators for human and ecological health. Environ Monit Assess 66(1):23–46

Catterall K, Zhao H, Pasco N, John R (2003) Development of a rapid ferricyanide-mediated assay for biochemical oxygen demand using a mixed microbial consortium. Anal Chem 75 (11):2584–2590

Chang IS, Jang JK, Gil GC, Kim M, Kim HJ, Cho BW, Kim BH (2004) Continuous determination of biochemical oxygen demand using microbial fuel cell type biosensor. Biosens Bioelectron 19(6):607–613

Charrier T, Durand MJ, Jouanneau S, Dion M, Pernetti M, Poncelet D, Thouand G (2011) A multi-channel bioluminescent bacterial biosensor for the on-line detection of metals and toxicity. Part I: design and optimization of bioluminescent bacterial strains. Anal Bioanal Chem 400(4):1051–1060

Checa SK, Zurbriggen MD, Soncini FC (2012) Bacterial signaling systems as platforms for rational design of new generations of biosensors. Curr Opin Biotechnol 23(5):766–772

Chee G-J, Nomura Y, Ikebukuro K, Karube I (2005) Development of photocatalytic biosensor for the evaluation of biochemical oxygen demand. Biosens Bioelectron 21(1):67–73

Choo-Smith L-P, Maquelin K, van Vreeswijk T, Bruining HA, Puppels GJ, Thi NAN, Kirschner C, Naumann D, Ami D, Villa AM et al (2001) Investigating microbial (micro)colony heterogeneity by vibrational spectroscopy. Appl Environ Microbiol 67(4):1461–1469
Dai C, Choi S (2013) Technology and applications of microbial biosensor. Open J. Appl. Biosens. 2(3):83–93
Diesel E, Schreiber M, van der Meer J (2009) Development of bacteria-based bioassays for arsenic detection in natural waters. Anal Bioanal Chem 394(3):687–693
Dudal Y, Holgado R, Knoth K, Debroux M (2006) A fluorescence-based microplate assay to quantify DOM-induced catabolic activity. Anal Bioanal Chem 384(1):175–179
Durand MJ, Thouand G, Dancheva-Ivanova T, Vachon P, DuBow M (2003) Specific detection of organotin compounds with a recombinant luminescent bacteria. Chemosphere 52(1):103–111
Durand MJ, Hua A, Jouanneau S, Cregut M, Thouand G (2015) Detection of metal and organometallic compounds with bioluminescent bacterial bioassays. In: Thouand G, Marks R (eds) Bioluminescence: fundamentals and applications in biotechnology, vol 3. Advances in biochemical engineering/biotechnology. Springer International Publishing, Berlin, pp 77–99
D'Souza SF (2001) Microbial biosensors. Biosens Bioelectron 16(6):337–353
Elad T, Benovich E, Magrisso S, Belkin S (2008) Toxicant identification by a luminescent bacterial bioreporter panel: application of pattern classification algorithms. Env Sci Technol 42 (22):8486–8491
Eltzov E, Marks R (2011) Whole-cell aquatic biosensors. Anal Bioanal Chem 400(4):895–913
Fai PB, Grant A (2010) An assessment of the potential of the microbial assay for risk assessment (MARA) for ecotoxicological testing. Ecotoxicology 19(8):1626–1633
Fan C, Hu Z, Riley LK, Purdy GA, Mustapha A, Lin M (2010) Detecting food- and waterborne viruses by surface-enhanced Raman spectroscopy. J Food Sci 75(5):M302–307
Fernández-Piñas F, Rodea-Palomares I, Leganés F, González-Pleiter M, Muñoz-Martín MA (2014) Evaluation of the ecotoxicity of pollutants with bioluminescent microorganisms. In: Thouand G, Marks R (eds) Bioluminescence: fundamentals and applications in biotechnology, vol 2. Advances in biochemical engineering/biotechnology. Springer, Berlin, pp 65–135
Gellert G (2000) Sensitivity and significance of luminescent bacteria in chronic toxicity testing based on growth and bioluminescence. Ecotoxicol Environ Saf 45(1):87–91
Gellert G, Stommel A, Trujillano AB (1999) Development of an optimal bacterial medium based on the growth inhibition assay with *Vibrio fischeri*. Chemosphere 39(3):467–476
Great Britain (1908) Royal commission on sewage disposal. Final report of the commissioners appointed to inquire and report what methods of treating and disposing of sewage (including any liquid from any factory or manufacturing process) may properly be adopted. General summary of conclusions and recommendations, Cornell University Library
Gu BM, Gil CG (2001) A multi-channel continuous toxicity monitoring system using recombinant bioluminescent bacteria for classification of toxicity. Biosens Bioelectron 16(9–12):661–666
Gueune H, Durand MJ, Thouand G, DuBow MS (2008) The ygaVP genes of *Escherichia coli* form a tributyltin-inducible operon. Appl Env Microbiol 74(6):1954–1958
Gueune H, Thouand G, Durand MJ (2009) A new bioassay for the inspection and identification of TBT-containing antifouling paint. Mar Pollut Bull 58(11):1734–1738
Guzzo A, Diorio C, DuBow MS (1991) Transcription of the *Escherichia coli* fliC gene is regulated by metal ions. Appl Env. Microbiol 57(8):2255–2259
Hakkila K, Green T, Leskinen P, Ivask A, Marks R, Virta M (2004) Detection of bioavailable heavy metals in EILATox-Oregon samples using whole-cell luminescent bacterial sensors in suspension or immobilized onto fibre-optic tips. J Appl Toxicol 24(5):333–342
Hua A, Gueuné H, Cregut M, Thouand G, Durand M-J (2015) Development of a bacterial bioassay for atrazine and cyanuric acid detection. Front Microbiol 6
Huang WE, Li M, Jarvis RM, Goodacre R, Banwart SA (2010) Shining light on the microbial world: the application of Raman microspectroscopy. In: Advances in applied microbiology, vol 70. Academic Press, New York, pp 153–186
Hynninen A, Tönismann K, Virta M (2010) Improving the sensitivity of bacterial bioreporters for heavy metals. Bioeng Bugs 1(2):132–138

ISO 5815-1:2003 (2003a) Water quality—determination of biochemical oxygen demand after n days (BODn)—Part 1: dilution and seeding method with allylthiourea addition
ISO 5815-2:2003 (2003b) Water quality—determination of biochemical oxygen demand after n days (BODn)—part 2: method for undiluted samples http://www.iso.org/iso/catalogue_detail.htm?csnumber=31091. Accessed 7 July 2016
Ivask A, Green T, Polyak B, Mor A, Kahru A, Virta M, Marks R (2007) Fibre-optic bacterial biosensors and their application for the analysis of bioavailable Hg and As in soils and sediments from Aznalcollar mining area in Spain. Biosens Bioelectron 22(7):1396–1402
Ivask A, Rolova T, Kahru A (2009) A suite of recombinant luminescent bacterial strains for the quantification of bioavailable heavy metals and toxicity testing. BMC Biotechnol 9:41
Jarvis RM, Goodacre R (2005) Genetic algorithm optimization for pre-processing and variable selection of spectroscopic data. Bioinformatics 21(7):860–868
Jennings VLK, Rayner-Brandes MH, Bird DJ (2001) Assessing chemical toxicity with the bioluminescent photobacterium (vibrio fischeri): a comparison of three commercial systems. Water Res 35(14):3448–3456
Jouanneau S, Durand M-J, Courcoux P, Blusseau T, Thouand G (2011) Improvement of the identification of four heavy metals in environmental samples by using predictive decision tree models coupled with a set of five bioluminescent bacteria. Environ Sci Technol 45(7):2925–2931
Jouanneau S, Recoules L, Durand MJ, Boukabache A, Picot V, Primault Y, Lakel A, Sengelin M, Barillon B, Thouand G (2014) Methods for assessing biochemical oxygen demand (BOD): a review. Water Res 49:62–82
Jung Y, Park C-B, Kim Y, Kim S, Pflugmacher S, Baik S (2015) Application of multi-species microbial bioassay to assess the effects of engineered nanoparticles in the aquatic environment: potential of a luminous microbial array for toxicity risk assessment (LumiMARA) on testing for surface-coated silver nanoparticles. Int J Environ Res Public Health 12(7):8172–8186
Kaur H, Kumar R, Babu JN, Mittal S (2015) Advances in arsenic biosensor development—a comprehensive review. Biosens Bioelectron 63:533–545
King JMH, DiGrazia PM, Applegate B, Burlage R, Sanseverino J, Dunbar P, Larimer F, Sayler GS (1990) Rapid, sensitive bioluminescent reporter technology for naphthalene exposure and biodegradation. Science 249(4970):778–781
Lasch P (2012) Spectral pre-processing for biomedical vibrational spectroscopy and microspectroscopic imaging. Chemom Intell Lab Syst 117:100–114
Lei Y, Chen W, Mulchandani A (2006) Microbial biosensors. Anal Chim Acta Mol Electron Anal Chem 568(1–2):200–210
Ma Z, Jacobsen FE, Giedroc DP (2009) Coordination chemistry of bacterial metal transport and sensing. Chem Rev 109(10):4644–4681
Marshall CP, Marshall AO (1922) The potential of Raman spectroscopy for the analysis of diagenetically transformed carotenoids. Philos Trans R Soc Lond Math Phys Eng Sci 2010 (368):3137–3144
Marshall CP, Edwards HG, Jehlicka J (2010) Understanding the application of Raman spectroscopy to the detection of traces of life. https://kuscholarworks.ku.edu/bitstream/handle/1808/17372/MarshallC_Astrobio_10%282%29229.pdf
Menz J, Schneider M, Kümmerer K (2013) Toxicity testing with luminescent bacteria—characterization of an automated method for the combined assessment of acute and chronic effects. Chemosphere 93(6):990–996
Merulla D, Buffi N, Beggah S, Truffer F, Geiser M, Renaud P, van der Meer JR (2013) Bioreporters and biosensors for arsenic detection. Biotechnological solutions for a world-wide pollution problem. Curr Opin Biotechnol 24(3):534–541
Merulla D, van der Meer JR (2016) Regulatable and modulable background expression control in prokaryotic synthetic circuits by auxiliary repressor binding sites. ACS Synth Biol 5(1):36–45
Nakazawa T, Inouye S, Nakazawa A (1980) Physical and functional mapping of RP4-TOL plasmid recombinants: analysis of insertion and deletion mutants. J Bacteriol 144(1):222–231

Neugebauer U, Schmid U, Baumann K, Holzgrabe U, Ziebuhr W, Kozitskaya S, Kiefer W, Schmitt M, Popp J (2006) Characterization of bacterial growth and the influence of antibiotics by means of UV resonance Raman spectroscopy. Biopolymers 82(4):306–311

Park M, Tsai S-L, Chen W (2013) Microbial biosensors: engineered microorganisms as the sensing machinery. Sensors 13(5):5777–5795

Pasco N, Baronian K, Jeffries C, Webber J, Hay J (2004) MICREDOX®—development of a ferricyanide-mediated rapid biochemical oxygen demand method using an immobilised *Proteus vulgaris* biocomponent. Biosens Bioelectron 20(3):524–532

Peinado MT, Mariscal A, Carnero-Varo M, Fernández-Crehuet J (2002) Correlation of two bioluminescence and one fluorogenic bioassay for the detection of toxic chemicals. Ecotoxicol Environ Saf 53(1):170–177

Pence I, Mahadevan-Jansen A (2016) Clinical instrumentation and applications of Raman spectroscopy. Chem Soc Rev 45(7):1958–1979

Ponomareva O, Arlyapov V, Alferov V, Reshetilov A (2011) Microbial biosensors for detection of biological oxygen demand (a review). Appl Biochem Microbiol 47(1):1–11

Radix P, Léonard M, Papantoniou C, Roman G, Saouter E, Gallotti-Schmitt S, Thiébaud H, Vasseur P (2000) Comparison of four chronic toxicity tests using algae, bacteria, and invertebrates assessed with sixteen chemicals. Ecotoxicol Environ Saf 47(2):186–194

Rae A, Stosch R, Klapetek P, Hight Walker AR, Roy D (2014a) State of the art Raman techniques for biological applications. Methods 68(2):338–347

Rae A, Stosch R, Klapetek P, Hight Walker AR, Roy D (2014b) State of the art Raman techniques for biological applications. Methods 68(2):338–347

Raud M, Kikas T (2013) Bioelectronic tongue and multivariate analysis: a next step in BOD measurements. Water Res 47(7):2555–2562

Raud M, Tenno T, Jõgi E, Kikas T (2012) Comparative study of semi-specific *Aeromonas hydrophila* and universal *Pseudomonas fluorescens* biosensors for BOD measurements in meat industry wastewaters. Enzyme Microb Technol 50(4–5):221–226

Roda A, Cevenini L, Michelini E, Branchini BR (2011a) A portable bioluminescence engineered cell-based biosensor for on-site applications. Biosens Bioelectron 26(8):3647–3653

Roda A, Roda B, Cevenini L, Michelini E, Mezzanotte L, Reschiglian P, Hakkila K, Virta M (2011b) Analytical strategies for improving the robustness and reproducibility of bioluminescent microbial bioreporters. Anal Bioanal Chem 401(1):201–211

Rutledge DN, Jouan-Rimbaud Bouveresse D (2013) Independent components analysis with the JADE algorithm. TrAC Trends Anal Chem 50:22–32

Schuster KC, Urlaub E, Gapes JR (2000) Single-cell analysis of bacteria by Raman microscopy: spectral information on the chemical composition of cells and on the heterogeneity in a culture. J Microbiol Methods 42(1):29–38

Siegfried K, Endes C, Bhuiyan AFMK, Kuppardt A, Mattusch J, van der Meer JR, Chatzinotas A, Harms H (2012) Field testing of arsenic in groundwater samples of Bangladesh using a test kit based on lyophilized bioreporter bacteria. Environ Sci Technol 46(6):3281–3287

Singer AC, Huang WE, Helm J, Thompson IP (2005) Insight into pollutant bioavailability and toxicity using Raman confocal microscopy. J Microbiol Methods 60(3):417–422

Su L, Jia W, Hou C, Lei Y (2011) Microbial biosensors: a review. Biosens Bioelectron 26 (5):1788–1799

Sun J-Z, Peter Kingori G, Si R-W, Zhai D-D, Liao Z-H, Sun D-Z, Zheng T, Yong Y-C (2015) Microbial fuel cell-based biosensors for environmental monitoring: a review. Water Sci Technol J Int Assoc Water Pollut Res 71(6):801–809

Tian H, Zhuang G, Ma A, Jing C (2012) Arsenic interception by cell wall of bacteria observed with surface-enhanced Raman scattering. J Microbiol Methods 89(3):153–158

Turner K, Raut N, Pasini P, Daunert S, Michelini E, Cevenini L, Mezzanotte L, Roda A (2010) Cell-based bioluminescent biosensors. In: Roda A (ed) Chemiluminescence and bioluminescence. Royal Society of Chemistry, Cambridge, pp 511–542

Vankeirsbilck T, Vercauteren A, Baeyens W, Van der Weken G, Verpoort F, Vergote G, Remon JP (2002) Applications of Raman spectroscopy in pharmaceutical analysis. TrAC Trends Anal Chem 21(12):869–877

Wadhia K (2008) ISTA13—international interlaboratory comparative evaluation of microbial assay for risk assessment (MARA). Environ Toxicol 23(5):626–633

Wadhia K, Dando T, Thompson KC (2007) Intra-laboratory evaluation of microbial assay for risk assessment (MARA) for potential application in the implementation of the water framework directive (WFD). J Environ Monit 9(9):953–958

Xu T, Close DM, Sayler GS, Ripp S (2013) Genetically modified whole-cell bioreporters for environmental assessment. Ecol Indic 28:125–141

Yagur-Kroll S, Belkin S (2014) Molecular manipulations for enhancing luminescent bioreporters performance in the detection of toxic chemicals. In: Thouand G, Marks R (eds) Bioluminescence: fundamentals and applications in biotechnology, vol 3. Advances in biochemical engineering/biotechnology. Springer, Berlin, pp 137–149

Yoshida N, Hoashi J, Morita T, McNiven SJ, Nakamura H, Karube I (2001) Improvement of a mediator-type biochemical oxygen demand sensor for on-site measurement. J Biotechnol 88 (3):269–275

Chapter 13
Microbial Biosensors for Metal(loid)s

Juan Carlos Gutiérrez, Francisco Amaro and Ana Martín-González

Abstract In this chapter we carry out an updated review on metal(loid)s biosensors using microorganisms as bioreceptor element of a classic biosensor or as a whole-cell biosensor. We analyze the potential advantages and possible disadvantages to use prokaryotic or eukaryotic microorganisms in metal(loid) biosensors. Likewise, the presence or absence of a cell wall in the microbial system can determine the degree of permeability of the target molecule to be detected. Sensitivity versus specificity of the biosensor is also discussed. We call attention on the necessity to carry out more bioassays using real environmental samples, and not only laboratory prepared once. A greater interest on designing biosensors using protozoa is also reclaimed, because these eukaryotic microorganism are much more sensitive to metal(loid)s than other microorganisms, and they share a higher degree of functional conservation with human genes than do other eukaryotic microbial models. Finally, a collection and analysis of the main metal(loid) microbial biosensors and genetic constructs potentially useful to design metal biosensors is reported.

Abbreviations

CB Classical or conventional biosensor/s
GFP Green fluorescent protein
MT Metallothionein
WCB Whole-cell biosensor/s

J.C. Gutiérrez (✉) · F. Amaro · A. Martín-González
Dpto. Microbiología-III, Facultad de Biología, Universidad Complutense (UCM), C/ José Antonio Novais, N° 12, 28040 Madrid, Spain
e-mail: juancar@bio.ucm.es

F. Amaro
e-mail: famaroto@ucm.es

A. Martín-González
e-mail: anamarti@bio.ucm.es

C. Cravo-Laureau et al. (eds.), *Microbial Ecotoxicology*,
DOI 10.1007/978-3-319-61795-4_13

13.1 Introduction to Metal(loid) Biosensors: Basic Concepts

Certain metal(loid)s (mainly those considered as "heavy metal(loid)s") are among the most abundant, toxic and persistent inorganic environmental pollutants (Hill 2004). Anthropogenic sources, mainly mining and industrial activities, have substantially increased the metal(loid)s content in the atmosphere and in may terrestrial and aquatic ecosystems (Peñuelas and Fillela 2002). This is the main reason to consider to these toxic compounds a priority in ecotoxicology, with the aim of minimizing the exposure of animals or humans. It is difficult to predict the global effects of increasing the different types of environmental pollutants, so there is an pressing need to develop screening methods for environmental monitoring. This necessity is for both, the early detection of environmental pollution by metal(loid)s and/or for testing the bioremediation process of a metal(loid) polluted ecosystem.

Low metal(loid) concentrations can be measured using molecular recognition or chemical analysis, such as absorption spectroscopy, mass spectroscopy, chromatography, polarography, among others. These techniques require qualified personnel and involve a high cost, and in addition, is not possible to carry out in situ analysis by using these techniques. On the other hand, critical ecotoxicological parameters such as bioavailability, toxicity and genotoxicity, can only be assayed using living cells. The most sensitive screening methods for detecting pollutants are those that incorporate biological components that are used as targets for an active substance or pollutant. In general, these screening techniques are known as biosensors or bioreporters. We can distinguish two types of biosensors: the classical or conventional biosensors (CB) and the whole-cell biosensors (WCB).

The CB can be defined as an integrated bioreceptor-physicochemical transducer device. This biosensor consists of three different elements: a bioreceptor or biological recognition element, which interacts with the pollutant molecules, a physicochemical transducer, which converts the biological response into a measurable physicochemical signal, and a microelectronic processor of this signal, which amplifies it and converts it into a numeric record (Fig. 13.1). The biological components can be macromolecules (such as enzymes, antibodies, nucleic acids, etc.) or whole-cells (prokaryotic or eukaryotic microorganisms or cells from multicellular organisms). At least, four main different types of transduction elements or transducers can be considered: electrochemical (potentiometric/amperometric), optical (spectrophotometry/fluorometry), piezoelectric or thermometric. Construction of these biosensors requires biological and physicochemical knowledge, which involves an interdisciplinary cooperation among different specialists, making construction more difficult and expensive. About the second type of biosensor (whole-cell biosensor), several authors have introduced the concept of the WCB as a very useful alternative to CB (Belkin 2003; Van der Meer and Belkin 2010). The main difference between both types of biosensors is that WCB use a whole prokaryotic or eukaryotic cell as a single reporter, incorporating both bioreceptor and transducer elements into the same cell (Fig. 13.1). This means that organisms used as WCB are, in general, experimentally modified to incorporate transducer capacity or increase their sensitivity. Another advantageous feature of these biosensors is the possibility to carry out both in situ or ex situ analysis.

When using WCB, two types of bioassays can be considered: *turn off* or *turn on* assays (Belkin 2003). *Turn off* assays are quite similar to general toxicological bioassays. In this case, the sample toxicity is estimated from the degree of inhibition of a cellular activity, such as growth inhibition, respiration rate, motility depletion, etc., or a specific reporter constitutive gene expression. In these bioassays, the toxic concentration is proportional to the measurement of any cellular function inhibition (Fig. 13.2). In *turn on* assays, a quantifiable molecular reporter is fused to a specific gene promoter, known to be activated by the chemical or environmental pollutant. Therefore, in this second type of bioassay, the sample toxicity is proportional to the gene expression of the reporter molecule (Fig. 13.2). These screening methods can be applied to detect the presence of both, any environmental pollutant causing general cellular stress or a specific pollutant (like metal(oid)s).

Turn off assays are more unspecific, because the signal decreases as a result of a broad range of cytotoxic effects, while WCB using *turn on* assays (based on an inducible gene expression) or CB (using specific molecules as bioreceptors), are usually more specific, as induction of the gene reporter, or interaction with the molecular bioreceptor, only takes place when the pollutant is present. The WCB specificity will therefore depend on the degree of the gene promoter specificity to be

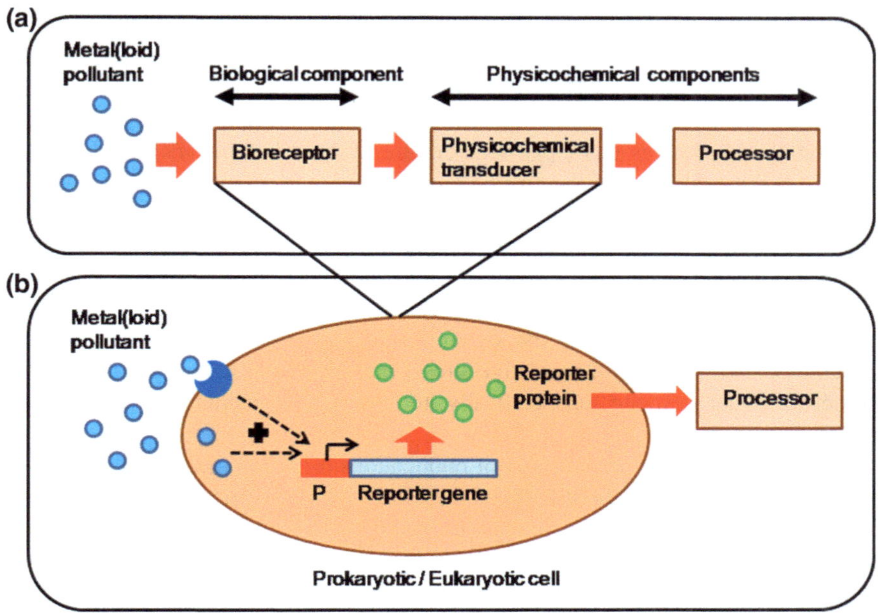

Fig. 13.1 Schematic representation of elements configuring classic and whole-cell biosensors. This figure is based on another previously published (Gutierrez et al. 2015)

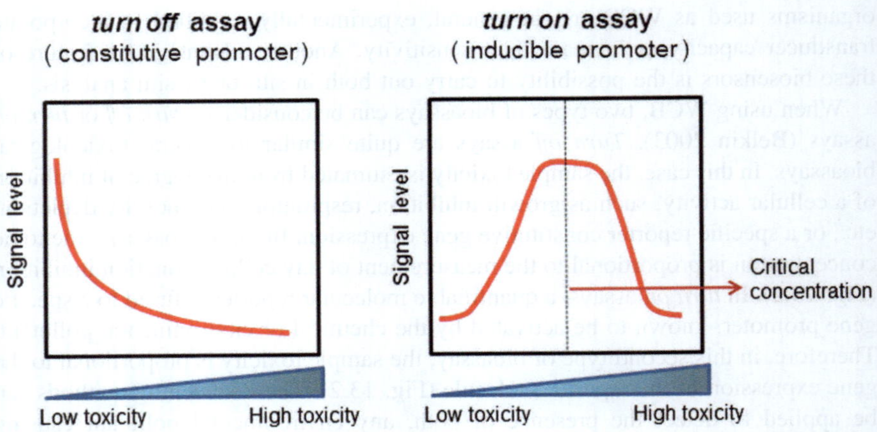

Fig. 13.2 *Turn off* and *turn on* bioassays. *Turn off* bioassays use constitutive promoters; therefore the signal level from the reporter gene decreases proportionally to toxic pollutant concentration. *Turn on* bioassays use inducible promoters; in this case the reporter signal level increases with the pollutant concentration. This reporter signal may reach a maximum value (critical concentration), after which decreases due to the increasing toxic concentration effect on the cell. The value of critical concentration will depend on the degree of cellular resistance to the pollutant. This figure is based on another previously published (Gutierrez et al. 2015)

activated by an exclusive pollutant or a chemically related group of pollutants. On the other hand, the CB specificity will depend on the specificity degree of the interaction between the bioreceptor and the pollutant. With respect to specificity, both *turn on* WCB and CB can be divided into effect- and compound-specific biosensors (Yagi 2007). Effect-specific biosensors respond to physicochemical environmental changes (e.g., pH, temperature or osmotic changes) or chemically diverse pollutants that give rise to a type of toxicity (e.g., oxidative stress or protein damage). On the other hand, compound-specific biosensors respond to only one type of pollutant or compounds with similar chemical features (e.g., metal(loid)s). For some specialists the before specific-based classification of CB or WCB, may be divided in three classes: (1) class-I biosensors that only respond to a specific or exclusive pollutant increasing the signal, (2) class-II biosensors that respond to a specific cellular stress (e.g., oxidative stress) increasing the signal, and (3) class-III biosensors that respond unspecifically to different pollutants or environment stressors.

In the last ten years, the number of publications reporting metal(loid) biosensors has been doubled, and about 85% of these are based on bacteria (Magrisso et al. 2008), while about 15% are based on eukaryotes (being yeasts the majority of them). Several reviews focused on general or specific aspects of biosensors to detect metal(loid)s have appeared in the recent years, such as Walmsley and Keenan (2000), Gu et al. (2004), Belkin (2003), Van der Meer et al. (2004), Kröger and Law (2005), Verma and Singh (2005), Yagi (2007), Magrisso et al. (2008), Van der

Meer and Belkin (2010), Gutierrez et al. (2015), Mehta et al. (2016). In this chapter, only biosensors (CB and WCB) using whole microorganisms or microbial macromolecules for detecting metal(loid)s present in environmental samples are considered.

13.2 Advantages and Disadvantages of Using Whole Microorganisms as Biological Recognition Components in CB and WCB

A majority of reported CB and WCB for metal(loid)s detection are based on prokaryotic or eukaryotic microorganisms (Verma and Singh 2005; Gutiérrez et al. 2015; Mehta et al. 2016). For an experimental point of view, is more easy to get a high microbial biomass than to reach the necessary amount of a specific purified macromolecule (enzyme, antibody, etc.) for getting the sufficiently quantifiable signal in any biosensor. So, this can be resolved using organisms with a high growth speed or short generation time, features that are almost exclusive of microorganisms. Microbial strains are cheaper than isolated enzymes, and the same enzyme used as biological component in a CB present more activity into the microbial cells owing to the optimal micro-environment provided by the cells (Verma and Singh 2005). Another advantages using microorganisms is that most of them can be easily manipulated and grown on a wide variety of different media or culture types. Recent advances in microbial genetic analyses and their genetic modification, and an increasing number of sequencing microbial genomes have facilitated the design and development of microbial biosensors with improved selectivity toward metal(loid)s or any other pollutants. In the case of WCB this technological capacity is essential due to necessity to introduce a transduction capacity into the cell. Furthermore, microorganisms are distributed all over the world, and occupy all known ecosystems, which constitutes a great advantages when the biosensor designer is looking for a particular microbial capacity to detect a specific environmental pollutant. For instance, the β-protobacterium *Ralstonia metallidurans* (formerly known as *Alcaligenes eutrophus*) is specifically adapted to ecosystems with a high content of metals, such as industrial and polluted biotopes or metallurgic wastes (Mergeay et al. 2003). From the knowledge on the mechanisms of metal(loid) resistant and their regulation obtained from this bacterium and other metal-resistant microorganisms, several types of metal biosensors have been designed (Corbisier et al. 1999; Leth et al. 2002) (Table 13.1).

Among eukaryotic microorganisms, there is the possibility to use microbial cells from three different taxonomic groups; fungi, microalgae and protozoa. The "eukaryotic" feature is particularly important because, in general, biosensors aimed at the detection of potential environmental toxic substances affecting other eukaryotic organism (including humans). The existence of a more similar metabolism, genome, and cellular organization in microbial eukaryotic biosensors with

Table 13.1 Microbial biosensors for metal(loid)s or microbial genetic constructs useful to design metal biosensors

Metal(loid)s[a]	Microorganism		Biosensor[b]		Reference
	Prokaryotic	Eukaryotic	CB	WCB	
Cd(II) > Ni(II) = Zn(II) > Cu(II)	*R. leguminosarum*			Bioluminescence (*turn off*)	Paton et al. (1997)
Cr(II)	*R. metallidurans*			Bioluminescence (*turn on*)	Corbisier et al. (1999)
As(III)	*S. aureus*			Bioluminescence (*turn off*)	Corbisier et al. (1993)
Cu(II), Pb(II), Cd(II)	*A. torulosa*		Optical (fluorescence)		Wong et al. (2013)
Cd(II)	*E. coli*		Electrochemical		Souiri et al. (2012)
Cd(II), Hg(II)	*E. coli*		Acoustic		Gammoudi et al. (2010)
Pb(II) > As(V) > Cd(II) > Cu(II) > Zn(II) > Hg (II)		*T. thermophila*		Bioluminescence (*turn on*)	Amaro et al. (2011)
Cd(II) > Hg(II) > Zn(II) > Cu(II) > Pb(II) > As(V)		*T. thermophila*		Bioluminescence (*turn on*)	Amaro et al. (2011)
Cd(II)		*T. thermophila*		Bioluminescence (*turn on*)	Amaro et al. (2014)
Cu(II)		*S. cerevisiae*		Electrochemical (*turn on*)	Tag et al. (2007)
Cu(II)	*R. metallidurans*		Optical (bioluminescence)		Leth et al. (2002)
Cu(II)	*T. chuii*[c]		Electrochemical (potentiometric)		Alpat et al. (2008)
Cr(VI)	*Thiobacillus* sp.[d]		Electrochemical		Oh et al. (2011)
Cd(II), Cu(II)	*A. torulosa*		Electrochemical (amperometric)		Chay et al. (2005)

(continued)

Table 13.1 (continued)

Metal(loid)s[a]	Microorganism		Biosensor[b]		Reference
	Prokaryotic	Eukaryotic	CB	WCB	
Cr(VI)	*A. ferrooxidans*		Electrochemical (amperometric)		Zlatev et al. (2006)
Cd(II)	*C. vulgaris*			Fluorescence (*turn off*)	Nguyen-Ngoc et al. (2009)
Cu(II)		*Circinella* sp.	Electrochemical		Alpat et al. (2008)
Hg(II)		*Chlorella* sp.	Electrochemical (amperometric)		Singh and Mittal (2012)
Cd(II)		*C. vulgaris*	Electrochemical (conductometric)		Chouteau et al. (2004)
Cu(II)	*D. chlorelloides*		Optical		Peña-Vázquez et al. (2010)
Pb(II)		*Phormidium* sp.	Electrochemical		Yüce et al. (2010a)
Cu(II)		*R. mucilaginosa*	Electrochemical		Yüce et al. (2010b)
Ni(II)	*B. sphaericus*		Electrochemical (amperometric)		Verma and Singh (2006)
Cu(II)		*S. cerevisiae*	Electrochemical (amperometric)		Lehmann et al. (2000)
Cu(II), Ag(I)		*S. cerevisiae*		Fluorescence (*turn on*)	Shetty et al. (2004)
Cu(II)		*S. cerevisiae*		Bioluminescence (*turn on*)	Roda et al. (2011)

(continued)

Table 13.1 (continued)

Metal(loid)s[a]	Microorganism		Biosensor[b]		Reference
	Prokaryotic	Eukaryotic	CB	WCB	
Cu(II), Ni(II)		*C. reinhardtii*	Electrochemical (amperometric)		Shitanda et al. (2005)
Hg(II)	*E. coli*			Bioluminescence (*turn on*)	Selifonova et al. (1993)
Cd(II), Pb(II), Sb(III)	*S. aureus*			Bioluminescence (*turn on*)	Tauriainen et al. (1998)
Cd(II), Sb(III), Zn(II), Sn(II)	*B. subtilis*			Bioluminescence (*turn on*)	Tauriainen et al. (1998)
As(V), Sb(III)	*E. coli*			Fluorescence (*turn on*)	Liao and Ou (2005)
As(V), Sb(III), As(III), Bi(III)	*S. aureus*			Colorimetric (*turn on*)	Ji and Silver (1992)
As(V), Cd(II), Sb(III)	*E. coli*			Bioluminescence (*turn on*)	Tauriainen et al. (1999)
Zn(II), Cd(II), Hg(II)	*Synechocystis* sp.			Bioluminescence (*turn on*)	Erbe et al. (1996)
Zn(II), Co(II)	*Synechocystis* sp.			Bioluminescence (*turn on*)	Peca et al. (2008)
Zn(II), Cd(II), Cr(VI), Hg(II), Pb(II)	*E. coli*			Bioluminescence (*turn on*)	Riether et al. (2001)
Cd(II), Hg(II), Zn(II)	*E. coli*			Bioluminescence (*turn on*)	Ivask et al. (2002)
Cd(II), Pb(II)	*E. coli*			Fluorescence (*turn on*)	Shetty et al. (2003)

(continued)

Table 13.1 (continued)

Metal(loid)s[a]	Microorganism		Biosensor[b]		Reference
	Prokaryotic	Eukaryotic	CB	WCB	
Hg(II), Cd(II)	*E. coli*			Bioluminescence (*turn on*)	Virta et al. (1995)
Hg(II)	*Vibrio anguillarum*			Bioluminescence (*turn on*)	Golding et al. (2002)
Hg(II), Cd(II)	*Pseudomonas fluorescens*			Bioluminescence (*turn on*)	Petänen et al. (2001)
Hg(II)	*E. coli*			Bioluminescence (*turn on*)	Ivask et al. (2001)
Hg(II), Cd(II), Zn(II)	*E. coli*			Bioluminescence (*turn on*)	Ivask et al. (2002)
Pb(II), Cd(II)	*E. coli*			Colorimetric (*turn on*)	Shetty et al. (2003)
Pb(II), Sb(III)	*E. coli*			Fluorescence (*turn on*)	Liao et al. (2006)
Cu(II), Ag(I)	*E. coli*			Bioluminescence (*turn on*)	Hakkila et al. (2004)
Cu(II)	*P. fluorescens*			Bioluminescence	Tom-Petersen et al. (2001)

[a]The order of sensitivity for tested metal(loid)s is reported, when it is available. [b]The transducer element of biosensors and the type of bioassay for WCB are indicated. [c]In this CB the no-living biomass of the alga *Tetraselmis chui* (biosortion-based CB) is used. [d]In this CB several sulfur-oxidant bacteria (including *Thiobacillus* sp.) are used

those organisms (plants and animals) undergoing a chemical pollution, makes the extrapolation and comparison of results more accurate and reliable.

13.2.1 *Bacteria-Based Biosensors*

To design CB or WCB, we can consider two basic types of bacteria; the naturally existing or wild-type and those genetically modified (anthropogenic origin). The first one, usually presents a peculiar natural characteristic which can be used for designing a metal(loid) biosensor, such as bioluminescence, a color or pigmentation, or any other feature that can be modified or altered after metal(loid) exposure. These biosensors are based on the inhibition or blocking of a natural bacterial feature, existing a proportional ratio between the metal toxicity and the bacterial signal decreasing (which should be measurable). These biosensor can be considered as *turn off* bioassays.

An example of whole bacteria-based CB is the one using the bioluminescent bacterium *Photobacterium phosphoreum* (Lee et al. 1992), immobilized on a cellulose nitrate membrane and used to detect chromium. Also *Anabaena torulosa* cyanobacterium cells embedded in a cellulose membrane have been used to detect Cu(II), Pb(II) and Cd(II) (Wong et al. 2013) (Table 13.1). The presence of these toxic ions reduce the photosynthetic activity changing the fluorescence quenching of these cells, and the release of photosynthetic oxygen is also inhibited under the metal presence and this oxygen emission reduction is detected by an oxygen electrode (Shing et al. 2008).

Likewise, in metal(loid) CB using enzymes isolated from microorganisms, toxic metals might inhibit the normal enzyme activity, exhibiting a direct correlation between the enzyme inhibition rate and metal toxicity. Consequently a reduction in enzyme activity can be read as a signal, which can be amplified to get the desired sensitivity level (Vel Krawczyk et al. 2000). Several enzymes, such as alkaline phosphatase, glucose oxidase or urease, among others, have been used to detect Cd (II), Pb(II), Zn(II), Ni(II) or Co(II) (Berezhetskyy et al. 2008), Cr(III), Hg(II), Ag(I), Cu(II), Cd(II), Pb(II), Fe(III), Co(II), Ni(II) or Zn(II) (Guascito et al. 2008; Samphao et al. 2012) and Cd(II) or Pb(II) (Ilangovan et al. 2006).

Genetically modified bacteria could be used in both *turn off* and *turn on* bioassays of CB or WCB. Several recombinant strains from both Gram-positive (*Staphylococcus aureus* and *Bacillus subtilis*) and Gram-negative (*Escherichia coli* and *Pseudomonas fluorescens*) bacteria were constructed to express a bioluminescence reporter gene (*lux* genes) to be used as metal(loid) WCB (Ivask et al. 2009). Both, *turn off* and *turn on* bioassays were carried out, and five strains to detect Cu(II) and Hg(II) were target metal specific, whereas eight other strains were induced by Cd(II), Hg(II), Zn(II) and Pb(II), so showing a lower metal specificity. The soil bacterium *R. metallidurans* has been also used as a WCB (*turn on*) to detect Cr(II) using *lux* reporter gen system (Corbisier et al. 1999) (Table 13.1). Another strain of *R. metallidurans* has been used as WCB for detection of Ni(II) and Co(II) in soil samples, after transformation with the megaplasmid pMOL1550 which contains the promoter of the *cnr* operon (resistance system to Ni(II) and Co (II) present in this bacterium) (Tibazarwa et al. 2000) fused to *lux* reporter gene system (Tibazarwa et al. 2001). The promoter from the *cadA* resistance determinant system for Cd(II) and Zn(II) of *Staphylococcus aureus* (Yoon et al. 1991) fused to the firefly luciferase reporter gene into the plasmid pT0024 has been used to design WCB in *S. aureus* and *B. subtilis* (Tauriainen et al. 1998) (Table 13.1). A WCB to detect Hg(II) has been also genetically constructed by using fusions of the Tn21 mercury resistance operon (*mer*) with *lux* reporter gene system from *Vibrio fischeri*, and the bacterium *E. coli* was used to design the WCB (Table 13.1). The *turn on* bioassays using this Hg-WCB was able to detect bioavailable mercury in water samples at a nM to μM concentration range (Selifonova et al. 1993). Although, they are not commented here, others different bacteria genetic constructs used to design CB or WCB are reported in Table 13.1.

13.2.2 Eukaryotic Microorganisms-Based Biosensors

In general, both CB or CWB using eukaryotic microorganisms are more scarce than those using prokaryotic one (Table 13.1). Probably, it is due to the greater cellular complexity of the eukaryotic cells and relative difficulty to work with them, but however they present several advantages in comparison with prokaryotic cells (see Sect. 13.2). Like in prokaryotic-based biosensors, we can distinguish between two basic types of eukaryotic microorganisms to be used to design both CB or WCB; wild-type and genetically modified microorganisms.

13.2.2.1 Microalgae-Based Biosensors

Microalgae are important in biosensor construction for aquatic (freshwater or marine) ecosystems applications (Kröger and Law 2005). Immobilized whole-cells of the microalga *Chlorella vulgaris* has been used to design a conductometric CB based on the inhibition of alkaline phosphatase activity in presence of Cd(II) ions (Chouteau et al. 2004) (Table 13.1). This same microalga was used as a WCB to detect Cd(II), in water suspension or immobilized in a translucent silica matrix. The Cd(II) toxicity affected the algal photosynthetic activity (*turn off* bioassay) resulting in a quenching of cellular fluorescence (Nguyen-Ngoc et al. 2009) (Table 13.1). For monitoring Cu(II) in water supplies the chlorophita *Dictyosphaerium chlorelloides* was used with an optic fiber coupled to the cellular flow or a microwell-plate reader (Peña-Vázquez et al. 2010) (Table 13.1). On the basis of flagellar motility of the microalga *Chlamydomonas reinhardtii* electrochemical biosensing systems for detecting Cu(II) or Ni(II) have been developed (Shitanda et al. 2005) (Table 13.1).

In many of these microorganisms, it is likely that the lack of usable genetic tools for bioengineering considerably limit the construction of improved WCB; in fact, they are used as wild type strains in both, as bioreceptor elements in CB or cells in WCB (*turn off* bioassays) (Table 13.1). Although, there is an exception related to the microalgal model *C. reinhardtii*, this suitable microalga model has not yet been genetically modified to design biosensors for environmental metal(loid) monitoring, excepting for sensing triazine and urea types herbicides (Lambreva et al. 2013), even though diverse studies have already been done on metal toxicity in this microorganism (Aksmann et al. 2014; De Schamphelaere et al. 2014). Particularly, De Schamphelaere et al. (2014) demonstrated inter-specific differences in Pb(II) sensitivity among three microalgae species (*Pseudokirchneriella subcapitata*, *Chlorella kesslerii* and *C. reinhardtii*), which should be taken in account when designing WCB for detecting this metal. Likewise, Aksmann et al. (2014) report a gene expression analysis of several antioxidant enzymes (superoxide dismutase, catalase and ascorbate peroxidase) under oxidative stress induced by Cd(II), together with an analysis of photosynthetic activity on this alga, which could be useful to select molecular elements to design biosensors for detecting Cd(II).

Taking in account that microalgae present enough qualities to be considered as good potential metal(loid) biosensors, we conclude that this biotechnological aspect has not been yet sufficiently exploited.

13.2.2.2 Filamentous Fungi and Yeasts-Based Biosensors

Both, filamentous fungi and yeasts are well-known eukaryotic microbial models that are widely used in toxicology, biotechnological and basic biological studies. Among them, the yeast *Saccharomyces cerevisiae* is the most widely used eukaryotic microorganism in very diverse biological areas, especially in genetic and bioengineering, and for this reason some authors (Walmsley and Keenan 2000) consider that it has certain advantages as a biosensor to be used with natural polluted environment samples. It is a robust eukaryotic microorganism with a considerably physicochemical tolerance to very diverse chemicals, and good genetic tools that make possible the construction of genetically modified yeasts showing optimized features to design better biosensors.

However, like bacteria, fungi and yeasts have a cell wall that protects the cell and acts as a selective barrier for very different molecules (including substrates used by the biosensor transducer system in WCB), which makes transducer signal emission more difficult. Therefore, in some occasions it is necessary to increase cell wall permeability before using these microorganisms as WCB or bioreceptor in CB, which constitutes an additional difficulty. Mutans with enhanced cell permeability can be used for this purpose (Terziyska et al. 2000; Walmsley and Keenan 2000).

As it also occurs with bacteria and microalgae, yeasts have been used almost exclusively as bioreceptor elements in CB (Table 13.1) (Baronian 2004). The pigmented yeast *Rhodotorula mucilaginosa* has been used to construct a microbial biosensor based on carbon paste for determination of Cu(II) (Yüce et al. 2010b) (Table 13.1). A similar construction was carried out on the filamentous fungus *Circinella* sp., consisting of concentrated whole cells on the carbon paste electrode surface for Cu(II) detection. This CB is based on the biosorption capacity of the fungus cell wall to Cu(II) ions (Alpat et al. 2008) (Table 13.1). A recombinant *S. cerevisiae* strain has been used to construct an amperometric CB (Lehmann et al. 2000) to detect Cu(II) (Table 13.1). A plasmid containing the copper inducible *cup1* gene promoter and the *E. coli lacZ* gene as a reporter gene was constructed, then this plasmid construct was introduced into *S. cerevisiae* and recombinant strain was immobilized with polyvinyl alcohol on a capillary membrane. If Cu(II) is present in the sample, this recombinant strain is able to utilize lactose as a carbon source, which leads to alterations in the oxygen consumption of the cell. So, changes in the oxygen concentration were quantified by an oxygen electrode (similar to Clark's oxygen electrode for glucose quantification) (Clark et al. 1953). Another amperometric CB using other different recombinant *S. cerevisiae* strain to detect Cu(II) was constructed also using *lacZ* reporter gene (Tag et al. 2007) (Table 13.1).

Few WCB have been designed using yeasts, some of them are: *S. cerevisiae* cells and GFP (green fluorescent protein) as a reporter protein was developed to detect Cu(II) ions (Shetty et al. 2004) (Table 13.1). The transcriptional activator protein AceI present in this yeast was used to control expression of the reporter gene *gfp* (encoding GFP). When Cu(II) ions are present, the AceI protein actives the *cup1* gene promoter located upstream from the *gfp* gene (*Pcup1::gfp*) into a plasmid, there by inducing GFP production. This system is selective for Cu(II) over other metals, except for Ag(II) (Shetty et al. 2004). Another similar *S. cerevisiae* WCB, also for Cu(II) detection, has been constructed using the same promoter (*cup1*) but a different reporter gene (luciferase), and showing a similar detection level for this metal (Roda et al. 2011) (Table 13.1).

From a microarray gene expression analysis carry out in the methylotrophic yeast *Hansenula polymorpha*, under Cd(II) treatment, several over-expressed genes were selected (Park et al. 2007). This analysis revealed that the promoter from the *seo1* gene (with an unknown cellular function), fused with the GFP gene, was the reporter construct with the highest GFP expression level with regard to other promoters tested. This reporter construct is not specific for Cd(II) because it is also inducible by As(III). Likewise, the *seo1* promoter from *S. cerevisiae* revealed that this is inducible by As(III) > Cd(II) > Hg(II), being also unspecific for Cd(II). These constructs could be useful to design metal WCB with these eukaryotic microorganisms.

13.2.2.3 Protozoa-Based Biosensors

Among protozoa, ciliates have been extensively used in ecotoxicological analysis (Gutierrez et al. 2008). Ciliates have, at least, two additional advantages with regard to other microorganisms. In first place, unlike bacteria, yeasts or microalgae, ciliates have not a cell wall in their vegetative stage. As it has been previously mentioned, a major limitation in using microorganisms with cell walls as WCB or bioreceptor in CB, is the diffusion of substrates or molecules through the cell wall, resulting in a lower signal emission or less effective cell response. To prevent this, cells have to be permeabilized by physicochemical or enzymatic methods. Furthermore, the presence of a cell wall could involves a not controlled unspecific metal(loid) biosortion process which would affect the real cellular response to the external metal concentration, when it is not used as a metal biosortion-based biosensor. Using ciliates might therefore avoid or diminish this serious problem, so the absence of a wall in these eukaryotic microorganisms results in greater sensitivity to environmental pollutants and a faster cell response (Martin-Gonzalez et al. 1999; Gutierrez et al. 2003). Secondly, ciliates are eukaryotic cells with a series of metabolic traits that are more similar to those of human cells than bacteria, microalgae, or even yeasts. After completing genome sequencing projects of two ciliate models such as *Tetrahymena thermophila* and *Paramecium tetraurelia* (Aury et al. 2006; Eisen et al. 2006), results shown that they share a higher degree of functional conservation with human genes than do other eukaryotic microbial models. Humans and *T. thermophila* share more ortholog genes with each other (about 2280) than are shared between humans and the yeast *S. cerevisiae* (Eisen et al. 2006). Likewise, the scores of *P. tetraurelia* proteins against human proteins are the highest with regard to the scores of yeast proteins to human proteins, suggesting that the *Paramecium* proteins are most similar to human proteins (Sperling et al. 2002). Therefore, this similarity with human biology makes it more reasonable to use these eukaryotic microorganisms in ecotoxicological studies (Gutierrez et al. 2008, 2011) or as biosensors to detect metal(loid) or organic pollutants. In addition, ciliates are cosmopolitan microorganisms living in aquatic or terrestrial ecosystems, and can be used as biosensors for monitoring pollutants in both habitats.

T. thermophila has five metallothionein gene isoforms. Two of these (*MTT1* and *MTT5*) are preferably over-expressed under Cd(II) or Pb(II), respectively, though they are also induced by other metals (Diaz et al. 2007; Gutierrez et al. 2011). Both genes, but mainly MTT5, respond quickly and strongly to metal stress, and their promoters have been used to design metal(loid) WCB. The first two WCB using ciliates, to detect metal pollution in soil and aquatic samples, were reported in 2011 (Amaro et al. 2011). These WCB (*turn on* bioassays) were designed using *MTT1* or *MTT5* gene promoters from *T. thermophila* and the firefly luciferase as reporter gene (Table 13.1), then these lineal constructions were introduced into nuclear genome by biolistic transformation. Validation of these WCB was carried out using artificial and natural (soil and aquatic) samples, including methods to detect false positives and negatives. A second type of *T. thermophila* WCB has been constructed with *MTT1* gene promoter and the GFP as a reporter molecule fused to *MTT1* or *MTT5* complete open reading frames into a plasmid (Amaro et al. 2014) (Table 13.1). A comparative analysis of both WCB revealed that: (1) in those using luciferase the minimal exposure time to obtain a detectable signal is $\approx$1 h, however for GFP-WCB an exposure of $\approx$2 h is necessary to have a stable signal, indicating a faster response in those with luciferase as reporter gene; (2) for the same MT promoter gene (*MTT1*), the minimum detectable Cd(II) concentration is lower in luciferase-WCB than GFP-WCB, so being luciferase-WCB more sensitive than

GFP-WCB; and (3) the bioluminescence emission from luciferase-WCB viable cells is up to 5 μM Cd(II), while cells with fluorescence emission (GFP-WCB) are viable up to 15 μM Cd(II). GFP-WCB are more resistant to Cd(II) than luciferase-WCB strains, because they have a higher copy number (plasmid constructs) of MTT1 or MTT5 genes (Amaro et al. 2014). Therefore, to detect low Cd (II) concentrations in polluted samples is better to use the luciferase-WCB strain, while for higher Cd(II) concentrations is more reasonable to use GFP-WCB strains.

At present, the only protozoa-based biosensors to detect metal(loid)s are those using the ciliate *T. thermophila* (Amaro et al. 2011, 2014), and although it has been only used as a metal(loid) WCB, in a next future these microorganism might be also used for monitoring other pollutants.

13.3 A Comparative Analysis Among Microbial Metal (Loid) Biosensors

Although the four types of microorganisms (bacteria, yeasts, microalgae or protozoa) can be used to design both CB and WCB for metal(loid) environmental monitoring, they present their advantages and disadvantages, from which the most significant will be discussed in the next sections.

13.3.1 To Have or not to Have Cell Wall: Advantage or Handicap?

For many prokaryotic and eukaryotic microorganisms to have a rigid wall is essential for survival, and, in general, if this structure is removed the cell dies. To use microorganisms with a cell wall to design metal biosensors has advantages and disadvantages. Microorganisms with a cell wall are more resistant to physico-chemical disturbances than those without cell wall, so to immobilize cells to design CB it is more easy using microorganisms with cell wall (bacteria, yeasts or microalgae). This is a good reason because these microorganisms are mostly used to design CB, where cells are in solution or immobilized into an inert matrix. But, another important point to be considered to design any biosensor is the cellular permeability capacity to target molecules (to be detected by the biosensor). Unlike ciliate protozoa, the presence of a cell wall in the rest of microorganisms (bacteria, yeasts or microalgae) may require a preliminary permeabilization process to facilitate the transit of the pollutant through the cell wall. This treatment might disturb the response of the cell to the pollutant modifying the level of the biological signal to be translated by the external transducer system (CB) or the cellular transducer (WCB). Likewise, to obtain the reporter signaling in substrate-dependent reporters, the substrate must cross the cell wall and reach the cytoplasm, where the enzymatic reaction takes place, or be added to the cells previously lysated. The substrate for eukaryotic luciferase, D-luciferin, is membrane-permeant only in its protonated form (pH 5); at neutral pH, it crosses the plasma membrane very slowly. For this reason, most luciferase-based bioassays are performed using cell extracts (Van der Meer et al. 2004) or permeabilized cells (Lagido et al. 2001). An additional problem to design metal(loid) biosensors using microorganisms with a cell wall, is the possible biosorption process carry out by the wall polymers. It could alter the response of the biosensor to the metal(loid) target to be detected and/or quantified, because a important part of metallic ions would be trapped by the microbial cell wall (biosorption).

As ciliated protozoa do not present a cell wall in their vegetative phase, they have a great advantage over other potential metal microbial biosensors, because there is no need for any preliminary permeabilization treatment or cellular lysis. For instance, in *T. thermophila* permeabilization pre-treatment or cellular lysis is not necessary when used as a WCB with luciferase as a reporter gene, because the luciferin crosses through the cell membrane. In this ciliate, luciferase activity can be measured as efficiently in intact viable cells as in permeabilized cells, and similar induction in vivo and in vitro was observed (Amaro et al. 2011).

13.3.2 Specificity Versus Sensitivity

The majority of microbial biosensors (CB and WCB) respond to two or more metal (loid)s, although some of them show greater specificity (Corbisier et al. 1999; Tom-Petersen et al. 2001; Ivask et al. 2009). It is not easy to find a gene promoter responding exclusively to one metal(loid), in fact metallothioneins (the main proteins binding metals) respond to several different metal(loid)s (de Francisco et al. 2016; Gutierrez et al. 2011). Likewise, the same bacterial operon responding to a specific metal stress has been used to design biosensors for other different metals (Tauriainen et al. 1998). In general, cells are ready to respond to diverse metal(loid) stresses using the same molecular protection system. Probably, the main reason of this is that natural or anthropogenic metal polluted ecosystems present very frequently a mixture of metal(loid)s rather than a single one (Fairbrother et al. 2007; Preston et al. 2000). Therefore, one valuable aim of environmental monitoring may be to determine the overall toxicity of a sample rather than the presence of a specific metal. Metal(loid) specificity is therefore not so important when designing a biosensor to be used for monitoring environmental metal pollution. However, several authors (Elad et al. 2008; Jouanneau et al. 2011) have tried to resolve this problem using a panel of luminescent bacteria as WCB with different stress-responsive gene promoters. These bacteria were treated with different toxic compounds (including heavy metals), and each toxic treatment activated different promoters. From this experimental approach these authors (Elad et al. 2008) were able to identify the toxic elements into the experimental sample within 30 min and with an error rate estimate that did not exceed 3% at a 95% confidence level. Later,

Jouanneau et al. (2011) based on a similar experimental approach, have elaborated predictive decision tree models and by using a specific software they can choose the best decision tree to identify the toxic metal(loid) from a four metal(loid) mixture. This method showed a high correlation (≈98%) for the metal(loid) identification. Although, these two experimental approaches represent good contributions to identify metallic elements from a polluted sample, the problem of the biosensor specificity is not still resolved.

On the other hand, the sensitivity level of the biosensor is really important when trying to detect metals present in very low concentrations, mainly those that are lower than the maximum allowable metal concentrations established by international commissions. A comparative analysis of the ranking of sensitivity values to different metal(loid)s among reported biosensors is summarized in Table 13.2, and described in the following points: (1) with regard to As(V), the *T. thermophila* MTT5Luc strain (WCB with the reporter construct *MTT5::LucFF*) is the biosensor with the highest sensitivity (25 nM) (Amaro et al. 2011) in comparison with other eukaryotic or prokaryotic-based biosensors; (2) for Zn(II) the ciliate *T. thermophila* MTT1Luc strain (with *MTT1::LucFF* construct) (Amaro et al. 2011) together the cyanobacterium *Synechococcus* sp. (with *smtB::luxCBDAE* construct) (Erbe et al. 1996) are the WCB with the highest sensitivity (0.5 μM), while among CB the synthetic phytochelatin-based capacitive biosensor (Bontidean et al. 2003) is the one showing the highest sensitivity (0.1 pM); (3) the *Escherichia coli*-based biosensor (with *cadC::gfp* construct) is the prokaryotic WCB with the highest

Table 13.2 Comparative analysis of the ranking of sensitivity values to different metal(loid)s among reported biosensors

Metal(loid)	Type	Bioreceptor[a]	Sensitivity[b]	Reference
As(V)	WCB	*T. thermophila*	25 nM	Amaro et al. (2011)
Zn(II)	WCB	*T. thermophila*	0.5 μM	Amaro et al. (2011)
	WCB	*Synechococcus* sp.	0.5 μM	Erbe et al. (1996)
	CB	Phytochelatin	0.1 pM	Bontidean et al. (2003)
Cd(II)	WCB	*E. coli*	0.1 nM	Liao and Ou (2005)
	WCB	*T. thermophila*	0.5 nM	Amaro et al. (2011)
	CB	Phytochelatin	0.1 pM	Bontidean et al. (2003)
Hg(II)	WCB	*E. coli*	1 fM	Virta et al. (1995)
	WCB	*T. thermophila*	0.25 nM	Amaro et al. (2011)
	CB	*E. coli*	1 pM	Gammoudi et al. (2010)
Pb(II)	WCB	*E. coli*	0.1 nM	Shetty et al. (2003)
	WCB	*T. thermophila*	50 nM	Amaro et al. (2011)
	CB	Phytochelatin	0.1 pM	Bontidean et al. (2003)
Cu(II)	WCB	*E. coli*	0.3 μM	Hakkila et al. (2004)
	WCB	*S. cerevisiae*	0.5 μM	Shetty et al. (2003)
	CB	Phytochelatin	0.1 pM	Bontidean et al. (2003)

[a]Microorganism or biomolecule. [b]Lowest detectable metal concentration

sensitivity to Cd(II) (0.1 nM) (Liao and Ou 2005), while, among eukaryotic WCB, the ciliate *T. thermophila* is that with the highest sensitivity (5 nM) for this metal (Amaro et al. 2011). And, among CB, the synthetic phytochelatin-based capacitive biosensor (Bontidean et al. 2003) shows the highest sensitivity value (0.1 pM) for Cd(II); (4) the *E. coli*-based WCB (with *merR::LucFF* construct) is that reporting the highest sensitivity (1 fM) for Hg(II) (Virta et al. 1995), while, among eukaryotic biosensors, *T. thermophila*-based WCB (Amaro et al. 2011) is the one showing the highest sensitivity (0.25 nM). Immobilized *E. coli* cells used to design an acoustic wave-based biosensor, seen to be the most sensitive for this metal (1 pM) (Gammoudi et al. 2010); (5) the WCB using the bacterium *E. coli* containing the *zntA::lacZ* construct (Shetty et al. 2003) presents the lowest detectable concentration for Pb(II) (0.1 nM), while *T. thermophila*, among eukaryotic WCB, presents the highest sensitivity (50 nM) (Amaro et al. 2011), likewise, the synthetic phytochelatin-based capacitive CB (Bontidean et al. 2003) is that reporting the highest sensitivity (0.1 pM) for Pb(II); and (6) with regard to Cu(II) ions, the *E. coli* strain with *copA::lucFF* construct (Hakkila et al. 2004) presents the highest sensitivity (0.3 μM), and among eukaryotic-based biosensors, a strain of the yeast *S. cerevisiae* with the *cup1::gfp* construct (Shetty et al. 2004), used as a WCB, has the highest sensitivity (0.5 μM) for Cu(II). Again, the synthetic phytochelatin-based capacitive CB (Bontidean et al. 2003) is that showing the most high sensitivity (0.1 pM) to this essential metal. Although these last authors indicate that this phytochelatin-based biosensor is able to detect metal ions in concentration range of 0.1 pM–10 mM, reporting an order of sensitivity (Zn > Cu > Hg ≫ Cd ≈ Pb), they do not indicate the concentration values for each metal. So, we cannot assure the real sensitivity values for each metal detected by this CB.

A summary of several features among different microorganisms which could affect (positively or negatively) the design of a metal(loid) biosensor is showed in Table 13.3.

Table 13.3 Comparative analysis of several features among different types of microorganisms which could affect (positively or negatively) the design of metal(loid) biosensors

Feature	Microorganism			
	Bacteria	Microalgae	Fungi/yeasts	Ciliates
Rapid growth	+++	+++	+++	+++
Easy manipulation	+++	+++	+++	+++
Genetic modification	+++	+	+++	+++
Metal(loid) sensitivity	++	++	++	+++
Cellular immobilization	+++	+++	+++	?
Used as WCB	+++	+	+++	+++
Presence of cell wall	+++	+++	+++	−

(+++): high; (++): low; (+): very low; (−): absent; (?): unknown. See the text for a more extensive explanation

13.4 Concluding Remarks

From this review on metal(loid) microbial biosensors the following general conclusions can be drawn:

(1) In general, microorganisms present more advantages than disadvantages to be used as CB or WCB to detect metal(loid)s from polluted ecosystems, in regard with other organisms. This is due to their easy cultivation and maintenance, their higher growth rate and genetic manipulation facilities. Likewise, eukaryotic microorganisms used as WCB have certain advantages over prokaryotic ones. Among them, the comparative analysis with multicellular organisms (including humans) is more reliable than using bacteria.

(2) The biotechnology for using microalgae as WCB is still underdeveloped, although these photosynthetic microorganisms have a great potential as CB or WCB based on genetic constructs involving photosynthesis genes. Likewise, ciliated protozoa also present a great potential to design both WCB or CB (using isolated molecular metal bioreceptors like metallothioneins).

(3) Microorganisms with cell wall (bacteria, fungi or microalgae) present a considerable disadvantage with regard to protozoa, because the presence of the cell wall could hinder the permeability of the pollutant or hold it by extracellular biosorption. Furthermore, using substrate-dependent reporters, the substrate must cross the cell wall to reach the cytoplasm, where the enzymatic reaction takes place. Therefore, sometimes a pre-treatment to increase cell permeabilization is necessary.

(4) The capacity for sensitivity of a metal biosensor is more important than its level of specificity to a metal. Because, in the real world the anthropogenic environmental metal pollution is generally by several metal(loid)s.

(5) At present, many CB and WCB using microorganisms have been designed to detect metal(loid)s in laboratory experiments, but, in general, very few of them have been validated using bioassays with real environmental samples. After rigorous bioassays using real metal polluted aquatic or terrestrial environmental samples, a lot of the biosensors reported to be specific to only one metal(loid) could be finally considered as non-specific. This is due to the presence of other unknown inorganic or organic components that can interact with the bioreceptor of the biosensor disturbing the response. This point is really important to build useful biosensors to detect metals in environmental samples.

(6) The future development of microbial WCB for environmental metal pollution monitoring could be considerably furthered by applying a synthetic biology approach. This would facilitate the design of WCB with multi-input systems based on two or more regulatory gene promoters in the same genetic construct, thereby increasing the capacity of the biosensor for detecting simultaneously several different metal(loid)s in the same polluted sample.

References

Aksmann A, Pokora W, Bascik-Remisiewicz A et al (2014) Time-dependent changes in antioxidative enzyme expression and photosynthetic activity of *Chlamydomonas reinhardtii* cells under acute exposure to cadmium and anthracene. Ecotoxicol Environ Saf 110C:31–40. doi:10.1016/j.ecoenv.2014.08.005

Alpat S, Alpat SK, Cadirci BH et al (2008) A novel microbial biosensor based on *Circinella* sp. modified carbon paste electrode and its voltammetric application. Sens Actuators 134:175–181. doi:10.1016/j.snb.2008.04.044

Amaro F, Turkewitz AP, Martin-Gonzalez A et al (2011) Whole-cell biosensors for detection of heavy metal ions in environmental samples based on metallothionein promoters from *Tetrahymena thermophila*. Microb Biotechnol 4:513–522. doi:10.1111/j.1751-7915.2011.00252.x

Amaro F, Turkewitz AP, Martin-Gonzalez A et al (2014) Functional GFP-metallothionein fusión protein from *Tetrahymena thermophila*: a potential whole-cell biosensor for monitoring heavy metal pollution and a cell model to study metallothionein overproduction effects. Biometals 27:195–205. doi:10.1007/s10534-014-9704-0

Aury JM, Jaillon O, Duret L et al (2006) Global trends of whole-genome duplications revealed by the ciliate *Paramecium tetraurelia*. Nature 444:171–178. doi:10.1038/nature05230

Baronian KHR (2004) The use of yeast and moulds as sensing elements in biosensors. Biosens Bioelectron 19:953–962. doi:10.1016/j.bios.2003.09.010

Belkin S (2003) Microbial whole-cell sensing systems of environmental pollutants. Curr Opin Microbiol 6:206–212

Berezhetskyy AL, Sosovska OF, Durrieu C et al (2008) Alkaline phosphatase conductometric biosensor for heavy-metal ions determination. IRBM 29:136–140

Bontidean I, Ahlqvist J, Chen W et al (2003) Novel synthetic phytochelatin-based capacitive biosensor for heavy metal ion detection. Biosens Bioeletron 18:547–553

Chay TC, Surif S, Heng LY (2005) A copper toxicity biosensor using immobilized cyanobacteria, *Anabaena torulosa*. Sens Lett 3:49–54

Chouteau C, Dzyadevych S, Chovelon JM et al (2004) Development of novel conductometric biosensors based on immobilised whole cell *Chlorella vulgaris* microalgae. Biosens Bioelectron 19:1089–1096. doi:10.1016/j.bios.2003.10.012

Clark LC, Wolf R, Granger D et al (1953) Continuous recording of blood oxygen tensions by polarography. J Appl Physiol 6:189–193

Corbisier P, Ji G, Nuyts G et al (1993) luxAB gene fusions with the arsenic and cadmium resistance operons of *Staphylococcus aureus* plasmid pI258. FEMS Microbiol Lett 110:231–238

Corbisier P, van der Lelie D, Borremans B et al (1999) Whole cell-and protein-based biosensors for the detection of bioavailable heavy metals in environmental samples. Anal Chim Acta 387:235–244

De Francisco P, Melgar LM, Diaz S et al (2016) The *Tetrahymena* metallothionein gene family: twenty-one new cDNAs, molecular characterization, phylogenetic study and comparative analysis of the gene expression under different abiotic stressors. BMC Genom 17:346. doi:10.1186/s12864-016-2658-6

De Schamphelaere KAC, Nys C, Janssen CR (2014) Toxicity of lead (Pb) to freshwater green algae: development and validation of a bioavailability model and inter-species sensitivity comparison. Aquat Toxicol 155:348–359. doi:10.1016/j.aquatox.2014.07.008

Diaz S, Amaro F, Rico D et al (2007) *Tetrahymena* metallothioneins fall into two subfamilies. PLoS ONE 2:e291. doi:10.1371/journal.pone.0000291

Eisen JA, Coyne RS, Wu M et al (2006) Macronuclear genome sequence of the ciliate Tetrahymena thermophile, a model eukaryote. PLoS Biol 4:e286. doi:10.1371/journal.pbio.0040286

Elad T, Benovich E, Magrisso S et al (2008) Toxicant identification by a luminescent bacterial bioreporter panel: application of pattern classification algorithms. Environ Sci Technol 42:8486–8491

Erbe JL, Adams AC, Taylor KB et al (1996) Cyanobacteria carrying an smt-lux transcriptional fusion as biosensors for the detection of heavy metal cations. J Ind Microbiol 17:80–83

Fairbrother A, Wenstel R, Sappington K et al (2007) Framework for metals risk assessment. Ecotoxicol Environ Saf 68:145–227

Gammoudi I, Tarbague H, Othmane A et al (2010) Love-wave bacteria-based sensor for the detection of heavy metal toxicity in liquid medium. Biosens Bioelectron 26:1723–1726

Golding GR, Kelly CA, Sparling R et al (2002) Evidence for facilitated update of Hg(II) by *Vibrio anguillarum* and *Escherichia coli* under anaerobic and aerobic conditions. Limnol Oceanogr 47:967–975

Gu MB, Mitchell RJ, Kim BC (2004) Whole-cell-based biosensors for environmental biomonitoring and application. Adv Biochem Eng Biotechnol 87:269–305. doi:10.1007/b13533

Guascito MR, Malitesta C, Mazzotta E et al (2008) Inhibitive determination of metal ions by an amperometric glucose oxidase biosensor: study of the effect of hydrogen peroxide decomposition. Sens Actuators B Chem 131:394–402

Gutierrez JC, Martin-Gonzalez A, Diaz S et al (2003) Ciliate as potential source of cellular and molecular biomarker/biosensors for heavy metal pollution. Eur J Protistol 39:461–467. doi:10.1078/0932-4739-00021

Gutierrez JC, Martin-Gonzalez A, Diaz S et al (2008) Ciliates as cellular tools to study the eukaryotic cell-heavy metal interactions. In: Brown SE, Welton WC (eds) Heavy metal pollution. Nova Science Publishers, New York, pp 1–44

Gutierrez JC, Amaro F, Diaz S et al (2011) Ciliate metallothioneins: unique microbial eukaryotic heavy-metal-binder molecules. J Biol Inorg Chem 16:1025–1034. doi:10.1007/s00775-011-0820-9

Gutierrez JC, Amaro F, Martin-Gonzalez A (2015) Heavy metal whole-cell biosensors using eukaryotic microorganisms: an updated critical review. Front Microbiol 6:1–8. doi:10.3389/fmicb.2015.00048

Hakkila K, Green T, Leskinen P et al (2004) Detection of bioavailable heavy metals in EILATox-Oregon samples using whole-cell luminescent bacterial sensors in suspension or immobilized onto fibreoptic tips. J Appl Toxicol 24:333–342

Hill MK (2004) Understanding environmental pollution. A primer. Cambridge University Press, Cambridge

Ilangovan R, Daniel D, Krastanov A et al (2006) Enzyme based biosensor for heavy metal ions determination. Biotechnol Equip 20:184–189

Ivask A, Hakkila K, Virta M (2001) Detection of organomercurials with sensor bacteria. Anal Chem 73:5168–5171

Ivask A, Virta M, Kahru A (2002) Construction and use of specific luminescent recombinant bacterial sensors for the assessment of bioavailable fraction of cadmium, zinc, mercury and chromium in the soil. Soil Biol Biochem 34:1439–1447

Ivask A, Rolova T, Kahru A (2009) A suite of recombinant luminescent bacterial strains for the quantification of bioavailable heavy metals and toxicity testing. BMC Biotechnol 9:41. doi:10.1186/1472-6750-9-41

Ji G, Silver S (1992) regulation and expression of the arsenic resistance operon from *Staphylococcus aureus* plasmid pI258. J Bacteriol 174:3684–3694

Jouanneau S, Durand M-J, Courcoux P et al (2011) Improvement of the identification of four heavy metals in environmental samples by using predictive decision tree models coupled with a set of five bioluminescent bacteria. Environ Sci Technol 45:2925–2931. doi:10.1021/es1031757

Kröger S, Law RJ (2005) Biosensors for marine applications. We all need the sea, but does the sea need biosensors? Biosens Bioelectron 20:1903–1913. doi:10.1016/j.bios.2004.08.036

Lagido C, Pettit J, Porter A et al (2001) Development and application of bioluminescent *Caenorhabditis elegans* as multicellular eukaryotic biosensors. FEBS Lett 493:36–39. doi:10.1016/S0014-5793(01)02271-2

Lambreva MD, Giardi MT, Rambaldi I et al (2013) A powerful molecular engineering tool provided efficient Chlamydomonas mutants as bio-sensing elements for herbicides detection. PLoS ONE 8:e61851. doi:10.1371/journal.pone.0061851

Lee S, Sode K, Nakanishi K et al (1992) A novel microbial sensor using luminous bacteria. Biosens Bioelectron 7:273–277

Lehmann M, Riedel K, Adler K et al (2000) Amperometric measurement of copper ions with a deputy substrate using a novel *Saccharomyces cerevisiae* sensor. Biosens Bioelectron 15:211–219

Leth S, Maltoni S, Simkus R et al (2002) Engineered bacteria based biosensors for monitoring bioavailable heavy metals. Electroanalysis 14:35–42. doi:10.1002/1521-4109(200201)14:1<35:AID-ELAN35>3.0.CO;2-W

Liao VH, Ou KL (2005) Development and testing of a green fluorescent protein-based bacterial biosensor for measuring bioavailable arsenic in contaminated groundwater samples. Environ Toxicol Chem 24:1624–1631

Liao VH, Chien MT, Tseng YY et al (2006) Development and testing of a Green fluorescent protein based bacterial biosensor for measuring bioavailable arsenic in contaminated groundwater saples. Environ Pollut 142:17–23

Magrisso S, Erel Y, Belkin S (2008) Microbial reporters of metal bioavailability. Microb Biotechnol 1:320–330. doi:10.1111/j.1751-7915.2008.00022.x

Martin-Gonzalez A, Diaz S, Jareño C et al (1999) The use of protists in ecotoxicology. Recent Res Dev Microbiol 3:93–111

Mehta J, Bhardwaj SK, Bhardwaj N et al (2016) Progress in the biosensing techniques for trace-level heavy metals. Biotechnol Adv 34:47–60. doi:10.1016/j.biotechadv.2015.12.0010734-9750

Mergeay M, Monchy S, Vallaeys T et al (2003) Ralstonia metallidurans, a bacterium specifically adapted to toxic metals: towards a catalogue of metal-responsive genes. FEMS Microbiol Rev 27:385–410. doi:10.1016/S0168-6445(03)00045-7

Nguyen-Ngoc H, Durrieu C, Tran-Minh C (2009) Synchronous-scan fluorescence of algal cells for toxicology assessment of heavy metals and herbicides. Ecotoxicol Environ Saf 72:316–320. doi:10.1016/j.ecoenv.2008.04.016

Oh SE, Hassan SH, Van Ginkel SW (2011) A novel biosensor for detecting toxicity in water using sulfur-oxidizing bacteria. Sens Actuators B Chem 154:17–21

Park JN, Shon MJ, Oh DB et al (2007) Identification of the cadmium-inducible *Hansenula polymorpha* SEO1 gene promoter by transcriptome analysis and its application to whole-cell heavy metal detection systems. Appl Environ Microbiol 73:5990–6000. doi:10.1128/AEM.00863-07

Paton GI, Palmer G, Burton M et al (1997) Development of an acute chronic ecotoxicology assay using *lux*-marked *Rhizobium leguminosarum* biovar *trifolii*. Lett App Microbiol 24:296–300

Peca L, Kos PB, Mate Z et al (2008) Construction of bioluminescent cyanobacterial reporter strains for detection of nickel, cobalt and zinc. FEMS Microbiol Lett 289:258–264

Peña-Vázquez E, Perez-Conde C, Costas E et al (2010) Development of a microalgal PAM test method for Cu(II) in waters: comparison of using spectrofluorometry. Ecotoxicology 19:1059–1065. doi:10.1007/s10646-010-0487-y

Peñuelas J, Fillela I (2002) Metal pollution in Spanish terrestrial ecosystems during the twentieth century. Chemosphere 46:501–505

Petänen T, Virta M, Karp M (2001) Construction and use of broad host range mercury and arsenite sensor plasmids in the soil bacterium *Pseudomonas fluorescens* OS8. Microb Ecol 41:360–368

Preston S, Coad N, Townend J et al (2000) Biosensing the acute toxicity of metal interactions: are they additive, synergistic or antagonistic? Environ Toxicol Chem 19:775–780

Roda A, Roda B, Cevenin IL et al (2011) Analytical strategies for improving the robustness and reproducibility of bioluminescent microbial bioreporters. Anal Bioanal Chem 401:201–211. doi:10.1007/s00216-011-5091-3

Riether KB, Dollard MA, Billard P (2001) Assessment of heavy metal bioavailability using *Escherichia coli* zntAp:lux and copAp:lux-based biosensors. Appl Microbiol Biotechnol 57:712–716

Samphao A, Rerkchai H, Jitcharoen J et al (2012) Indirect determination of mercury by inhibition of glucose oxidase immobilizated on carbon paste electrode. Int J Electrochem 7:1001–1010

Selifonova O, Burlage R, Barkay T (1993) Bioluminescent sensors for detection of bioavailable Hg(II) in the environment. Appl Environ Microbiol 59:3083–3090

Souiri M, Gammoudi I, Mora L et al (2012) A novel 3-D nano-assembly bacteria based biosensor for enhanced detection of heavy metal pollutants. J Environ Sci Eng A 1:924–935

Shetty RS, Deao SK, Shah P et al (2003) Luminescen-based whole cell sensing systems for cadmium and lead using genetically engineered bacteria. Anal Bioanal Chem 376:11–17

Shetty RS, Deo SK, Liu Y et al (2004) Fluorescence-based sensing system for copper using genetically engineered living yeast cells. Biotechnol Bioeng 88:664–670

Singh J, Mittal SK (2012) *Chlorella* sp. based biosensor for selective determination of mercury in presence of silver ions. Sens Actuators B Chem 165:48–52

Shing WL, Surif S, Heng LY (2008) Toxicity biosensor for the evaluation of cadmium toxicity based on photosynthetic behaviour of cyanobacteria *Anabaena torulosa*. Asian J Biochem 3:162–168

Shitanda I, Takada K, Sakai Y et al (2005) Amperometric biosensing systems based on motility and gravitaxis of flagellate algae for aquatic risk assessment. Anal Chem 77:6715–6718. doi:10.1021/ac050894b

Sperling L, Dessen P, Zagulski M et al (2002) Random sequencing of paramecium somatic DNA. Eukaryot Cell 1:341–352. doi:10.1128/EC.1.3.341-352.2002

Tag K, Riedel K, Bauer HJ et al (2007) Amperometric detection of Cu^{2+} by yeast biosensors using flow injection analysis (FIA). Sens Act B Chem 122:403–409

Tauriainen S, Karp M, Chang W et al (1998) Luminescent bacterial sensor for cadmium and lead. Biosens Bioelectron 13:931–938

Tauriainen S, Virta M, Chang E et al (1999) Measurement of firefly luciferase reporter gene activity from cells and lysates using *Escherichia coli* arsenite and mercury sensors. Anal Biochem 272:191–198

Terziyska A, Waltschewa L, Venkov P (2000) A new sensitive test based on yeast cells for studying environmental pollution. Environ Pollut 109:43–52. doi:10.1016/S0269-749(99)00237-7

Tibazarwa C, Wuertz S, Mergeay M et al (2000) Regulation of the *cnr* cobalt and nickel resistance determinant of *Ralstonia eutropha* (*Alcaligenes eutrophus*) CH34. J Bacteriol 182:1399–1409

Tibazarwa C, Corbisier P, Mench M et al (2001) A microbial biosensor to predict bioavailable nickel in soil and its transfer to plants. Environ Pollut 113:19–26

Tom-Petersen A, Hosbond C, Nybroe O (2001) Identification of copper-induced genes in *Pseudomonas fluorescens* and use of a reporter strain to monitor bioavailable copper in soil. FEMS Microbiol Ecol 38:59–65. doi:10.1111/j.1574-6941.2001.tb00882.x

Van der Meer JR, Belkin S (2010) Where microbiology meets microengineering: design and applications of reporter bacteria. Nat Rev Microbiol 8:511–522

Van der Meer JR, Tropel D, Jasper M (2004) Iluminating the detection chain of bacterial bioreporters. Environ Microbiol 6:1005–1020. doi:10.1111/j.1462-2920.2004.00655.x

Vel Krawczyk TK, Moszczynska M, Trojanowicz M (2000) Inhibitive determination of mercury and other metal ions by potentiometric urea biosensor. Biosens Bioelectron 15:681–691

Verma N, Singh M (2005) Biosensors for heavy metals. Biometals 18:121–129. doi:10.1007/s.10534-004-5787-3

Verma N, Singh M (2006) A *Bacillus sphaericus* based biosensor for monitoring nickel ions in industrial effluents and foods. J Anal Methods Chem 83427:1–4. doi:10.1155/JAMMC/2006/83427

Virta M, Lampinen J, Karp M (1995) A luminescence based mercury biosensor. Anal Chem 67:667–669

Walmsley RM, Keenan P (2000) The eukaryotic alternative: advantages of using yeasts in place of bacteria in microbial biosensor development. Biotechnol Bioprocess Eng 5:387–394. doi:10.1007/BF02931936

Wong L S, Lee Y H, Surif S (2013) Whole cell biosensor using *Anabaena torulosa* with optical transduction for environmental toxicity evaluation. J Sens 567272. doi:10.1155/2013/567272

Yagi K (2007) Applications of whole-cell bacterial sensors in biotechnology and environmental science. Appl Microbiol Biotechnol 73:1251–1258. doi:10.1007/s00253-006-0718-6

Yoon KP, Misra TK, Silver S (1991) Regulation of the *cadA* cadmium resistance determinant of *Staphylococcus aureus* plasmid pI258. J Bacteriol 173:7643–7649

Yüce M, Nazir H, Dönmez G (2010a) An advanced investigation on a new algal sensor determining Pb(II) ions from aqueous media. Biosens Bioelecron 26:321–326

Yüce M, Nazir H, Dönmez G (2010b) A voltammetric *Rhodotorula mucilaginosa* modified microbial biosensor for Cu(II) determination. Bioelectrochemistry 79:66–70

Zlatev R, Magnin JP, Ozil P et al (2006) Bacterial sensors based on *Acidithiobacillus ferrooxidans*: part II. Cr (VI) determination. Biosens Bioelectron 21:1501–1506

Part III
Perspectives

Chapter 14
Microbial Ecotoxicology: Looking to the Future

Sergi Sabater

Abstract Chemicals pose a certain risk to microbial communities, but effects may be accumulative or synergistic with other stressors also affecting microbes in freshwater ecosystems. Precisely, a main challenge faced by microbial ecotoxicologists is to discern the effects of chemicals from those caused by co-occurring stressors on microbial communities, and this requires the ability to use and integrate new and not so new techniques to accurately describe their responses and scaling them up to the ecosystem.

Keywords Biofilm · Contaminant · Microbial community · Species · Modeling · Omics · Mesocosm · Environmental conditions

14.1 Introduction

A possible manner of prospecting the challenges in a discipline is by understanding its roots. The retrospective can set the basis from where to speculate on future developments.

The first part of the term "microbial ecotoxicology" sets the interest on the microbes as a large fraction of the ecosystem's biota. Microbes not only are affected by contaminants but also perform essential functions in freshwater ecosystems. The variety of functions they perform include transducing solar energy into organic matter, transform, decompose and mineralize organic matter, and contribute to the retention and transformation of materials (including pollutants) in the freshwater ecosystem. In itself, the "microbial" approach envisaged in *microbial ecotoxicology* should not sustain a long distance from other disciplines using the "microbial" perspective. All these share the common interest for microorganisms, and in the

S. Sabater (✉)
Catalan Institute for Water Research (ICRA), GRECO-IEA,
University of Girona, Girona, Spain
e-mail: ssabater@icra.cat

C. Cravo-Laureau et al. (eds.), *Microbial Ecotoxicology*,
DOI 10.1007/978-3-319-61795-4_14

case of the microbial ecotoxicology by focusing on the effects and responses of microbes and the processes they catalyze to pollutants.

The second part of the term "microbial ecotoxicology" is in its turn a composite of two different terms, corresponding to the disciplines "ecology" and "toxicology". In the past, these two were separated and evolved separately, mutually ignoring each other. Ecology aimed at understanding the reasons why natural systems function, and why there was an order in the distribution of species, populations and communities. Toxicology has specially focused on understanding the adverse effects of chemicals, strongly determined to quantify the dose (concentrations) of chemicals respective to their effects on living organisms. The application of ecological concepts to the pollutants response on the biological communities has been rare, though with important exceptions. The theoretical framework of pollution—induced community tolerance (PICT) described that a toxicant could exert a selection pressure towards a more tolerant community, when applied at a sufficient concentration for an adequate duration (Blanck et al. 1988). The biological community therefore experienced a progressive compositional or physiological shift in the overall tolerance to the toxicant, leading to a causal explanation of long-term effects of chemicals on the community. This concept also recognized that tolerance to one or several stressors was modified by external (environmental) factors, and that cross-effects between toxicants and other environmental stressors might occur. This breakthrough concept opened the perspective towards a rising awareness of the necessity to incorporate the ecological perspective into the toxicological machinery. This is already the perspective of modern microbial ecotoxicology. In the way stated by Ghiglione et al. (2016), microbial ecotoxicology is related to taxonomic and functional microbial biodiversity that in its turn is important to support ecosystem functions and ensure their stability and recovery.

Considering the impact of chemicals on freshwater ecosystems requires an understanding of their origin in a myriad of environmental perturbations associated to land uses (e.g., urbanization, agriculture) and hydrological modifications (e.g., dams, dikes, and channelization). Approximately 300 million tons of synthetic compounds are used annually in industrial and consumer products, some reaching freshwater ecosystems via wastewater effluents and diffuse sources (Schwarzenbach et al. 2006). Many of these chemicals pose a risk to microbial communities, and effects in some cases may be accumulative or synergistic. Still, chemicals are only one of the many types of stressors that exist and affect microbes in freshwater ecosystems (Segner et al. 2014). It is clear, therefore, that the challenge faced by microbial ecotoxicologists is to discern the effects of chemicals from those caused by co-occurring stressors, altogether affecting the structure (biodiversity) and functioning (Burton et al. 2012; Segner et al. 2014) of microbial communities. Such a challenge can be decomposed in: (i) the recognition of the role of microbial communities, and not only that of isolated species, as receptors of pollutants, (ii) the ability to use and integrate new and not so new techniques to accurately describe the responses, and (iii) the ability to implement these in the context of occurrence of multiple stressors.

14.2 The Challenges

The previously enumerated challenges are critically appraised in this section, providing their state-of-the-art as well as some of the many gaps they still exhibit.

A. ***Considering multiple stressors and not only isolated impacts of toxicants on microbial communities.***

Persistent chemical pollution has been seen as a prevalent driver in freshwater ecosystems (Malaj et al. 2014), particularly in systems highly impacted by human activity. Organic microcontaminants and heavy metals constitute complex mixtures characteristic of specific land uses, i.e., extensive agriculture, industrial activities, or human conurbations (Posthuma et al. 2008). The composition and concentration of these mixtures varies between periods (Petrovic et al. 2011), and is highly responsive of the dilution capacity of the system (Barceló and Sabater 2010). These pollutants co-occur with nutrients in excess, or with abundant dissolved organic matter, especially in systems heavily impacted by industrial or urban effluents (Hatt et al. 2004). These stressors have higher or lower associated energy, and might cause a more intense effect, in relation to their frequency of occurrence (Fig. 14.1). Chemicals are amongst those showing lower effective energy if we compare them with the one associated with nutrients, biological interactions or hydrogeomorphological alterations (Fig. 14.1). As a derivative of this reality, we should expect that effects of chemicals would be the least, though by no means irrelevant to the fate of microbial communities. In many impaired systems, a high co-occurring number of stressors with varying modes of action (MoA) exert complex effects on biological communities (Segner et al. 2014), and chemicals are part of this.

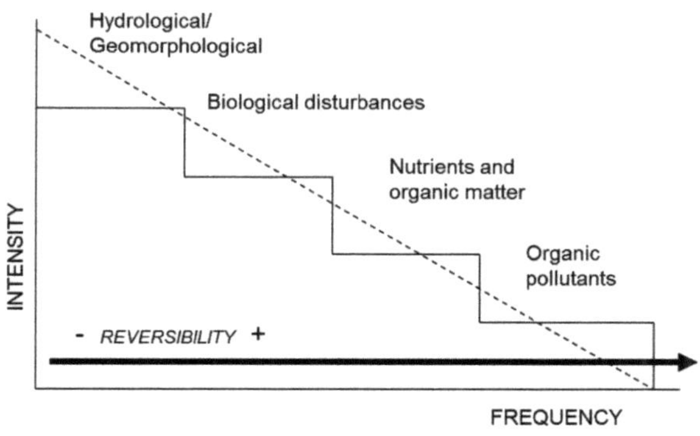

Fig. 14.1 Relationship between intensities and frequencies in the occurrence of stressors in the environment, and the expected responses on the biota. The ones with higher energy are the least frequent, and vice versa. When the associated intensity becomes lower, the ability of the organisms to cope with the stressor is higher, and reversibility increases. The effect of pollutants is amongst those with lower energy, but effects may be complicated because of their chronic influence

Recognizing the relevance of multiple stressors is therefore acknowledging the reality of environmental conditions occurring in impaired systems. Organisms are net receivers of influences at multiple spatial and temporal scales, and their ultimate response defines the carrying capacity of a system (Posthuma et al. 2014). Appreciating the relevance of multiple stressor occurrences is not new, but the approach to solve their complexity requires of refreshed views. Early in the history of the use of indicator organisms, it was already accepted that biological communities were modulated by dissolved organic matter (evaluated by means of the DOC or TOC), as well as by nutrients in excess, or by heavy metal pollution (Margalef 1960; Goodnight 1956). We should therefore not only recognize the occurrence and complex responses of biological communities, but also work on the attribution of potential effects of stressors on the microbial communities, with their direct and indirect effects, as well as on the real consequences for freshwater ecosystems.

Evaluating potential patterns of response of biological communities to multiple stressors requires of a combination of field-based evidences and laboratory experiments. Observations and experiments should be tailored to encompass the potential effect of stressors with different exergies (Fig. 14.2). Higher associated energy or increasing complexity of multiple stressors might produce rising effects, and shifts between transitional states (Scheffer et al. 2012). It is true, however, that the potential effects of stressors can be somehow compensated by increasing biological complexity (Fig. 14.2). Biodiversity and ecosystem function are hand-by-hand with each other (Hooper et al. 2005). It has been shown that the number of species required to sustain ecosystem functions increases with the number of processes considered (Hector and Bagchi 2007). This so-called *functional redundancy* indicates that several species perform the same function, but also that a given species may perform more than one function at the same time. Microbial communities may

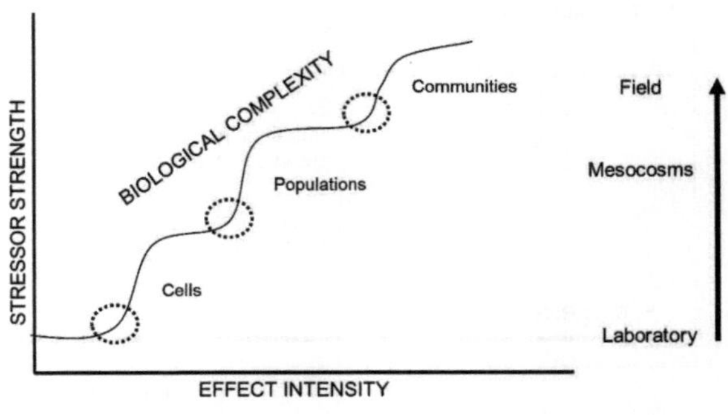

Fig. 14.2 Direct relationship exists between the stressor 'strength (or intensity) and the corresponding effect in microbial organisms. Each scale of organization (cells, populations, communities) is affected by a given stressor or combination of stressors until a threshold, and this threshold increases with biological complexity. Hence, communities have mechanisms that allow withstanding stressors of higher strength than those able to break the resistance of populations, and so forth. The scales of observation (*right hand arrow*) need to match to biological responses in order to reliably detect the stressors 'effects

have high functional redundancy because of their high diversity, high dispersal capacity, physiological versatility and horizontal gene transfer (Peter et al. 2011), making them highly adaptable to occurring stressors.

Experiments, in the laboratory (e.g. using micro- or mesocosms), as well as manipulative in the field (e.g. translocation), provide the necessary testing of responses to specific stressors, though at different scales and complexity. Extensive field studies allow characterizing the patterns of physical, chemical and biological variables in a set of sites, and therefore define the distribution patterns of organisms according to the driving pressures occurring there. Field-based evidence and laboratory experiments need to be seen as complementary, and optimally, need to be performed conjointly. The two have recognized strengths and weaknesses.

Implications derived from field studies are correlational, but may have sufficient statistical power to indicate emerging patterns on the structure of ecological data (Legendre and Legendre 1998). Laboratory experiments might shed light on the relevance of the factors tested (Hall et al. 1999; Sabater et al. 2002), but are simplifications of the reality since they replace the multiplicity of factors operating at the same time with a few. While these experiments have higher statistical power, their results cannot be directly transferred to the environmental reality. Field and laboratory-based studies combine different scales of space and time, and when used in combination may solve complex questions of multiple stressor effects.

Using mesocosms (artificial streams) and ANOVA design, Corcoll et al. (2015) determined the effects of pharmaceutical compounds on biofilm communities when submitted to flow intermittency. The experiment showed that moderate concentrations of pharmaceuticals affected both the algal biomass and bacterial taxa richness, and that effects of flow intermittency were analogous but comparatively much higher. The combined effect of water flow interruption and pharmaceutical inputs affected more negatively the algae, while bacteria were more resistant. This experiment therefore qualified the respective contribution and interaction strength of flow interruption and pharmaceutical occurrence on stream biofilms, in a way that was not achievable in the field, but also with the uncertainty associated to the inherent simplification. The use of mesocosms can be therefore understood as an intermediate step between pure laboratory experiments and field observations. As said, findings do not imply that results can be transposed to real, environmental situations, but indeed open a better understanding on the underpinning mechanisms.

In one field-based study, Ponsatí et al. (2016) observed that relatively high concentrations (reaching a total sum of micrograms) of organic microcontaminants co-occurred with high DIN and DOC concentrations (at the milligram level) in polluted sites. The conjoint presence of pharmaceutical compounds, industrial organic compounds and herbicides, but also DOC and DIN, affected biofilms (bacteria and algae) in situations of high and low flows. Defining which group of stressors was the most relevant was approached by means of variance partitioning analysis. This analysis showed that the organic compounds were significantly but poorly related to the observed biofilm responses, and that the interaction between organic pollutants and nutrients had a higher share of the variance. This study also showed that chemical pollution and hydrological patterns interacted with one another. The type of pollutants reaching the biofilm and the different composition and biomass of biofilms differed during high flow and low flow conditions. During high water flow, biofilms were thinner, more active and potentially more reactive to chemical influences, and the reverse occurred during low flow conditions.

Assembled field-derived evidence show that several sorts of variables affect the biota in impaired rivers, often coinciding in space and time, and following a rather general mechanism: the most sensitive species are affected, even becoming locally extinct, while other tolerant species are favored (Blanck and Wängberg 1988). The most affected organisms were those jointly exposed to the presence of micropollutants and other stressors (Allan et al. 2013; Coors and De Meester 2008). The overall decrease in diversity of biological communities (including microbes) has been associated with increasing nutrient concentrations (Johnson and Hering 2009) as well as to the continuous inputs of pharmaceutical products and other contaminants (Burkart and Kolpin 1993; Allan 2004).

The use of manipulative experiments in the field might be seen as complementary to smaller-scale experiments in the laboratory. Proia et al. (2013) used translocation experiments between sites with increasing concentrations of nutrients, organic matter, and organic microcontaminants. The study observed that biofilms translocated from lower to higher pollution sites were those showing most evident changes. The translocated community quickly reacted to the mixture of changes associated to increasing in conductivity and rising concentrations of analgesics and barbiturics.

The use of these different approaches shows that environmental stressors may reinforce the effect of organic micropollutants, or vice versa (Segner et al. 2014). Stressors occurring at multiple spatial and temporal scales define a so-called "stressor space" where the net receivers are the biological communities, and where synergies could produce much higher effects than the ones attributed solely to organic microcontaminants or to inorganic nutrients. Whatever the causes, it is obvious that the simultaneous occurrence of multiple stressors challenges the carrying capacity of ecosystems (Posthuma et al. 2014) by affecting their biodiversity and basic functions. Understanding the real risks of biological communities requires quantifying the effects of multiple stressors.

We can derive from the above that organic microcontaminants cannot be contemplated aside from other pollutants such as nutrients, or TOC, which may interfere with the potential toxicity of micropollutants in freshwaters. Hydrology, particularly the dilution capacity of a system, is also highly relevant to understand the interaction between contaminants and microbial communities. Some recent evidences indicate that the conjoint effect of land-use descriptors, hydrological irregularity and TOC and nutrients may be much higher than the correlational effects of organic microcontaminants on biofilms and invertebrates (Sabater et al. 2016). Still, many knowledge gaps exist in the definition of the responsibility of micropollutants in the composition and activity of microbial biota in freshwaters. The relationship of toxic effects of contaminants with the nutrient condition of microbial communities is rather unknown. Whether the physical stress associated to excess light, hydraulics, may buffer or increase the effects of pollutants may also be a matter of research. Finally, the occurrence of new challenges such as the effect of antibiotic resistance genes, or the response to disinfection-by-products, once reaching the freshwaters, deserve attentive research in an environmental-based perspective.

B. ***Considering microbial communities as an objective.***

One of the main challenges faced by microbial ecotoxicology is accounting for the complexity of microbial communities. Species have different roles and functions, and the complexity in a community does not match the simple accumulation of species. Autotrophs and heterotrophs co-exist in a highly complex entity, their roles sometimes being co-operative and others being competitive, altogether showing that species play different roles and may respond differently to stressors.

Community ecotoxicology is a shift of paradigm compared to the classical approach of using species tests for chemicals. While population ecotoxicology is mainly based on laboratory assays of single species, community ecotoxicology uses the summary of responses of each of the species making part of the community, and their positive and negative feed-backs (Fig. 14.3). The reproducibility of the response may therefore decrease if we compare it with that obtained when using monospecies tests, but on the other hand realism increases; increasing realism is at the expenses of reliability (Sabater et al. 2007).

Microbial communities occurring in freshwaters are organized with autotrophs (basically algae and cyanobacteria) and heterotrophs (bacteria, fungi), and the occurrence of an EPS matrix for those associated to a substratum. In the case of algae, chemical toxicity depends on the mode of action of each toxicant and the time and concentration of exposure, and in many cases on the sensitivity of organisms. Diatoms and green algae have different sensitivities to copper; the pedunculated-forming diatom *Gomphonema gracile* was enhanced by continued exposure to copper, while the chain-forming diatom *Fragilaria capucina* was

Fig. 14.3 Cell or population ecotoxicology is mainly based on laboratory assays of single species, while community ecotoxicology encompasses the responses of each of the components together with their positive and negative feed-backs (Fig. 14.3). Reliability of the response may therefore decrease from the cell to the community, but realism increases; realism may be increased at the expenses of the reliability and reproducibility of the assay

inhibited, and filaments of the green alga *Mougeotia* showed malformations (Guasch et al. 2002). Organic herbicides such as diuron, atrazine, prometryn and/or isoproturon have relevant effects on algal photosynthesis (e.g. Ricart et al. 2009). Herbicides can also affect nutrient (NO_3, NO_2, and Si) uptake by algae (Debenest et al. 2010). Some heavy metals such as copper or mercury directly act on the photosynthetic process (Juneau et al. 2001), while others such as arsenic affect their nutrient uptake, particularly phosphorus (Anderson and Bruland 1991; Rodriguez Castro et al. 2015).

Bacteria are more sensitive to pharmaceutical products reaching the environment, especially antibiotics. The persistent release of large amounts of these products might alter the population dynamics of microorganisms, including selection of resistance (Martínez 2008). Still, sensitivity to antibiotics differs between species. It has been observed that different bacterial strains may be selected as a function of their resistance to antibiotics and the co-occurrence of other environmental stressors (Tlili et al. 2010). The effects may not only concern the established bacterial communities, but might represent a selective pressure during the first phases of biofilm development, when bacteria are early colonizers and cover the mineral surfaces with their polysaccharide glycocalix (Bärlocher and Murdoch 1989).

Pollutants directly affect algal or bacterial metabolism with a direct MoA, but may also produce effects beyond their primary target. The bactericide triclosan (Proia et al. 2013), pharmaceuticals such as β-blockers (Bonnineau et al. 2010) or antibiotics (Liu et al. 2011) affect algal photosynthesis. Diuron effects on bacterial communities can be a result of indirect effects of the herbicide on heterotrophs (Ricart et al. 2009), but pollutants may also be used as a carbon and energy source by some bacteria (Dantas et al. 2008). Microbial communities face complex trophic and metabolic interactions (Freeman and Lock 1995; Rier el al. 2002), where both direct and indirect effects may occur. Interactions do not necessarily result in additive or synergistic effects, but they can also buffer stressor impacts promoting genetic, phenotypic or community composition adjustments (Segner et al. 2014). Stressors may cause the rearrangement of the species composition (or of its overall physiological functioning) by replacing stressor-sensitive species by stressor-tolerant others (Blanck et al. 1988). When this complexity is transferred to the scenario's implications for many co-occurring chemicals in polluted systems, we might face a myriad of direct and indirect sources of stress for bacteria and algae, with the perspective of a highly complex response.

The EPS matrix in itself also complicates the prediction of responses to chemicals. The EPS is a potential barrier for heavy metals and hydrophobic contaminants, and acts as a modifier of the response to toxicants. It acts as site for adsorption and can impede the diffusion and penetration of the toxicants. Admiraal et al. (1999) in a study of a polluted stream related the limited toxicity of zinc with the biofilm thickness and its EPS, a result stressing the role of EPS on the overall response of a biofilm to a toxic substance.

Communities differ in their composition and architecture at very small scales (Flemming and Wingender 2010), but heterogeneity is a characteristic of higher spatial scales in a fluvial or lacustrine habitat. The transition from the vial to the ecosystem requires incorporation of other sources of variation existing in a freshwater ecosystem, such as the different substrata as well as the associated values in light, water velocity and resources availability. These may differ spatially at the habitat spatial scale, and provide different opportunities for the autotrophs and heterotrophs, acting as a selection driver for their distribution in patches (Tornés and Sabater 2010). Including this heterogeneity is part of the necessary realism of the ecosystem but it also complicates standardization, even if such a shortcoming can partially be balanced by the use of artificial substrata (Kutka and Richards 1996).

All in all, the challenge to provide a standard account of the response of complex communities is far from being resolved and requires co-operative work involving different scientific skills. Microbiologists, ecologists, ecotoxicologists, and chemists need to work together to improve the common knowledge on the effects of multiple stressors on microbial communities. Though this is a long way off, reaching standardization and regulating its use is the best way to promote the extensive use of natural microbial communities as a test unit in ecotoxicology.

C. ***Using integrative techniques for generalization and scaling up***

From the cell to the community

Classical methods in microbial ecology have long been used and applied to ecotoxicology. These include microscopical techniques, and molecular techniques for the determination of community composition, as well as measurements of microbial respiration, photosynthesis, or decomposition (Sabater et al. 2007; Morin et al. 2010; Tlili et al. 2011; Lupini et al. 2011). The use of these techniques has allowed detailed accounts of single stressor or multiple stressor effects on microbial communities (e.g. Barthès et al. 2015; Proia et al. 2013; Pesce et al. 2011), by providing added evidences of microbial responses at multiple structural and functional scales.

The arrival of the—omics techniques has allowed analysis in unprecedented detail at subcellular and community scales. Genomic techniques have allowed sequencing, assembling and analysis of the function and structure of genomes (the complete set of DNA within a single cell or of a community). Genomics can provide detailed accounts (at the Phylum, order, family, genera or OTUs level) of the community composition. Transcriptomic techniques may identify genes differentially expressed as a response to different treatments. Gene identification and their relative expression or repression can be used as a tool to determine the early responses of organisms to stressors (Bier et al. 2015). Metabolomic techniques identify metabolite end-products of cellular processes in organisms or communities, and finally provide a metabolic profile that is a snapshot of their physiology. The use of aggregated metabolic fingerprints provides quantitative estimates which may result valuable for establishing adverse effects of chemicals (Riedl et al. 2015).

These latest advances in next-generation sequencing methods are revolutionizing the way microbial research and its relationship to stressors is conducted. The large number of new molecular tools is, no doubt, a significant advance, but this should not preclude the use of more established techniques which can still provide useful and complementary information at the population and community scale. Coupling the use of these techniques, and using them through progressively complex spatial scales, could be the key of success to providing reliable and environmentally relevant diagnoses. Such integration between these different sets of techniques may provide the necessary amount of evidences to understand the response of microbial communities to contaminants.

From the community to the ecosystem

Upscaling responses from the laboratory to the ecosystem requires the use of modelling techniques. These may be used to generalize and conceptualize the responses of microbial communities to pollutants and other stressors. However, bacterial and algal communities are highly complex and do show characteristic ways to respond to the occurring stressors, and this complicates modeling patterns and transferring results to higher spatial scales. Different groups of organisms have particular life cycles, habitat preferences, and environmental requirements. Therefore it is difficult to use a single model that may account for the algal and bacterial responses to stressors.

Empirical and mechanistic models involving microbial organisms have been used for a variety of purposes in many circumstances. Empirical models have been used to predict the biomass dynamics of algae and bacteria, as well as their relationships to environmental factors (e.g. Bird and Duarte 1984; Hardenbicker et al. 2015). This type of models also allowed the presence of pollutants to be related to the microbial biomass (Federle et al. 1986). Mechanistic models are based on pre-defined conservation principles, and the ability to provide detailed information. Copin et al. (2016) developed a model to predict the cell density inhibition of a pulse exposure of the alga *Scenedesmus vacuolatus* to herbicides. Graba et al. (2013) defined a mechanistic model of stream epilithic biofilm that combined the biomass dynamics, the autogenic detachment, and the estimated feeding activity of biofilm-dwelling invertebrates. These modelling exercises are useful examples on the path to follow, that is, the one incorporating multiple elements of real communities together with their environmental constraints, to understand the mechanisms governing microbial communities submitted to stress. Still, advancing on the coupling of empirical and mechanistic models could help to clearly establish how stressors affect microbial organisms. This combination was recently used to track dispersion of fecal organisms (Safaie et al. 2016). Empirical modelling in that study was used to quantify the relationships between environmental measurements and known concentrations of fecal organisms, while mechanistic modelling allowed defining simulated contaminant paths associated to the complex hydrologic patterns in the area. Enforcing cooperative modeling in such a way could result in improved models to better describe the complex dynamics of microbial organisms in response to multiple stressors.

14.3 Conclusion

Chemicals are part of the multiple stressors that pose a risk to microbial communities, with cumulative, antagonistic or synergistic effects. Microbial ecotoxicologists have to consider the role of microbial communities because of their central place in the ecosystem and the implications their response to stressors might have for essential ecosystem functions. Doing so, requires integrating techniques at multiple scales (subcellular, populations, communities), and combine laboratory techniques with field observations and modelling tools.

References

Admiraal W, Blanck H, Buckert-de Jong M, Guasch H, Ivorra N, Lehmann V, Nyström BAH, Paulsson M, Sabater S (1999) Short-term toxicity of zinc to microbenthic algae and bacteria in a metal polluted stream. Water Res 33(9):1989–1996

Allan JD (2004) Landscapes and riverscapes: the influence of land use on stream ecosystems. Annu Rev Ecol Syst 35:257–284

Allan JD, Mcintyre PB., Smith SDP, Halpern BS, Boyer GL, Buchsbaum A, Burton JGA, Campbell LM, Chadderton WL, Ciborowski JJH, Doran PJ, Ederk T, Infante DM, Johnson LB, Joseph CA, Marino AL, Prusevich A, Read JG, Rose JB, Rutherford ES, Sowa SP, Steinman AD (2013) Joint analysis of stressors and ecosystem services to enhance restoration effectiveness. Proc Nat Acad Sci 110:372–377

Anderson LC, Bruland KW (1991) Biogeochemistry of arsenic in natural waters: the importance of methylated species. Environ Sci Technol 25:420–427

Barceló D, Sabater S (2010) Water quality and assessment under scarcity. Prospects and challenges in Mediterranean watersheds. J Hydrol 383:1–4

Bärlocher F, Murdoch JH (1989) Hyporheic biofilms—a potential food source for interstitial animals. Hydrobiologia 184:61–67

Barthès A, Ten-Hage L, Lamy A, Rols JL, Leflaive J (2015) Resilience of aggregated microbial communities subjected to drought–small-scale studies. Microb Ecol 70:9–20

Bier RL, Bernhardt ES, Boot CM, Graham EB, Hall EK, Lennon JT and Wallenstein MD (2015) Linking microbial community structure and microbial processes: an empirical and conceptual overview. FEMS Microbiol Ecol 91(10):fiv113

Bird DF, Duarte CM (1984) Empirical relationship between bacterial abundance and chlorophyll concentration in fresh and marine waters. Can J Fish Aquat Sci 41:1015–1023

Blanck H, Wängberg S-A (1988) Induced community tolerance in marine periphyton established under arsenate stress. Can J Fish Aquat Sci 45:1816–1819

Blanck H, Wängberg S-A, Molander S (1988) Pollution-induced community tolerance—a new ecotoxicological tool. In: Cairns J Jr, Pratt JR (eds) Functional testing of aquatic biota for estimating hazards of chemicals. ASTM STP 988 American Society for testing materials, Philadelphia, pp 219–230

Bonnineau C, Guasch H, Proia L, Ricart M, Geiszinger A, Romaní AM, Sabater S (2010) Fluvial biofilms: a pertinent tool to assess b-blockers toxicity. Aquat Toxicol 96:225–233

Burkart MR, Kolpin DW (1993) Hydrologic and land-use factors associated with herbicides and nitrate in near-surface aquifers. J Environ Qual 22:646–656

Burton GA, De Zwart D, Diamond J, Dyer SD, Kapo KE, Liess M, Posthuma L (2012) Making ecosystem reality checks the status quo. Environ Sci Technol 31:459–468

Coors A, De Meester L (2008) Synergistic, antagonistic and additive effects of multiple stressors: predation threat, parasitism and pesticide exposure in Daphnia magna. J Appl Ecol 45:1820–1828
Copin P-J, Perronet L, Chèvre N (2016) Modelling the effect of exposing algae to pulses of S-metolachlor: how to include a delay to the onset of the effect and in the recovery. Sci Total Environ 541:257–267
Corcoll N, Casellas M, Huerta B, Guasch H, Acuña V, Rodriguez-Mozaz S, Serra-Compte A, Barceló D, Sabater S (2015) Effects of flow intermittency and pharmaceutical exposure on the structure and metabolism of stream biofilms. Sci Total Environ 503–504:159–170
Dantas G, Sommer MO, Oluwasegun RD, Church GM (2008) Bacteria subsisting on antibiotics. Science 320:100–103
Debenest T, Silvestre J, Coste M, Pinelli E (2010) Effects of pesticides on freshwater diatoms. In: Reviews of environmental contamination and toxicology. Springer, pp 87–103
Federle TW, Dobbins DC, Thornton-Manning JR, Jones DD (1986) Microbial biomass, activity, and community structure in subsurface soils. Ground Water 24:365–374
Flemming HC, Wingender J (2010) The biofilm matrix. Nat Rev Microbiol 8:623–633
Freeman C, Lock MA (1995) The biofilm polysaccharide matrix: a buffer against changing organic substrate supply? Limnol Oceanogr 40:273–278
Ghiglione J-F, Martin-Laurent F, Pesce S (2016) Microbial ecotoxicology: an emerging discipline facing contemporary environmental threats. Environ Sci Pollut Res 23:3981–3983
Goodnight CJ (1956) The use of aquatic invertebrates as indicators of stream pollution. Trans Am Microscopical Soc 92:1–13
Graba M, Sauvage S, Moulin FY, Urrea G, Sabater S, Sanchez-Perez JM (2013) Interaction between local hydrodynamics and algal community in epilithic biofilm. Water Res 47:2153–2163
Guasch H, Paulsson M, Sabater S (2002) Effect of copper on algal communities from oligotrophic calcareous streams. J Phycol 38:241–248
Hall LW, Anderson RD, Kilian J, Tierney DP (1999) Concurrent exposure assessments of atrazine and metolachlor in the mainstem, major tributaries and small streams of the Chesapeake Bay watershed: indicators of ecological risk. Environ Monit Assess 59:155–190
Hardenbicker P, Weitere M, Ritz S, Schöll F, Fischer H (2015) Longitudinal plankton dynamics in the Rivers Rhine and Elbe. River Res Appl. doi:10.1002/rra.2977
Hatt BE, Fletcher TD, Walsh CJ, Taylor SL (2004) The influence of urban density and drainage infrastructure on the concentration and loads of pollutants in small streams. Environ Manage 34:112–124.
Hector A, Bagchi R (2007) Biodiversity and ecosystem multifunctionality. Nature 448:188–190
Hooper DU, Chapin IFS, Ewel JJ, Hector A, Inchausti P, Lavorel S, Lawton JH, Lodge DM, Loreau M, Naeem S, Schmid B, Setälä H, Symstad AJ, Vandermeer J, Wardle DA (2005) Effects of biodiversity on ecosystem functioning: a consensus of current knowledge. Ecol Monogr 75:3–35
Johnson RK, Hering D (2009) Response of taxonomic groups in streams to gradients in resource and habitat characteristics. J Appl Ecol 46:175–186
Kutka FJ, Richards C (1996) Relating diatom assemblage structure to stream habitat quality. J North Am Benthological Soc 15:469–480
Legendre P, Legendre L (1998) Numerical Ecology. Elsevier BV, Amsterdam, The Netherlands
Liu B, Liu W, Nie X, Guan Ch, Yang Y, Wang Z, Liao W (2011) Growth response and toxic effects of three antibiotics on Selenastrum capricornutum evaluated by photosynthetic rate and chlorophyll biosynthesis. J Environ Sci 23:1558–1563
Lupini G, Proia L, Di Maio M, Amalfitano S, Fazi S (2011) CARD-FISH and CLSM to assess successional changes of the bacterial community in freshwater biofilms. J Microbiol Methods 86:248–251
Malaj E, Von Der Ohe PC, Groted M, Kühne R, Mondy CP, Usseglio-Polatera P, Brack W, Schäferb RB (2014) Organic chemicals jeopardize the health of freshwater ecosystems on the continental scale. Proc Nat Acad Sci 111:9549–9554

Margalef R (1960) Ideas for a synthetic approach to the ecology of running waters. Int Rev gesamten Hydrobiol 45:133–153
Martínez JL (2008) Antibiotics and antibiotic resistance genes in natural environments. Science 321:365–367
Morin S, Proia L, Ricart M, Bonnineau C, Geiszinger A, Ricciardi F, Guasch H, Romani A, Sabater S (2010) Effects of a bactericide on the structure and survival of benthic diatom communities. Vie et milieu 60:109–116
Pesce S, Morin S, Lissalde S, Montuelle B, Mazzella N (2011) Combining polar organic chemical integrative samplers (POCIS) with toxicity testing to evaluate pesticide mixture effects on natural phototrophic biofilms. Environ Pollut 159:735–741
Peter H, Ylla I, Gudasz C, Romaní AM, Sabater S, Tranvik LJ (2011) Multifunctionality in microbial biofilms is constrained by diversity. PLoS ONE 6(8):e23225
Petrovic M, Ginebreda A, Acuña V, Batalla RJ, Elosegi A, Guasch H, De Alda ML, Marce R, Muñoz I, Navarro-Ortega A, Navarro E, Vericat D, Sabater S, Barceló D (2011) Combined scenarios of chemical and ecological quality under water scarcity in Mediterranean rivers. TrAC Trends Anal Chem 30:1269–1278
Ponsatí L, Corcoll N, Petrovic M, Picó Y, Ginebreda A, Tornés E, Guasch H, Barceló D, Sabater S (2016) Multiple-stressor effects on river biofilms under different hydrological conditions. Freshw Biol. doi:10.1111/fwb.12764
Posthuma L, Bjørn A, Zijp MC, Birkved M, Diamond ML, Hauschild MZ, Huijbregts MAJ, Mulder C, Van De Meent D (2014) Beyond safe operating space: finding chemical footprinting feasible. Environ Sci Technol 48:6057–6059
Posthuma L, Eijsackers H, Koelmans JPA, Vijver MG (2008) Ecological effects of diffuse mixed pollution are site-specific and require higher-tier risk assessment to improve site management decisions: a discussion paper. Sci Total Environ 406:503–517
Proia L, Osorio V, Soley S, Köck-Schulmeyer M, Pérez S, Barceló D, Romaní A, Sabater S (2013) Effects of pesticides and pharmaceuticals on biofilms in a highly impacted river. Environ Pollut 178:220–228
Ricart M, Barceló D, Geiszinger A, Guasch H, Alda MLPD, Romaní AM, Vidal G, Villagrasa M, Sabater S (2009) Effects of low concentrations of the phenylurea herbicide diuron on biofilm algae and bacteria. Chemosphere 76:1392–1401
Riedl J, Schreiber R, Otto M, Heilmeier H, Altenburger R, Schmitt-Jansen M (2015) Metabolic Effect level index links multivariate metabolic fingerprints to ecotoxicological effect assessment. Environ Sci Technol 49:8096–8104
Rodriguez Castro MC, Urrea G, Guasch H (2015) Influence of the interaction between phosphate and arsenate on periphytons growth and its nutrient uptake capacity. Sci Total Environ 503–504:122–132
Sabater S, Navarro E, Guasch H (2002) Effects of copper on algal communities at different current velocities. J Appl Phycol 14(5):391–398
Sabater S, Guasch H, Ricart M, Vidal G, Klünder C, Schmitt-Jansen M (2007) Monitoring the effect of chemicals on biological communities. The biofilm as an interface. Anal Bioanal Chem 387:1425–1434
Sabater S, Barceló D, De Castro-Català N, Ginebreda A, Kuzmanovic M, Petrovic M, Picó Y, Ponsatí L, Tornés E, Muñoz I (2016) Shared effects of organic microcontaminants and environmental stressors on biofilms and invertebrates in impaired rivers. Environ Pollut 210:303–314
Safaie A, Wendzel A, Ge Z, Nevers MB, Whitman RL, Corsi SR, Phanikumar MS (2016) Comparative evaluation of statistical and mechanistic models of *Escherichia coli* at beaches in Southern Lake Michigan. Environ Sci Technol 50:2442–2449
Scheffer M, Carpenter SR, Lenton TM, Bascompte J, Brock W, Dakos V, Van De Koppel J, Van De Leemput IA, Levin SA, Van Nes EH, Pascual M, Vandermeer J (2012) Anticipating Critical Transitions. Science 338:344–348
Schwarzenbach RP, Escher BI, Fenner K, Hofstetter TB, Johnson CA, Von Gunten U, Wehrli B (2006) Global hydrological cycles and world water resources. Science 313:1072–1077

Segner H, Schmitt-Jansen M, Sabater S (2014) Assesing the impact of multiple stressors on aquatic biota: the receptor's sice matters. Environ Sci Technol 48:7690–7696

Tlili A, Bérard A, Roulier J-L, Volat B, Montuelle B (2010) PO_4 dependence of the tolerance of autotrophic and heterotrophic biofilm communities to copper and diuron. Aquat Toxicol 98:165–177

Tlili A, Marechal M, Montuelle B, Volat B, Dorigo U, Bérard A (2011) Use of the MicroResp method to assess pollution-induced community tolerance to metals for lotic biofilms. Environ Pollut 159:18–24

Tornés E, Sabater S (2010) Variable discharge alters habitat suitability for benthic algae and cyanobacteria in a forested Mediterranean stream. Mar Freshw Res 61:441–450

Index

C. Cravo-Laureau et al. (eds.), *Microbial Ecotoxicology*,
DOI 10.1007/978-3-319-61795-4

MIX
Papier aus verantwortungsvollen Quellen
Paper from responsible sources
FSC® C105338

If you have any concerns about our products,
you can contact us on
ProductSafety@springernature.com

In case Publisher is established outside the EU,
the EU authorized representative is:
Springer Nature Customer Service Center GmbH
Europaplatz 3, 69115 Heidelberg, Germany

Printed by Libri Plureos GmbH
in Hamburg, Germany